VOLUME SIX HUNDRED AND SIXTY EIGHT

METHODS IN ENZYMOLOGY

Coenzyme B_{12} Enzymes Part A

METHODS IN ENZYMOLOGY

VOLUME SIX HUNDRED AND SIXTY EIGHT

METHODS IN ENZYMOLOGY

Coenzyme B_{12} Enzymes Part A

Edited by

E. NEIL G. MARSH

Professor of Chemistry and Biological Chemistry, Department of Chemistry, University of Michigan, United States

ELSEVIER

ACADEMIC PRESS

An imprint of Elsevier

Academic Press is an imprint of Elsevier
50 Hampshire Street, 5th Floor, Cambridge, MA 02139, United States
525 B Street, Suite 1650, San Diego, CA 92101, United States
The Boulevard, Langford Lane, Kidlington, Oxford OX5 1GB, United Kingdom
125 London Wall, London, EC2Y 5AS, United Kingdom

First edition 2022

ISBN: 978-0-323-91591-5
ISSN: 0076-6879

For information on all Academic Press publications
visit our website at https://www.elsevier.com/books-and-journals

Publisher: Zoe Kruze
Developmental Editor: Federico Paulo S. Mendoza
Production Project Manager: James Selvam
Cover Designer: Christian Bilbow

Typeset by STRAIVE, India

Contents

Contributors

Ruma Banerjee
Department of Biological Chemistry, University of Michigan Medical School, Ann Arbor, MI, United States

Birgitta Beatrix
Gene Center, Department of Biochemistry, Ludwig Maximilian University, München, Germany

Wolfgang Buckel
Department of Biology, Philipps University, Marburg, Germany

Flavia G. Costa
Department of Microbiology, University of Georgia, Athens, GA, United States

Evelyne Deery
School of Biosciences, University of Kent, Canterbury, Kent, United Kingdom

Jorge C. Escalante-Semerena
Department of Microbiology, University of Georgia, Athens, GA, United States

Karl Fisher
School of Chemistry, University of Manchester, Manchester, United Kingdom

Harsha Gouda
Department of Biological Chemistry, University of Michigan Medical School, Ann Arbor, MI, United States

Zachary F. Hallberg
Department of Plant & Microbial Biology, University of California, Berkeley, CA, United States

Tom Halliwell
School of Chemistry, University of Manchester, Manchester, United Kingdom

Samantha J. Hardman
Manchester Institute of Biotechnology, Department of Chemistry, University of Manchester, Manchester, United Kingdom

Amrita B. Hazra
Department of Biology; Department of Chemistry, Indian Institute of Science Education and Research Pune, Pune, India

Derren J. Heyes
Manchester Institute of Biotechnology, Department of Chemistry, University of Manchester, Manchester, United Kingdom

Markos Koutmos
Department of Chemistry, Program in Biophysics, Program in Chemical Biology, University of Michigan, Ann Arbor, MI, United States

Bernhard Kräutler
Institute of Organic Chemistry and Center of Molecular Biosciences, University of Innsbruck, Innsbruck, Austria

Andrew D. Lawrence
School of Biosciences, University of Kent, Canterbury, Kent, United Kingdom

David Leys
School of Chemistry, University of Manchester, Manchester, United Kingdom

Zhu Li
Department of Biological Chemistry, University of Michigan Medical School, Ann Arbor, MI, United States

Lahiru Malalasekara
Department of Microbiology, University of Georgia, Athens, GA, United States

Romila Mascarenhas
Department of Biological Chemistry, University of Michigan Medical School, Ann Arbor, MI, United States

Yamini Mathur
Department of Biology, Indian Institute of Science Education and Research Pune, Pune, India

Kenny C. Mok
Department of Plant & Microbial Biology, University of California, Berkeley, CA, United States

Koichi Mori
Graduate School of Natural Science and Technology, Okayama University, Tsushima-naka, Kita-ku, Okayama, Japan

Shubhadra Pillay
Department of Biological Chemistry, University of Michigan Medical School, Ann Arbor, MI, United States

Harshwardhan Poddar
Manchester Institute of Biotechnology, Department of Chemistry, University of Manchester, Manchester, United Kingdom

Stephen E.J. Rigby
School of Chemistry, University of Manchester, Manchester, United Kingdom

Markus Ruetz
Department of Biological Chemistry, University of Michigan Medical School, Ann Arbor, MI, United States

Michiyo Sakuma
Manchester Institute of Biotechnology, Department of Chemistry, University of Manchester, Manchester, United Kingdom

Nigel S. Scrutton
Manchester Institute of Biotechnology, Department of Chemistry, University of Manchester, Manchester, United Kingdom

Naoki Shibata
Graduate School of Life Science, University of Hyogo, 3-2-1 Koto, Kamigori-cho, Ako-gun, Hyogo, Japan

Michiko E. Taga
Department of Plant & Microbial Biology, University of California, Berkeley, CA, United States

Takamasa Tobimatsu
Graduate School of Natural Science and Technology, Okayama University, Tsushima-naka, Kita-ku, Okayama, Japan

Tetsuo Toraya
Graduate School of Natural Science and Technology, Okayama University, Tsushima-naka, Kita-ku, Okayama, Japan

Aniket R. Vartak
Department of Biology, Indian Institute of Science Education and Research Pune, Pune, India

Elizabeth A. Villa
Department of Microbiology, University of Georgia, Athens, GA, United States

Martin J. Warren
School of Biosciences, University of Kent, Canterbury, Kent; Quadram Institute Bioscience, Norwich Research Park, Norwich, United Kingdom

Mamoru Yamanishi
Graduate School of Natural Science and Technology, Okayama University, Tsushima-naka, Kita-ku, Okayama, Japan

Madeline Yaw
Department of Biological Chemistry, University of Michigan Medical School, Ann Arbor, MI, United States

Oskar Zelder
Department of Biology, Philipps University, Marburg; Industrial Biotechnology I, BASF SE, Ludwigshafen am Rhein, Germany

Shaowei Zhang
Manchester Institute of Biotechnology, Department of Chemistry, University of Manchester, Manchester, United Kingdom

Preface

Vitamin B_{12} (cyanocobalamin) is the precursor to two biologically active forms of coenzyme B_{12}: adenosylcobalamin (AdoCbl) and methylcobalamin (MeCbl). Its structure comprises a highly modified tetrapyrrole macrocycle that chelates a cobalt ion; a further axial ligand to cobalt is provided by a nucleotide-derived ligand (most commonly dimethylbenzimidazole) that is tethered to the macrocycle. This leaves the sixth coordination site open to catalyze biological reactions, which take advantage of cobalt's ability to access the +1, +2, and +3 oxidation states. Coenzyme B_{12} has been called nature's most beautiful cofactor. Whether one shares this view or not, it is certainly one of the most structurally complex cofactors; its biosynthesis requires over 20 steps starting from uroporphyrinogen III, the common intermediate from which other tetrapyrroles such as heme and chlorophyll are derived.

Vitamin B_{12} is the precursor to the catalytically active cofactors for two enzymes in animals. Methylmalonyl-CoA mutase uses AdoCbl to catalyze the rearrangement of methylmalonyl-CoA to succinyl-CoA in the metabolism of odd-chain fatty acids, while methionine synthase uses MeCbl as the intermediate methyl donor in the transfer of a methyl group from methyl-tetrahydrofolate to homocysteine to produce methionine. These are the only known uses of this complex molecule in animals, although both AdoCbl- and MeCbl-dependent enzymes are found far more extensively in bacteria and archaea. By contrast, B_{12} appears to be absent from plants.

Vitamin B_{12} was discovered through efforts to cure pernicious anemia—a disease that we now know is generally caused by defects in the cobalamin uptake system in afflicted individuals. In 1926, Minot and Murphy showed that a diet high in liver and kidney, and low in fat, cured patients of their anemia noting "*prompt, rapid and distinct remission of their anemia and a marked symptomatic improvement, except for pronounced disorders due to spinal cord degeneration.*" It was this observation that would lead, some 96 years later, to this volume! Although liver extracts enriched in "antipernicious anemia factor" became an effective treatment for the disease during the 1920s and 1930s, it was not until 1947 that the compound was isolated in crystalline form. This herculean task, given the very small quantities present in liver, was achieved independently by Folkers' group at Merck in the United States and Smith's group at Glaxo in the United Kingdom.

In 1948, with its structure and mode of action still unknown, the crystalline "antipernicious anemia factor" was renamed vitamin B_{12}—"B" connotes only nutritional significance and the number "12" had not been used previously in vitamin nomenclature.

The structure of vitamin B_{12} was solved by X-ray crystallography in Hodgkin's laboratory in 1955—at the time, it was the largest molecular structure to be solved and a landmark in structural biology. However, the structure gave no immediate clue to the biochemical function of B_{12}; that would have to wait until 1958 for the discovery by Barker's laboratory of the first B_{12} enzyme, glutamate mutase. When the crystal structure of the active coenzyme was determined in 1960, again by Hodgkin, an adenosyl moiety was found to replace the cyanide ligand (cyanide is likely an artifact of the purification procedure) as the sixth ligand. Most unexpectedly, this ligand was bonded to cobalt through the 5′-carbon of adenosine. Shortly afterward, Wood and Lester demonstrated that methylcobalamin was involved in the biosynthesis of methionine. With the discovery of these two active forms of vitamin B_{12}, the stage was set to explore this new class of metalloenzymes that catalyze unusual reactions enabled by the unique chemistry of the B_{12} cobalt-carbon bond.

Homolysis of the AdoCbl cobalt-carbon bond serves to generate the highly reactive 5′-deoxyadenosyl radical. In this respect, AdoCbl-dependent enzymes parallel radical SAM enzymes, which form the same radical by reductive cleavage of *S*-adenosylmethionine. AdoCbl-dependent enzymes likely evolved after radical SAM enzymes, probably as a response to a more aerobic environment—all radical SAM enzymes are very oxygen-sensitive whereas AdoCbl-dependent enzymes are much less so. Historically though, radical SAM enzymes were discovered after B_{12} enzymes, prompting Barker to dub them "a poor man's coenzyme B_{12}." In subsequent decades this sobriquet has reverberated throughout the literature—and has often been challenged. In one respect, however, Barker was the most percipient; for just as there are many more poor people than rich people, there are now many more radical SAM enzymes known than B_{12} enzymes! In the six decades since glutamate mutase was discovered, only about a dozen AdoCbl enzymes have been discovered, all of which catalyze mechanistically similar rearrangements or dehydration/deamination reactions. By contrast, there are now around 100 different radical SAM enzymes known. These have been found in all kingdoms of life and catalyze an astonishingly wide array of reactions.

Despite their relative rarity, AdoCbl-dependent enzymes have fascinated enzymologists since their discovery because they catalyze unusual rearrangement reactions with no obvious counterpart in conventional organic chemistry. Elucidating the mechanisms by which AdoCbl-dependent enzymes catalyze these reactions has provided fundamental insights into how enzymes harness highly reactive free radicals for catalysis, thereby catalyzing reactions that would be hard to achieve by other means. What we have learned from studying AdoCbl-dependent enzymes has provided the foundational knowledge with which to tackle the much larger (and much more technically challenging) family of radical SAM enzymes.

With this brief historical introduction as context, I turn to this volume of *Methods in Enzymology* that propels B_{12} enzymology firmly into the 21st century. First, I express my profound thanks to all the authors who have so generously contributed their expertise and time to produce the chapters collected here. The response from the B_{12} community has been remarkable! Although I originally conceived this collection of methods-focused papers as a single volume, the enthusiasm of the contributors has been such that it has expanded into two volumes. Within these volumes, I have loosely organized the chapters into four sections.

Section I deals with enzymes involved in the biosynthesis of cobalamin and methods for the biosynthesis of cobalamin analogs. The structural complexity of coenzyme B_{12} ensures that the enzymes involved in its biosynthesis are both interesting and diverse! This section includes chapters on the enzymes involved in the construction of the nucleotide "tail" of the coenzyme, which in many microbes varies from the canonical dimethylbenzimidazole ligand, and the enzymes involved in transforming the vitamin form of B_{12} to the enzymatically active form, AdoCbl. In animals, the uptake, transport, and installation of B_{12} into methylmalonyl-CoA mutase and methionine synthase require a dedicated set of chaperone enzymes that are also discussed. Lastly, the synthesis and biological properties of "antivitamin B_{12}" analogs are discussed—these synthetic compounds have potential as mechanistic probes and therapeutic agents.

Section II groups together chapters that focus on the purification and characterization of enzymes that use either AdoCbl or MeCbl as cofactors. The chapters include discussions of enzymes that have been known from the early days, including diol dehydratase and ethanolamine ammonia-lyase, glutamate mutase and 2-methyleneglutarate mutase, and the human enzymes methylmalonyl-CoA mutase and methionine synthase. Also included are more recently recognized functions of B_{12}. Thus, B_{12} is used as a redox cofactor by a class of reductive dehalogenases that initiate the degradation of halogenated hydrocarbons by various anaerobic microorganisms.

And the most recently recognized role for AdoCbl is as a light sensor. AdoCbl has long been known to be light-sensitive, but this property has now been recognized to have a biological role with the discovery of the bacterial light-sensing transcription factor CarH.

Section III is devoted to the emerging group of enzymes that, excitingly for both B_{12} and radical SAM enzymologists, use both radical SAM and cobalamin chemistry. Many of these enzymes catalyze reactions in which methyl groups are added to nonacidic positions on the substrate, e.g., carbon and phosphorus. These methylations proceed by radical mechanisms in which radicals are generated from SAM and the methyl group is supplied by MeCbl. However, these enzymes have proved especially challenging to study due to the difficulty of recombinantly expressing enzymes that have both the iron-sulfur cluster and the cobalamin correctly installed. The chapters here focus on practical methods for expressing, purifying, and characterizing a number of these fascinating enzymes.

Section IV focuses on methods for studying the mechanisms of B_{12} enzymes. These chapters cover a wide range of techniques and describe how they have been used to elucidate the mechanisms of various B_{12} enzymes. In many cases, these techniques and approaches should be readily adaptable to study the mechanisms of other B_{12} enzymes, and related enzymes such as radical SAM enzymes. The rich electronic spectrum of the B_{12} corrin ring is sensitive to the oxidation and coordination state of cobalt. These properties make UV-visible spectroscopy and magnetically coupled circular dichroism particularly useful techniques to study the reactivity of the coenzyme, especially when combined with stopped flow. The mechanisms of many B_{12} enzymes involve radical intermediates, which naturally makes EPR spectroscopy a valuable tool to investigate them. The transfer of hydrogen between the substrate and coenzyme is a hallmark of AdoCbl-dependent reactions; as such, isotope effect measurements are an informative tool. Many of these techniques can be adapted to apply them at very short timescales (in some cases on the picosecond timescale), making it possible to precisely dissect individual steps in the mechanism. Last but not least, computational methods continue to increase in the power and sophistication with which they can model enzyme reactions, and B_{12} enzymes are no exception.

In closing, I hope this collection of chapters will serve as a valuable resource for the current community of scientists working on B_{12} and, most importantly, those entering the field in the future.

E. Neil G. Marsh

SECTION I

B_{12} Biosynthesis and B_{12} Chaperones

CHAPTER ONE

Biosynthesis of cobamides: Methods for the detection, analysis and production of cobamides and biosynthetic intermediates

Evelyne Deery[a], Andrew D. Lawrence[a], and Martin J. Warren[a,b,*]
[a]School of Biosciences, University of Kent, Canterbury, Kent, United Kingdom
[b]Quadram Institute Bioscience, Norwich Research Park, Norwich, United Kingdom
*Corresponding author: e-mail address: martin.warren@quadram.ac.uk

Contents

Abstract

Vitamin B_{12}, cobalamin, belongs to the broader cobamide family whose members are characterized by the presence of a cobalt-containing corrinoid ring. The ability to detect, isolate and characterize cobamides and their biosynthetic intermediates is an important

Methods in Enzymology, Volume 668
ISSN 0076-6879
https://doi.org/10.1016/bs.mie.2022.01.013

prerequisite when attempting to study the synthesis of this remarkable group of compounds that play diverse roles across the three kingdoms of life. The synthesis of cobamides is restricted to only certain prokaryotes and their structural complexity entails an equally complex synthesis orchestrated through a multi-step biochemical pathway. In this chapter, we have outlined methods that we have found extremely helpful in the characterization of the biochemical pathway, including a plate microbiological assay, a corrinoid affinity extraction method, LCMS characterization and a multigene cloning strategy.

1. Introduction

Vitamin B_{12}, or cobalamin, is a cobalt-containing corrinoid that belongs to the broader cobamide family (Costa, Deery, Warren, & Escalante-Semerena, 2020; Kennedy & Taga, 2020). There is considerable interest in the structure and function of cobamides and in the roles this family of molecules play across the three kingdoms of life (Banerjee & Ragsdale, 2003; Bridwell-Rabb & Drennan, 2017; Krautler, 2005). As cobamides are made exclusively by only a subset of prokaryotes, there is also interest in how these enigmatic cofactors and coenzymes are trafficked to organisms that require them as a nutrient but cannot make the molecules themselves (Lawrence et al., 2018; Roth, Lawrence, & Bobik, 1996). Cobalamin was the first cobamide to be identified, isolated, purified and characterized (Hodgkin et al., 1956; Rickes, Brink, Koniuszy, Wood, & Folkers, 1948; Smith, 1948) but it soon became apparent that there are many other related molecules that differed primarily in the nature of the base found in the lower nucleotide loop (Fig. 1) (Costa et al., 2020; Kennedy & Taga, 2020). Humans, and possibly all cobamide-requiring eukaryotes, appear to have opted exclusively for cobalamin as their nutrient of choice (Smith, Warren, & Refsum, 2018) from the more than 15 variant forms that represent the broader functional cobamide family (Kennedy & Taga, 2020), although some of the cobalamin analogs may be functional with eukaryotic enzymes (Sokolovskaya et al., 2021).

The biosynthesis of cobalamin has been studied in several bacteria (Blanche et al., 1995; McGoldrick, Deery, Warren, & Heathcote, 2002; Raux et al., 1996; Raux, Lanois, Rambach, Warren, & Thermes, 1998; Raux, Lanois, Warren, Rambach, & Thermes, 1998; Roessner, Huang, Warren, Raux, & Scott, 2002; Santos et al., 2008). As a modified tetrapyrrole, the molecule is based on the uroporphyrinogen III macrocyclic template that is similarly used

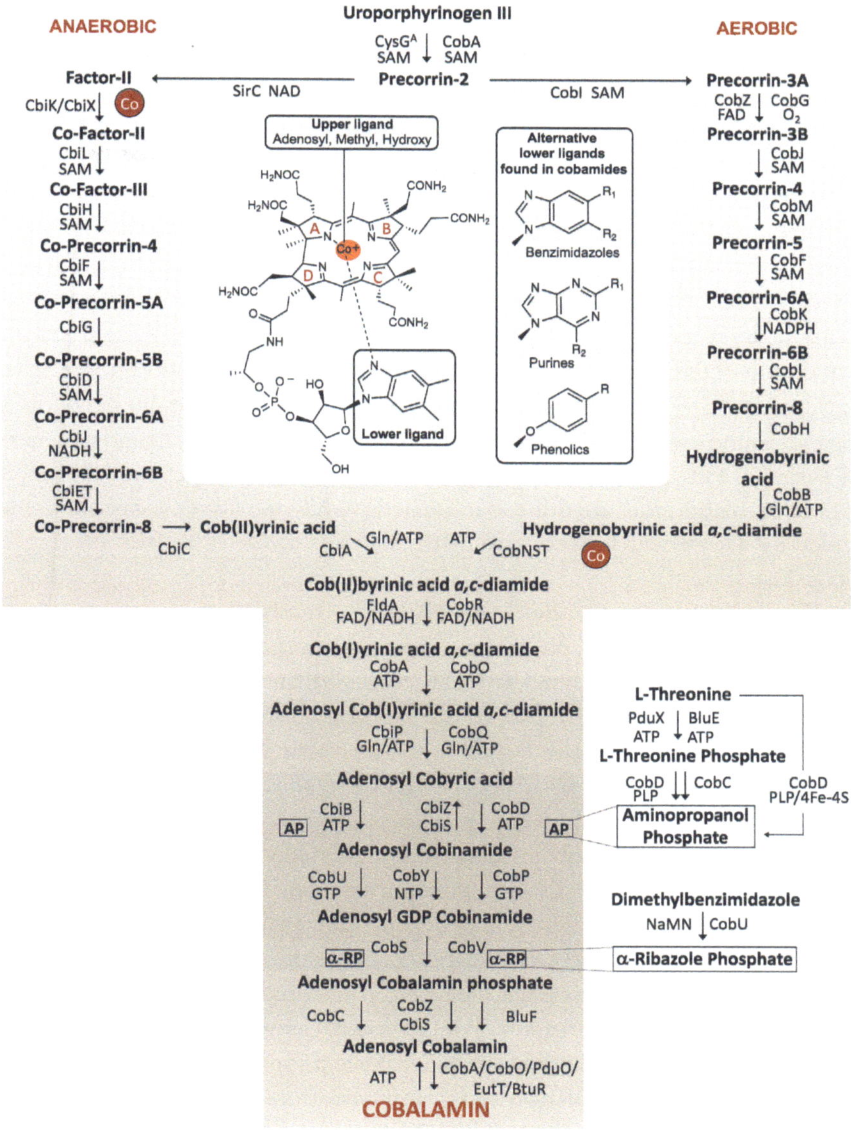

Fig. 1 Structure of cobamides and the pathways for the biosynthesis of cobalamin. The figure shows the structure of cobalamin, with the arrangement of the four pyrrole-derived rings (A–D) that give rise to the unique corrin structure. The upper ligand is normally found as an adenosyl, methyl or hydroxo group. The lower ligand in cobalamin is a dimethylbenzimidazole base attached to a nucleotide loop that is linked to ring D of the corrin ring. Alternative (analog) forms of cobalamins are found in nature where the dimethylbenzimidazole is substituted with alternative groups such as purines or phenolics as shown. The figure also outlines the two elucidated biosynthetic routes for the synthesis of cobalamin that have been described. The pathway to the left involves the early insertion of cobalt and is referred to as the cobalt-early or anaerobic route, whilst the route to the right is the cobalt-late or aerobic pathway. The biosynthetic intermediates and enzymes involved in the biosynthesis are highlighted.

as the starting point for the synthesis of related molecules that include heme, chlorophyll, coenzyme F_{430}, siroheme and heme d_1 (Bryant, Hunter, & Warren, 2020). However, in comparison to these other modified tetrapyrroles the structure of cobalamin is much more complex and this is reflected in a more complex biochemical pathway (Warren, Raux, Schubert, & Escalante-Semerena, 2002). The complexity of cobalamin reflects the fact that the corrin ring has undergone a unique ring-contraction, through extrusion of the meso-carbon position that joins rings A–D. A further level of complexity is added by the high level of methylation of the ring system and the amidation of the peripheral side chains (Fig. 1). Finally, in contrast to the other modified tetrapyrroles, cobalamin additionally provides both upper and lower ligands to the centrally chelated metal ion, where the upper ß-ligand generates the unique Co-carbon bond that underpins many of the biochemical properties of cobalamin that are utilized in catalysis, while the lower α-ligand is provided by a lower nucleotide loop that is attached to the side chain on ring D (Fig. 1).

Typically, the biosynthesis of cobalamin involves in excess of twenty steps for its complete biogenesis from uroporphyrinogen III (Fig. 1) (Costa et al., 2020). There are also two similar yet genetically distinct pathways for cobalamin biosynthesis found in bacteria, that differ largely on the timing on when the central cobalt ion is inserted (Warren et al., 2002). The cobalt-early pathway is often referred to as the anaerobic pathway and is characterized by the insertion of cobalt before ring contraction takes place; in this case cobalt insertion is mediated via a comparatively small type II chelatase in an ATP-independent fashion. By way of contrast, the cobalt-late pathway is often referred to as the aerobic pathway, as there is a step requiring molecular oxygen; here cobalt is inserted only after ring contraction is completed by a large type I enzyme complex that has an absolute requirement for ATP (Brindley, Raux, Leech, Schubert, & Warren 2003).

In this chapter we describe how we have adopted, adapted and developed methods for the isolation and characterization of a range of cobamides and their biosynthetic intermediates. The aerobic cobalt-late pathway generates more air-tolerant intermediates, and this has been used successfully for engineering biological approaches to generate useful intermediates (Fig. 2) (McGoldrick et al., 2005). The coupling of engineering biological and chemical approaches, for instance, has been used for the generation of non-natural metal isosteres of cobalamin, the so-called metbalamins (Kieninger et al., 2019, 2020; Widner et al., 2016).

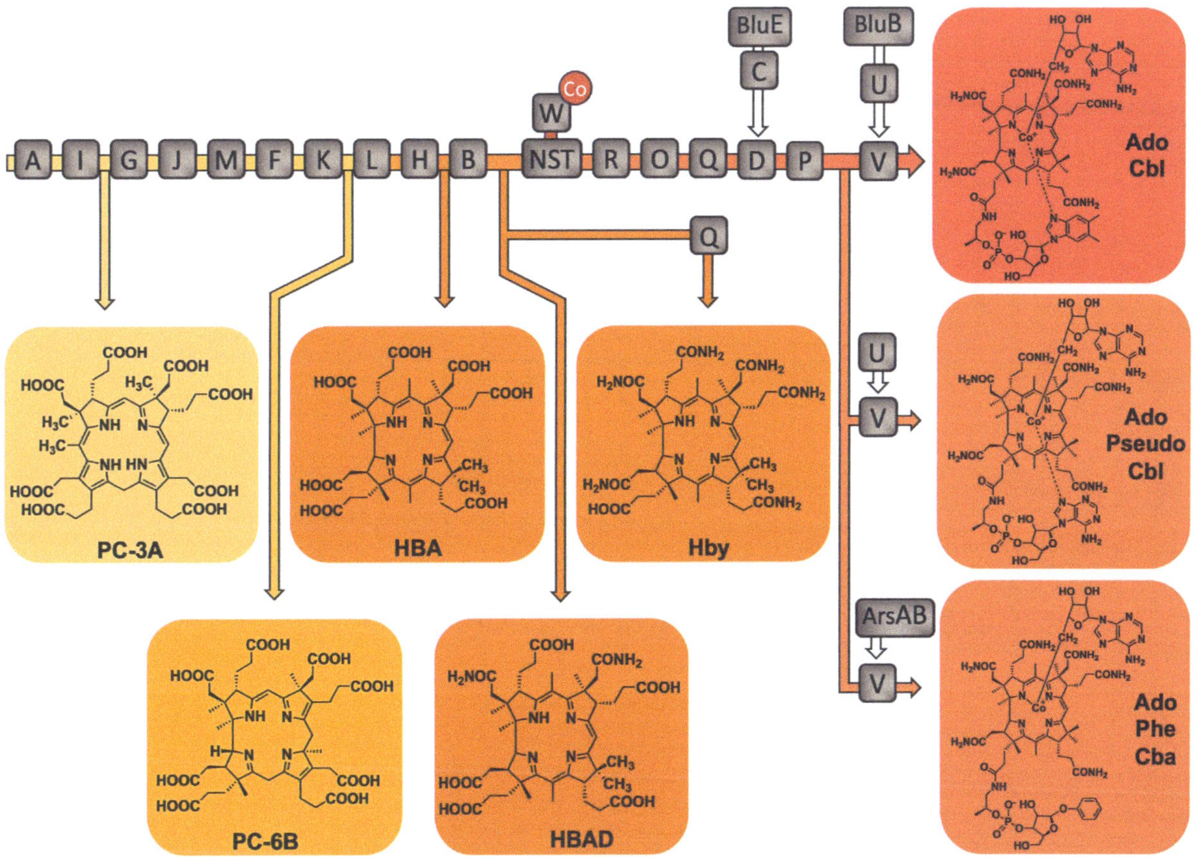

Fig. 2 See figure legend on next page.

2. Extraction, isolation and identification of cobamides

Methods for the isolation and identification of cobamides have varied and developed over the years. However, a major sticking point with the detection and analysis has always been the generally low natural abundance of cobamides, which has meant that the isolation of these compounds has been technically challenging. One liter of bacterial culture may only produce 50 μg of cobamide and this can be very difficult to detect let alone purify. Herein, we outline some approaches that are currently employed to identify and quantify cobamides.

2.1 The microbiological assay

The presence and quantification of cobamides can be estimated through the use of suitable microbiological bioassays, which can detect very small quantities of the nutrient. Indeed, it was the development of the microbiological assay for vitamin B_{12} by Mary Shaw Shorb (Shorb, 1948) that hastened the isolation and purification of the nutrient and variations of this assay are still in current use (Anderson, 1964; Kelleher, Walshe, Scott, & O'Broin, 1987). The bioassay approach was used also for many years to estimate the amount of cobalamin in human blood and was shown to be a very sensitive and reproducible method (Kelleher et al., 1987). The assay relies on the use of cobalamin auxotrophs, bacterial strains that are dependent upon cobalamin for growth. The strain used by Shorb was *Lactobacillus lactis* Dornier, a strain that requires cobalamin as a coenzyme for its ribonucleotide reductase. However, many other bacterial strains have subsequently been used for the assay, where growth of the strain is dependent upon the available nutrient to complement B_{12}-dependent enzymes such as ribonucleotide reductase or methionine synthase.

Fig. 2 Cobalt-late pathway highlighting how combinations of enzymes produce specific intermediates and alternative cobamide forms. The figure shows the order of gene products associated with the cobalt-late or aerobic cobalamin biosynthetic pathway and highlights certain biosynthetic intermediates and products. The combination of CobA and CobI generates precorrin-3A (PC-3A), whilst the enzymes CobAIGJMFK produces precorrin-6B (PC-6B). The further inclusion of CobL and H fabricates hydrogenobyrinic acid (HBA) and the addition of CobB makes hydrogenobyrinic acid *a, c*-diamide (HBAD). The pathway can be derailed by adding CobQ out of turn to produce hydrogenobyric acid (Hby). However, inclusion of the remaining pathway enzymes engenders the correct production of adenosylcobalamin (Ado-Cbl), although in some organisms cobalamin analogues are produced such as adenosyl-pseudocobalamin (Ado-Pseudo-Cbl) and adenosyl-phenylcobamide (Ado-Phe-Cba).

2.2 Affinity isolation of cobamides

Whilst microbial bioassays help in the detection of the nutrient, the laborious methods of extracting small quantities of the nutrient from complex mixtures of cellular material have previously meant that this is a tedious process involving many chromatographic steps. However, the identification of vitamin B_{12} binding proteins, such as human transcobalamin (Wuerges et al., 2006) and bacterial uptake components such as BtuF (Borths, Locher, Lee, & Rees, 2002) and BtuG (Wexler et al., 2018), provides a mechanism for a simpler extraction of the nutrient using high affinity binding columns. This approach was used successfully to extract a range of cobamides from fecal material (Allen & Stabler, 2008). The ability to produce these protein recombinantly and in comparatively high yields means that simpler and more rapid procedures have evolved for cobamide isolation.

2.3 LCMS as a tool to analyze cobamide

Finally, the advent of highly sensitive liquid chromatography mass spectrometry (LCMS) techniques means that cobamide forms can be much more easily identified. This identification is based on the mass-to-charge ratio, so that known masses of cobamides can be readily identified from the spectrum, and the composition of the cobamide further confirmed through MS/MS analysis after fragmentation. Early extraction and analysis of cobamides was usually performed after exposing the cobamide to cyanide in order to convert the cobamide into the cyano-form and thereby remove the upper ligand. This had the advantage of isolating and analyzing just a single form of the cobamide rather than a mixture of the adenosyl, hydroxo and methyl forms. The more rapid isolation techniques coupled with analysis by LCMS means that it is possible to detect the various upper ligand forms of the nutrient, albeit that the extracts need to be handled in reduced light conditions given the light-sensitive nature of the adenosyl and methyl forms. When attempting to identify the type of cobamides present in a sample it is often difficult to rely on MS alone as many of the possible molecules have very similar masses. This task increases in complexity if there is additional variation in the upper or β-ligand present in the sample or other natural variants such as the norcobamides. In these cases, identification requires the analysis of the MS/MS fragmentation pattern. The observed fragmentation pattern is usually relatively simple due to the stability of the corrin ring. The most common fragments typically arise from loss of the upper ligand and fragmentation of the nucleotide loop.

3. Isolation of biosynthetic intermediates of cobalamin biosynthesis

As highlighted above, there are two related but distinct pathways for the biosynthesis of cobalamin, requiring many enzymes to complete the process. Both pathways start from uroporphyrinogen III and proceed via precorrin-2, at which point the pathways diverge only to re-join at a much later stage with the production of cobyrinic acid *a,c*-diamide (Fig. 1). The final steps of the pathway are then shared. Many of the pathway intermediates up to the production of cobyrinic acid are very unstable and extremely oxygen sensitive. These various intermediates can be produced by mixing the relevant enzymes as a cocktail together with the required substrates and incubating them under strictly anaerobic conditions in a glove box (Spencer, Stolowich, Min, Roessner, & Scott, 1994). This approach has allowed the characterization of many intermediates in the pathway but does require specialized skills in preparing anaerobic environments and experience in handling the air-sensitive reagents and products.

3.1 Recombinant strains for isolation of cobamides and biosynthetic intermediates

The use of recombinant DNA approaches has allowed the development of specific strains of *E. coli* that can make cobalamin through cloning of the complete B_{12} pathway (Fang et al., 2018; Ko et al., 2014; McGoldrick et al., 2005; Raux et al., 1996; Raux, Lanois, Rambach, et al., 1998). Similarly, it is possible to construct strains that make intermediates, and those intermediates that are air-stable can accumulate within these strains to reasonable levels (Deery et al., 2012; Kieninger et al., 2019; McGoldrick et al., 2005; Widner et al., 2016). For instance, within the cobalt-pathway we have found that co-expression of the biosynthetic genes *cobAIGJMFKLH* leads to the production of hydrogenobyrinic acid, a vibrant orange-colored ring-contracted compound that is very stable (Fig. 2) (McGoldrick et al., 2005). Expression of subsequent enzymes allow the isolation of hydrogenobyrinic acid *a c*-diamide and hydrogenobyric acid (Fig. 2) (Kieninger, Deery, et al., 2019; Widner et al., 2016). These compounds have also been used for the synthesis of various metal isosteres of cobalamin through the chemical insertion of a range of alternative metal ions such as rhodium, zinc and nickel (Kieninger et al., 2020; Kieninger, Baker, et al., 2019; Widner et al., 2016).

3.2 Enzyme trap approach for isolation of intermediates

We have found that it is possible to isolate some of the more air-sensitive intermediates by isolating them as bound enzyme-product complexes (Deery et al., 2012). This enzyme-trap approach is comparatively simple and allows the terminal enzyme within a partial pathway to be isolated with its product attached. Many of the enzymes in the cobalamin pathway appear to hold onto their products in the absence of the proceeding enzyme as it appears the pathway operates through substrate channeling where the product of one enzyme is passed directly to the following enzyme within the biosynthetic sequence (Deery et al., 2012; Moore et al., 2013). A example of this is the isolation and purification of precorrin-6B, which was trapped as the enzyme-product complex representing CobK attached to precorrin-6B (Deery et al., 2012; Gu, Sushko, Deery, Warren, & Pickersgill, 2015).

3.3 Production of intermediates in an anaerobic chamber with purified enzymes—Precorrin-3A

Precorrin-3A is a relatively stable intermediate that can be generated under anaerobic conditions and stored there for several weeks although it will degrade when exposed to air. The compound is a useful starting point for further mechanistic studies along the pathway. We have developed a relatively simple system for the production of precorrin-3A using purified enzymes. It involves the co-production of the enzymes for uroporphyrinogen III synthesis from 5-aminolevelunic acid (HemB, C, D) together with CobA which converts uroporphyrinogen III into precorrin-2. This is accomplished by using a plasmid that co-produces all of these enzymes with N-terminal His-tags, allowing them to be isolated simultaneously on a Nickel-column. The CobI enzyme, which converts precorrin-2 into precorrin-3A, is not as stable as the other enzymes and is purified separately and then added to the cocktail. The production of precorrin-3A is then initiated by the addition of 5-aminolevulinic acid and S-adenosylmethionine. After 18–20 h of incubation in the dark in an oxygen-depleted environment, the yellow-colored precorrin-3A is ready to be purified for assaying the next enzyme along the pathway, CobG.

4. Protocols

All strains and plasmids used in these protocols are listed in Table 1.

Table 1 Strain and plasmid list

AR3612 Δ*metE* Δ*cysG*	Reference (Raux et al., 1996)
AR2680 Δ*metE* Δ*cbiB*	Reference (Raux, Lanois, Rambach, et al., 1998)
BL21 star DE3 pLysS	Invitrogen
pET3a	Novagen, www.merckmillipore.com/GB/en/product/pET-3a-DNA-Novagen,EMD_BIO-69418
pET14b	Novagen, www.merckmillipore.com/GB/en/product/pET-3a-DNA-Novagen,EMD_BIO-69418
pETcoco2	Novagen, www.merckmillipore.com/GB/en/product/pET-3a-DNA-Novagen,EMD_BIO-69418
pET3a-cobA	Reference (Deery et al., 2012)
pET3a-cobAI	Reference (Deery et al., 2012)
pET3a-*cobAIGJMFKLH*	Reference (Deery et al., 2012)
pET14b-*E. coli* btuF	Reference (Lawrence et al., 2018)

4.1 Bioassay plate method for estimation of cobalamin concentration using a *Salmonella* reporter strain

This agar-plate cobamide-reporter microbiological assay is based on a system that was used to monitor recombinant cobalamin production in *E. coli* (Raux et al., 1996).

1.1. Streak the relevant *Salmonella* indicator strain (either AR2680 (Δ*metE* & Δ*cbiB*) or AR3612 (Δ*metE* & Δ*cysG*)) on an LB plate and grown overnight at 37 °C. These strains rely on their vitamin B_{12}-dependent methionine synthase (MetH) for growth. Details of the genotypes are given in Fig. 3.

1.2. Re-streak the *Salmonella* strain on an M9 plate containing methionine (and cysteine for AR3612 only), such that the plate will be completely covered with *Salmonella* the following day, and grow overnight at 37 °C.

1.3. Scrape the bacteria from the overnight minimum medium plate and suspend the cells in 1 mL of 0.9% NaCl (w/v) in an Eppendorf tube. Spin the cells for 2 min at 8000 rpm and resuspend them twice more in the same solution in order to remove any traces of methionine.

1.4. Transfer the cells after their washes to 300 mL of reconstituted minimum medium agar that has been cooled to 47/48 °C (containing cysteine for AR3612 only). Once set and dried the plates can be use immediately or kept in the fridge for a maximum of up to 3 weeks.

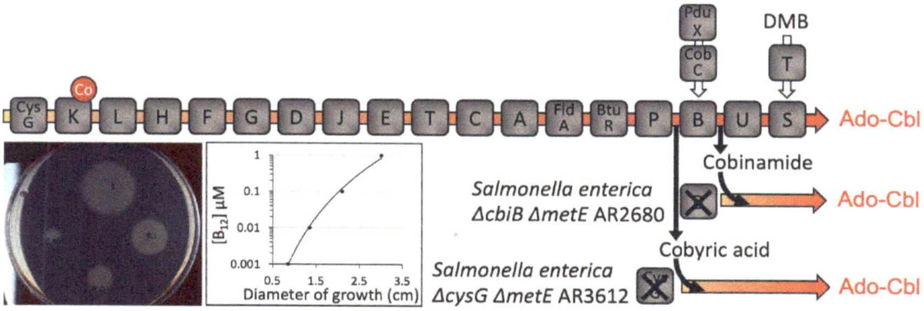

Fig. 3 Microbiological plate assay for estimation of cobamide levels in crude samples. This microbiological plate assay represents a simple and comparatively quick method to allow the detection and quantitative estimation of cobamides within any number of samples. It uses one of two *Salmonella enterica* reporter strains, that are both deleted in *metE* (cobalamin-independent methionine synthase) and are also inactivated in different parts of the cobalamin biosynthetic pathway. These strains are thus dependent on their *metH* gene to produce a cobalamin-dependent methionine synthase and are thus cobamide auxotrophs. *S. enterica* has the ability to salvage and utilize cobalamin breakdown components, but the earliest intermediate that can be transported into the cell is cobyric acid. *S. enterica strain* AR2680 is deleted in *cbiB* and hence will grow in the presence of cobinamide and later intermediates, whereas strain AR3612 can grow on cobyric acid and later intermediates. The assay works by suspending a culture of the reporter strain in minimal media agar, effectively making it dormant. Growth can be stimulated by the application of trace quantities of cobamide, where growth circles originate from the application point. The diameter of the growth circle can be correlated on an exponential scale to the concentration of cobamide that is applied to the plate.

1.5. To generate a standard curve, add 10 μL from known vitamin B_{12} solutions at 1; 0.1; 0.01 and 0.001 μM concentrations to specific areas on the plate. The presence of vitamin B_{12} will stimulate growth of the dormant *Salmonella* in the agar around the application point. After leaving the plate to incubate overnight at 37 °C, circles of growth will be observed around the application point. The diameter of the growth circle is related in an exponential fashion to the concentration of the vitamin B_{12} standard (Fig. 3). The standard curve can be plotted on a logarithmic scale (Fig. 3). The curve will change slightly, with less growth as the plate ages and becomes more dehydrated.

1.6. Finally, add 10 μL samples from cobamide solutions and compare the level of growth to the standard curve. If the samples have too much cobalamin they will need to be diluted until they fall within the standard curve.

Tip: This procedure provides a relatively quick method for the identification and quantification of cobamide within a solution and/or to detect in which fractions it is present. More accurate analysis can be achieved by running this as a liquid assay but the plate method is a simple way of comparing multiple samples on a single plate.

Tip: The *Salmonella* strains can similarly be used to quantify some of the pathway intermediates. The strain AR2680 will grow in the presence of cobinamide and later intermediates, while AR3612 will grow additionally in the presence of the previous intermediate, cobyric acid. The latter strain requires cysteine to be added to the minimal medium since it has a mutation in *cysG*, which is involved not only in cobalamin biosynthesis but also siroheme, a cofactor involved in cysteine biogenesis (Fig. 3). Despite having a deletion in the first gene of the B_{12} pathway from uroporphrinogen III, the first intermediate that can be taken up by the *Salmonella* Δ*metE* Δ*cysG* is cobyric acid. Previous intermediates are not recognized by the corrinoid transport import system or are too unstable. This method does not distinguish between the different functional cobamides.

4.2 Rapid isolation of cobamide through affinity chromatography

1.1. Obtain an *E. coli* strain that overproduces BtuF, such as BL21 star (DE3)(pLysS)(pET14b-*E. coli* btuF). Ensure that the gene is modified so that the periplasmic signal sequence of the encoded protein is removed, but that an N-terminal His-tag encoding sequence is introduced.

1.2. Grow 1 L of the strain in the presence of appropriate antibiotics in LB media or 2YT broth at 37 °C and induce with IPTG. Upon induction continue growth at 18 °C overnight. Harvest the bacteria and resuspend in 30 mL of binding buffer (20 mM HEPES pH 7.5, 500 mM NaCl and 5 mM imidazole). The pellets can be stored in the freezer for several months (−80 °C better than −20 °C).

1.3. Sonicate the resuspended bacterial pellet to lyse the cells and centrifuge 18,000 rpm for 20 min at 4 °C to remove the cell debris. The supernatant is next loaded onto a small column containing 5 mL of charged nickel chelating resin (Fig. 4). The column is washed with 25 mL of binding buffer (20 mM HEPES pH 7.5, 500 mM NaCl and 5 mM imidazole) followed by 10 mL of wash buffer (20 mM HEPES pH 7.5, 500 mM NaCl and 60 mM imidazole) to remove non-specific binding proteins and finally the His-BtuF bound resin is re-equilibrated with 10 mL of lower salt binding buffer (20 mM HEPES pH 7.5 100 mM NaCl and 5 mM imidazole).

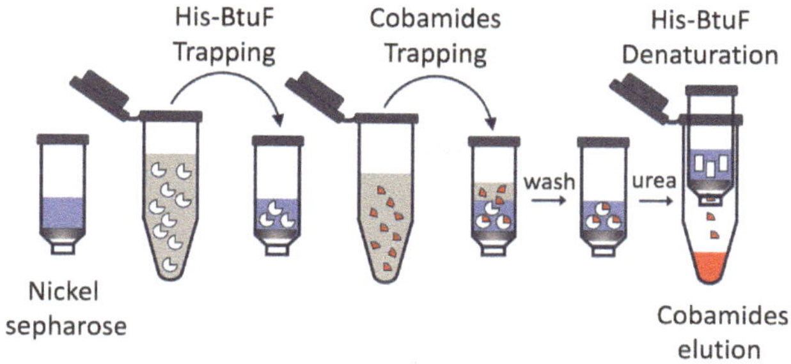

Fig. 4 Diagrammatic description for assembly of a cobamide affinity column for the isolation of cobamide from crude samples. The diagram highlights how a nickel-chelating sepharose can be used to bind a His-tagged version of the *E. coli* periplasmic cobalamin-binding protein BtuF. This column can be used as a cobalamin affinity column, using the bound BtuF to extract cobamides from a crude mixture that is passed through it. After washing the column, the cobamides are released by adding a solution of 8M urea, which denatures the BtuF but remains bound to the column through the His-tag, with the release of the previously bound cobamides.

1.4. For the isolation of small quantities of cobamide, the resin is transferred into a small beaker and gently stirred. Next, 0.5 to 1 mL of resin is poured into the desired number of mini-columns. In order to quantify the cobamide, ensure one column is saturated with vitamin B_{12} in order to know the maximum binding capacity of the resin in the column.

1.5. To isolate cobamide from a crude sample, the extract is loaded onto the mini-nickel chelating His-BtuF bound resin (Fig. 4). The column is washed twice with 1 mL of 20 mM HEPES pH 7.5, 100 mM NaCl. Purified cobamide is then eluted by applying 1 mL of 20 mM HEPES pH 7.5, 100 mM NaCl, containing 8 M urea. In the presence of 8 M urea, His-BtuF is denatured and cobalamin is released (Fig. 4). The His-BtuF remains bound to the nickel chelating resin due to the poly-histidine tag.

1.6. The amount of cobamide can be estimated by recording the UV/vis spectrum of the isolated material and the concentration worked out using the respective extinction coefficients, such as hydroxocobalamin at 357 nm is 21,400 M^{-1} cm^{-1} and cyanocobalamin at 361 nm is 27,500 M^{-1} cm^{-1} (Chemaly, 2008). The sample can be further analyzed by LCMS as described in Section 4.3 below.

1.7. This approach can be scaled up for the isolation of larger amounts of cobamide. To remove the urea from the cobalamin solution the cobamide can be separated from the urea on a LiChroprep RP18

(25–40 μM) (Merck) column. The LiChroprep is first suspended in methanol and equilibrated with 0.1% TFA (v/v). After passing the cobamide/urea solution over the column it is washed with 0.1% TFA and eluted in 50% methanol. The resulting pink solution is dried and can later be stored or resuspended in water.

Tip: In solution, the apo His-BtuF protein (29 kDa) is not very stable, but is very stable when bound to the Nickel resin and can be kept in the fridge for over a month.

Tip: To assist the isolation of the cobamide, whether from food material, bacterial extracts or feces, it is often best to homogenize the material in buffer and prior to heating to 80 °C for 10 mins. After centrifugation the soluble extract can be applied to the His-BtuF bound column. Beware that cobyric acid is heat sensitive.

Tip: to obtain an accurate cobamide concentration without using potassium cyanide, the samples are left under a bright light for 5 min to convert the cobamides to their hydroxo form.

Tip: We have not checked that the *E. coli* BtuF will bind all cobamides with the efficiency. There is a chance that specific cobalamin variants do not bind with the avidity as cobalamin. However, in our hands the BtuF-affinity approach does seem to bind most of the analogues that we have come across.

4.3 LCMS analysis of cobamide

1.1. Cobamide-containing samples can be analyzed by HPLC–MS (Allen & Stabler, 2008) and it is best to use samples that have been purified as described in Section 4.2 since this provides material that is relatively pure. The system has been developed for use with an ACE 5 AQ 150 x 2.1 mm column attached to an Agilent 1100 series HPLC coupled to a Bruker micrOTOF-Q II mass spectrometer (ESI +).

1.2. Mobile phase A consists of 0.1% (v/v) TFA in water and mobile phase B is acetonitrile. The HPLC column is equilibrated at 5% B and the column oven is set to 30 °C.

1.3. Samples are either filtered (0.2 μm) or spun in a 1.5 mL Eppendorf tube in a microcentrifuge at 13,000 rpm for 10 min prior to transfer to autosampler vials to remove small particulates. Amber glass vials are preferred when working with light sensitive cobamides.

1.4. After an initial hold at 5% B for 3 min, the concentration of B is increased to 90% over 30 min, held for 5 min and subsequently decreased back to 5% over 10 min. An equilibration time of 10 min is allowed before the next injection. HPLC chromatograms are typically monitored at 350 nm where mixed corrinoids are being analyzed (Fig. 5A).

1.5. The eluent from the HPLC is diverted to waste for the first 5 mins of the run before switching to the source of the mass spectrometer for the rest of the acquisition time.

1.6. MS data is collected in positive ion mode with a scan range of *m/z* 100–2000 and fragmentation is achieved through collision induced disassociation (Fig. 5B). The Auto MS/MS function typically performs well.

Tip: The MS is optimized for maximum sensitivity around 700 *m/z*. This works well for biosynthetic intermediates as well as cobamides whose prominent ion tends to be double charged.

Tip: If the sample contains a mixture of cobamides or biosynthetic intermediates that vary in the nature of the upper and lower ligands, then MS fragmentation analysis is crucial to establishing molecular identity.

Tip: TFA is routinely used in mobile phases but this can suppress ionization and sensitivity maybe increased using 0.1% (v/v) formic acid or a volatile buffer such as 10 mM ammonium formate (pH 3.5).

Tip: relative intensity of the fragment ions is dependent on the fragmentation voltage applied during CID.

4.4 Generation of strains for the production of cobamides, stable and air-sensitive intermediates

1.1. The process involves a multi-gene cloning strategy to construct plasmids with an ordered and specific arrangement of genes that reflect the biosynthetic pathway (McGoldrick et al., 2005). The genes are organized using a strategy called Link and Lock, where genes are prepared with specific restriction enzyme sites at the 5' and 3' ends (Fig. 6). This method describes how such constructs can be made. The relevant genes either have to be synthesized or amplified using primers that contain these specific sites. This method outlines how such a multi-gene plasmid can be prepared in the plasmid pET3a.

1.2. The first gene in the pathway, cobA, is amplified with a 5' primer containing an NdeI site (that overlaps with ATG initiation codon) and a 3' primer containing SpeI and BamHI sites. The amplified gene is cloned into pET3a using the NdeI and BamHI sites after appropriate restriction enzyme digestion and ligation to generate pET3a-cobA (Fig. 6).

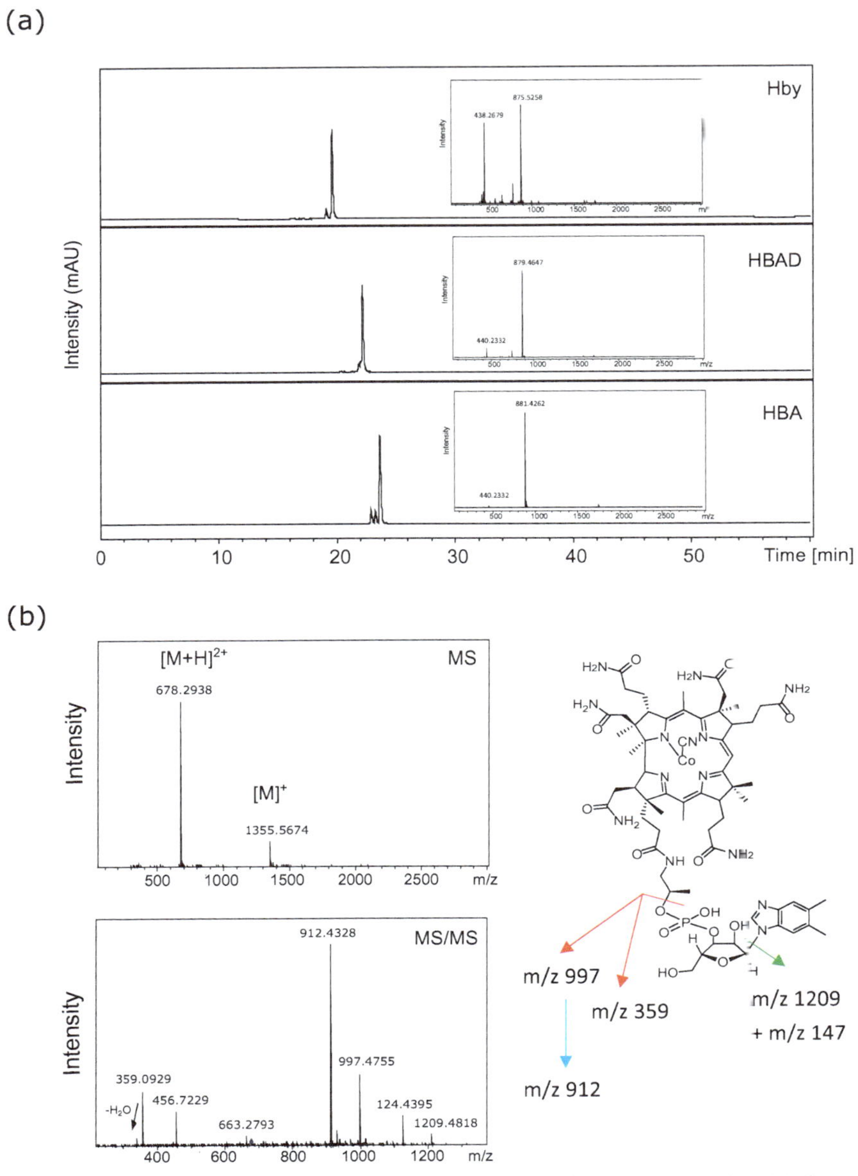

Fig. 5 Examples of the use of LCMS for the identification of corrinoid samples. (A) Shows the liquid chromatography (LC) traces of hydrogenbyric acid (Hby), hydrogenobyrinic acid *a,c*-diamide (HBAD) and hydrogenobyrinic acid (HBA), with the corresponding mass and uv/vis spectra of the main peak. (B) Shows the mass spectrum and MS/MS fragmentation of cyanocobalamin. The main fragments are correlated to the breakdown of the lower nucleotide loop of cobalamin as shown.

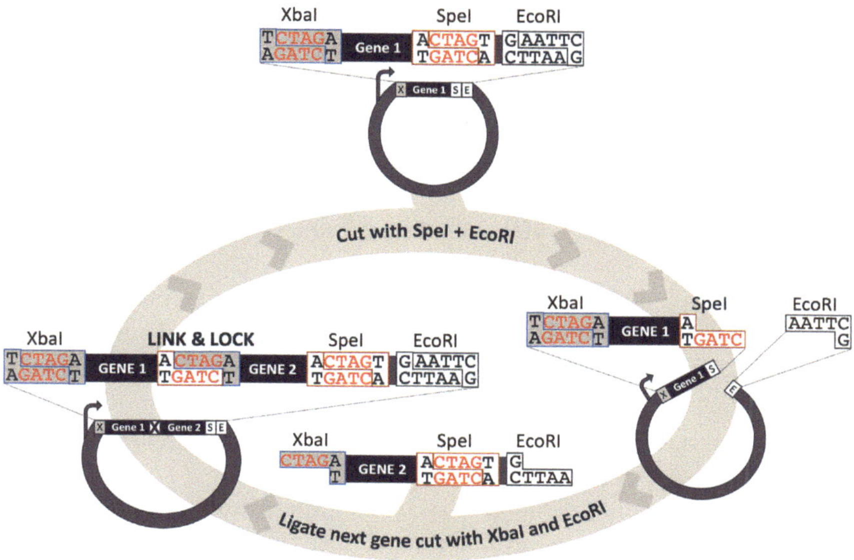

Fig. 6 Link and Lock strategy for the organized multi-gene assembly of the cobalamin pathway. The figure outlines how the Link and Lock cloning procedure allows for the multi-gene cloning of complete pathways. The methodology relies on the ability to amplify genes with specific restriction enzyme sites that are introduced via primers that are used in the amplification process. The genes can then be sub-cloned down-stream of each by using these restriction enzyme sites, which have compatible sticky ends that, when ligated together, are then lost. The first gene is cloned into a suitable plasmid and contains a down-stream SpeI and EcoRI site. After the plasmid is cleaved with both SpeI and EcoRI it is ligated with the second gene that has been excised with XbaI and EcoRI. The XbaI and SpeI overhangs are compatible allowing for the integration of the second gene downstream of the first. This can be repeated many times.

1.3. The second gene in the pathway, cobI, is also cloned into pET3a after amplification with a 5' primer containing the NdeI site and a 5' primer containing a SpeI site. Next, the pET3a-cobA plasmid is digested with NdeI and SpeI and the exuded cobA gene is discarded. The empty plasmid is then used to clone the cobI gene after it has been restricted with NdeI and SpeI (Fig. 6). This generates pET3a-cobI.

1.4. Building the multi-gene expression plasmid involves restricting the pET3a-cobA plasmid with SpeI and EcoRI, a process that releases a small nucleotide fragment that is discarded. The pET3a-cobI plasmid is restricted with XbaI and EcoRI and the cobI gene fragment, together with the ribosome binding site, is isolated and cloned into the restricted pET3a-cobA plasmid. As SpeI and XbaI have the same cohesive ends they can ligate into each other, but after ligation both sites are destroyed. This generates a new plasmid, pET3a-cobA-cobI, that still retains SpeI and EcoRI sites for repeat cloning (Fig. 6). In this way large multi-gene plasmid constructs are produced. Cloning of genes in this system to include cobAIGJMFKLH will generate an *E. coli* strain, after transformation, that has the ability to produce hydrogenobyrinic acid (Fig. 2). Further intermediates are made with cloning of subsequent genes.

1.5. For the plasmid to be used for enzyme-trap experiments in the isolation of air-sensitive intermediates as enzyme-product complexes, the terminal gene sequence needs to encode a protein with an N-terminal His-tag. Hence, the initial cloning of the terminal gene should be performed in pET14b, which will allow the gene to be subcloned with the encoded His-tag segment.

Tip: For genes that naturally contain the SpeI restriction enzyme sites both NheI or AvrII can be used instead since these are similarly compatible with XbaI.

Tip: It is possible to assemble large fragments of DNA within pET3a, but above 20 kb the plasmid becomes a bit unstable. At this stage it is better to switch to a lower copy number plasmid such as pETcoco.

5. Studying cobamide biosynthesis—Conclusions

The study of cobamide biogenesis requires a range of techniques and approaches. Within this chapter we have outlined some of the approaches that we have found to be very helpful when trying to understand more about how this family of remarkable molecules are made. Genome studies have revealed that there are a number of variations between our established view of cobalamin biosynthesis, suggesting that there may be some variations with the enzymes required for the process (Magnusdottir, Ravcheev, de Crecy-Lagard, & Thiele, 2015; Rodionov, Vitreschak, Mironov, & Gelfand, 2003; Shelton et al., 2019). This is exemplified by the use of CobZ for ring contraction in the cobalt-late pathway found in *Rhodobacter capsulatus* rather than CobG (McGoldrick et al., 2005). Some organisms appear to have neither CobG nor CobZ, again suggesting that alternative enzymes may exist. To probe the biosynthesis of cobalamin in these microorganisms there is a need for further studies.

There is growing interest in the biosynthesis of cobalamin analogues that contain alternative bases within the lower nucleotide loop (Kennedy & Taga, 2020). It has yet to be established if these variant forms are synthesized *de novo* or whether they are generated from remodeling of the cobalamin framework. This is all part of a bigger question as to why such a diverse range of cobamides have evolved and the role that these play in more complex microbial communities (Sokolovskaya, Shelton, & Taga, 2020).

Studies on cobamides have always been hampered by difficulties in being able to identify and isolate the various forms, largely because they are produced in such small quantities. Similarly, our understanding of the biosynthetic process and mechanisms involved has been impeded by the technical challenges of pathways that produce very small quantities of intermediates. Recently though, significant progress to be made through the combination of recombinant DNA technology, which allows biosynthetic enzymes to be overproduced and used to synthesize previously difficult-to-isolate intermediates, coupled with enhanced and highly sensitive analytical techniques such as LCMS. The integration of the biosynthetic genes into *E. coli* strains allows a range of stable intermediates and complete cobamides to be generated in much higher yields. The techniques and methods described in this chapter can be readily acquired by other groups and utilized to help develop the area further.

References

Allen, R. H., & Stabler, S. P. (2008). Identification and quantitation of cobalamin and cobalamin analogues in human feces. *The American Journal of Clinical Nutrition, 87*(5), 1324–1335.

Anderson, B. B. (1964). Investigations into the Euglena method for the assay of the vitamin B12 in serum. *Journal of Clinical Pathology, 17*, 14–26.

Banerjee, R., & Ragsdale, S. W. (2003). The many faces of vitamin B12: catalysis by cobalamin-dependent enzymes. *Annual Review of Biochemistry, 72*, 209–247.

Blanche, F., Cameron, B., Crouzet, J., Debussche, L., Thibaut, D., Vuilhorgne, M., et al. (1995). Vitamin-B12—How the problem of its biosynthesis was solved. *Angewandte Chemie. International Edition, 34*(4), 383–411.

Borths, E. L., Locher, K. P., Lee, A. T., & Rees, D. C. (2002). The structure of Escherichia coli BtuF and binding to its cognate ATP binding cassette transporter. *Proceedings of the National Academy of Sciences of the United States of America, 99*(26), 16642–16647.

Bridwell-Rabb, J., & Drennan, C. L. (2017). Vitamin B12 in the spotlight again. *Current Opinion in Chemical Biology, 37*, 63–70.

Brindley, A. A., Raux, E., Leech, H. K., Schubert, H. L., & Warren, M. J. (2003). A story of chelatase evolution: identification and characterization of a small 13-15-kDa "ancestral" cobaltochelatase (CbiXS) in the archaea. *The Journal of Biological Chemistry, 278*(25), 22388–22395.

Bryant, D. A., Hunter, C. N., & Warren, M. J. (2020). Biosynthesis of the modified tetrapyrroles-the pigments of life. *The Journal of Biological Chemistry, 295*(20), 6888–6925.

Chemaly, S. M. (2008). Cobalamins and the spectrochemical series. *Dalton Transactions, 42*, 5766–5773.

Costa, F. G., Deery, E., Warren, M. J., & Escalante-Semerena, J. C. (2020). New insights into the biosynthesis of cobamides and their use. In H.-W. Lui, & T. P. Begley (Eds.), *Comprehensive natural products III* Elsevier. In press.

Deery, E., Schroeder, S., Lawrence, A. D., Taylor, S. L., Seyedarabi, A., Waterman, J., et al. (2012). An enzyme-trap approach allows isolation of intermediates in cobalamin biosynthesis. *Nature Chemical Biology*, *8*(11), 933–940.
Fang, H., Li, D., Kang, J., Jiang, P., Sun, J., & Zhang, D. (2018). Metabolic engineering of Escherichia coli for de novo biosynthesis of vitamin B12. *Nature Communications*, *9*(1), 4917.
Gu, S., Sushko, O., Deery, E., Warren, M. J., & Pickersgill, R. W. (2015). Crystal structure of CobK reveals strand-swapping between Rossmann-fold domains and molecular basis of the reduced precorrin product trap. *Scientific Reports*, *5*, 16943.
Hodgkin, D. C., Kamper, J., Mackay, M., Pickworth, J., Trueblood, K. N., & White, J. G. (1956). Structure of vitamin B12. *Nature*, *178*(4524), 64–66.
Kelleher, B. P., Walshe, K. G., Scott, J. M., & O'Broin, S. D. (1987). Microbiological assay for vitamin B12 with use of a colistin-sulfate-resistant organism. *Clinical Chemistry*, *33*(1), 52–54.
Kennedy, K. J., & Taga, M. E. (2020). Cobamides. *Current Biology*, *30*(2), R55–R56.
Kieninger, C., Baker, J. A., Podewitz, M., Wurst, K., Jockusch, S., Lawrence, A. D., et al. (2019). Zinc substitution of cobalt in vitamin B12: Zincobyric acid and zincobalamin as luminescent structural B12-mimics. *Angewandte Chemie (International Ed. in English)*, *58*(41), 14568–14572.
Kieninger, C., Deery, E., Lawrence, A., Podewitz, M., Wurst, K., Nemoto-Smith, E., et al. (2019). The hydrogenobyric acid structure reveals the corrin ligand as an entatic state module empowering B12-cofactors for catalysis. *Angewandte Chemie (International Ed. in English)*.
Kieninger, C., Wurst, K., Podewitz, M., Stanley, M., Deery, E., Lawrence, A., et al. (2020). Replacement of the cobalt-center of vitamin B12 by Nickel—Nibalamin and nibyric acid prepared from metal-free B12-ligands hydrogenobalamin and hydrogenobyric acid. *Angewandte Chemie (International Ed. in English)*, *59*(45), 20129–20136.
Ko, Y., Ashok, S., Ainala, S. K., Sankaranarayanan, M., Chun, A. Y., Jung, G. Y., et al. (2014). Coenzyme B12 can be produced by engineered Escherichia coli under both anaerobic and aerobic conditions. *Biotechnology Journal*, *9*(12), 1526–1535.
Krautler, B. (2005). Vitamin B12: Chemistry and biochemistry. *Biochemical Society Transactions*, *33*(Pt. 4), 806–810.
Lawrence, A. D., Nemoto-Smith, E., Deery, E., Baker, J. A., Schroeder, S., Brown, D. G., et al. (2018). Construction of fluorescent analogs to follow the uptake and distribution of cobalamin (vitamin B12) in bacteria, worms, and plants. *Cell Chemical Biology*, *25*(8), 941–951 e6.
Magnusdottir, S., Ravcheev, D., de Crecy-Lagard, V., & Thiele, I. (2015). Systematic genome assessment of B-vitamin biosynthesis suggests co-operation among gut microbes. *Frontiers in Genetics*, *6*, 148.
McGoldrick, H., Deery, E., Warren, M., & Heathcote, P. (2002). Cobalamin (vitamin B(12)) biosynthesis in Rhodobacter capsulatus. *Biochemical Society Transactions*, *30*(4), 646–648.
McGoldrick, H. M., Roessner, C. A., Raux, E., Lawrence, A. D., McLean, K. J., Munro, A. W., et al. (2005). Identification and characterization of a novel vitamin B12 (cobalamin) biosynthetic enzyme (CobZ) from Rhodobacter capsulatus, containing flavin, heme, and Fe-S cofactors. *The Journal of Biological Chemistry*, *280*(2), 1086–1094.
Moore, S. J., Lawrence, A. D., Biedendieck, R., Deery, E., Frank, S., Howard, M. J., et al. (2013). Elucidation of the anaerobic pathway for the corrin component of cobalamin (vitamin B12). *Proceedings of the National Academy of Sciences of the United States of America*, *110*(37), 14906–14911.
Raux, E., Lanois, A., Levillayer, F., Warren, M. J., Brody, E., Rambach, A., et al. (1996). Salmonella typhimurium cobalamin (vitamin B12) biosynthetic genes: functional studies in S. typhimurium and Escherichia coli. *Journal of Bacteriology*, *178*(3), 753–767.

Raux, E., Lanois, A., Rambach, A., Warren, M. J., & Thermes, C. (1998). Cobalamin (vitamin B12) biosynthesis: functional characterization of the Bacillus megaterium cbi genes required to convert uroporphyrinogen III into cobyrinic acid a,c-diamide. *The Biochemical Journal*, *335*(Pt. 1), 167–173.
Raux, E., Lanois, A., Warren, M. J., Rambach, A., & Thermes, C. (1998). Cobalamin (vitamin B12) biosynthesis: Identification and characterization of a Bacillus megaterium cobI operon. *The Biochemical Journal*, *335*(Pt. 1), 159–166.
Rickes, E. L., Brink, N. G., Koniuszy, F. R., Wood, T. R., & Folkers, K. (1948). Crystalline vitamin B12. *Science*, *107*(2781), 396–397.
Rodionov, D. A., Vitreschak, A. G., Mironov, A. A., & Gelfand, M. S. (2003). Comparative genomics of the vitamin B12 metabolism and regulation in prokaryotes. *The Journal of Biological Chemistry*, *278*(42), 41148–41159.
Roessner, C. A., Huang, K. X., Warren, M. J., Raux, E., & Scott, A. I. (2002). Isolation and characterization of 14 additional genes specifying the anaerobic biosynthesis of cobalamin (vitamin B12) in Propionibacterium freudenreichii (P. shermanii). *Microbiology*, *148*(Pt. 6), 1845–1853.
Roth, J. R., Lawrence, J. G., & Bobik, T. A. (1996). Cobalamin (coenzyme B12): Synthesis and biological significance. *Annual Review of Microbiology*, *50*, 137–181.
Santos, F., Vera, J. L., van der Heijden, R., Valdez, G., de Vos, W. M., Sesma, F., et al. (2008). The complete coenzyme B12 biosynthesis gene cluster of Lactobacillus reuteri CRL1098. *Microbiology (Reading)*, *154*(Pt. 1), 81–93.
Shelton, A. N., Seth, E. C., Mok, K. C., Han, A. W., Jackson, S. N., Haft, D. R., et al. (2019). Uneven distribution of cobamide biosynthesis and dependence in bacteria predicted by comparative genomics. *The ISME Journal*, *13*(3), 789–804.
Shorb, M. S. (1948). Activity of vitamin B12 for the growth of Lactobacillus lactis. *Science*, *107*(2781), 397–398.
Smith, E. L. (1948). Purification of anti-pernicious anaemia factors from liver. *Nature*, *161*(4095), 638.
Smith, A. D., Warren, M. J., & Refsum, H. (2018). Vitamin B12. *Advances in Food and Nutrition Research*, *83*, 215–279.
Sokolovskaya, O. M., Plessl, T., Bailey, H., Mackinnon, S., Baumgartner, M. R., Yue, W. W., et al. (2021). Naturally occurring cobalamin (B12) analogs can function as cofactors for human methylmalonyl-CoA mutase. *Biochimie*, *183*, 35–43.
Sokolovskaya, O. M., Shelton, A. N., & Taga, M. E. (2020). Sharing vitamins: Cobamides unveil microbial interactions. *Science*, *369*(6499).
Spencer, J. B., Stolowich, N. J., Min, C. H., Roessner, C. A., & Scott, A. I. (1994). Genetically-engineered synthesis of hydrogenobyrinic acid. *Abstracts of Papers of the American Chemical Society*, *208*, 90-BIOL.
Warren, M. J., Raux, E., Schubert, H. L., & Escalante-Semerena, J. C. (2002). The biosynthesis of adenosylcobalamin (vitamin B12). *Natural Product Reports*, *19*(4), 390–412.
Wexler, A. G., Schofield, W. B., Degnan, P. H., Folta-Stogniew, E., Barry, N. A., & Goodman, A. L. (2018). Human gut Bacteroides capture vitamin B12 via cell surface-exposed lipoproteins. *eLife*, 7.
Widner, F. J., Lawrence, A. D., Deery, E., Heldt, D., Frank, S., Gruber, K., et al. (2016). Total synthesis, structure, and biological activity of adenosylrhodibalamin, the non-natural rhodium homologue of coenzyme B12. *Angewandte Chemie (International Ed. in English)*, *55*(37), 11281–11286.
Wuerges, J., Garau, G., Geremia, S., Fedosov, S. N., Petersen, T. E., & Randaccio, L. (2006). Structural basis for mammalian vitamin B12 transport by transcobalamin. *Proceedings of the National Academy of Sciences of the United States of America*, *103*(12), 4386–4391.

CHAPTER TWO

Guardian of cobamide diversity: Probing the role of CobT in lower ligand activation in the biosynthesis of vitamin B_{12} and other cobamide cofactors

Yamini Mathur[a], Aniket R. Vartak[a], and Amrita B. Hazra[a,b,*]
[a]Department of Biology, Indian Institute of Science Education and Research Pune, Pune, India
[b]Department of Chemistry, Indian Institute of Science Education and Research Pune, Pune, India
[*]Corresponding author: e-mail address: amrita@iiserpune.ac.in

Contents

Methods in Enzymology, Volume 668
ISSN 0076-6879
https://doi.org/10.1016/bs.mie.2022.01.001

Abstract

Enzymes catalyze a wide variety of reactions with exquisite precision under crowded conditions within cellular environments. When encountered with a choice of small molecules in their vicinity, even though most enzymes continue to be specific about the substrate they pick, some others are able to accept a range of substrates and subsequently produce a variety of products. The biosynthesis of Vitamin B_{12}, an essential nutrient required by humans involves a multi-substrate α-phosphoribosyltransferase enzyme CobT that activates the lower ligand of B_{12}. Vitamin B_{12} is a member of the cobamide family of cofactors which share a common tetrapyrrolic corrin scaffold with a centrally coordinated cobalt ion, and an upper and a lower ligand. The structural difference between B_{12} and other cobamides mainly arises from variations in the lower ligand, which is attached to the activated corrin ring by CobT and other downstream enzymes. In this chapter, we describe the steps involved in identifying and reconstituting the activity of new CobT homologs by deriving lessons from those previously characterized. We then highlight biochemical techniques to study the unique properties of these homologs. Finally, we describe a pairwise substrate competition assay to rank CobT substrate preference, a general method that can be applied for the study of other multi-substrate enzymes. Overall, the analysis with CobT provides insights into the range of cobamides that can be synthesized by an organism or a community, complementing efforts to predict cobamide diversity from complex metagenomic data.

Abbreviations

IPTG Isopropyl β-D-1-thiogalactopyranoside
RPM revolutions per minutes
Cba cobamide
PFAM protein family
HMM hidden Markov models
BLAST basic local alignment search tool
HPLC high performance liquid chromatography
LC-MS liquid chromatography mass spectrometry
NMR nuclear magnetic resonance
ERETIC electronic reference to access *in vivo* concentrations
Ni-NTA nickel-nitrilotriacetic acid

1. Introduction

Vitamin B_{12} (cobalamin) is an essential nutrient required by humans for the biosynthesis of methionine and metabolism of methylmalonyl-

CoA to succinyl-CoA (Banerjee, 1997; Roth, Lawrence, & Bobik, 1996). Interestingly, B_{12} is only one of the over 30 naturally produced cobamides used by enzymes as cofactors to catalyze complex reactions in biological systems (Banerjee & Ragsdale, 2003; Giedyk, Goliszewska, & Gryko, 2015). Despite their prevalent requirement in protists, algae, and animals, only a small subset of archaea and bacteria are known to produce cobamides in the natural world (Rodionov, Vitreschak, Mironov, & Gelfand, 2003; Shelton et al., 2018; Zhang, Rodionov, Gelfand, & Gladyshev, 2009).

The biosynthesis of cobamide cofactors follows a modular scheme wherein the three structural components—(i) a tetrapyrrolic corrin ring containing a cobalt ion, (ii) the β-axial/upper ligand, and (iii) a nucleotide loop containing the α-axial/lower ligand—are assembled in a step-wise manner (Fig. 1A) (Mattes, Escalante-Semerena, Deery, & Warren, 2017; Moore & Warren, 2012). The corrin ring synthesis follows the canonical tetrapyrrole biosynthesis pathway with two distinguishing steps—(a) a ring contraction and (b) the insertion of the cobalt ion. The upper ligand, a methyl group or a 5′-deoxyadenosine moiety, is installed on to the central cobalt ion and is directly involved in catalyzing methyl transfer or radical-based rearrangement reactions, respectively (Banerjee, 1997; Brown, 2005). Building the nucleotide loop entails attaching an aminopropanol tail that ends with a phosphoribosylated lower ligand on the corrin ring (Cameron et al., 1991; Maggio-Hall & Escalante-Semerena, 1999). The identity of the lower ligand varies—typically, they are derivatives of benzimidazoles, purines, and phenols (Fig. 1B, C).

Cobamide cofactor diversity arises mainly from two structural features—(i) the absence of a methyl group in the nucleotide loop (these molecules are called norcobamides) (Kräutler et al., 2003) and (ii) a variety of chemically distinct lower ligands. For the purpose of this chapter, we focus on the latter. The lower ligand in vitamin B_{12} is 5,6-dimethylbenzimidazole (DMB), whereas pseudocobalamin (pB_{12}) and phenolylcobamide ([Phe]Cba) are identified by adenine and phenol as their lower ligands, respectively (Fig. 1). Microbe-rich environmental samples are often found to contain a variety of cobamides. For example, extracts from the bovine rumen are abundant in cobalamin and 5-hydroxybenzimidazolylcobamide ([5-OHBza]Cba), human fecal samples predominantly contain 2-methyladeninylcobamide, and a culture from trichloroethylene-enriched groundwater contains high proportions of [5-OHBza]Cba and phenolylcobamide ([Phe]Cba) (Allen & Stabler, 2008; Girard, Santschi, Stabler, & Allen, 2009; Men et al., 2015).

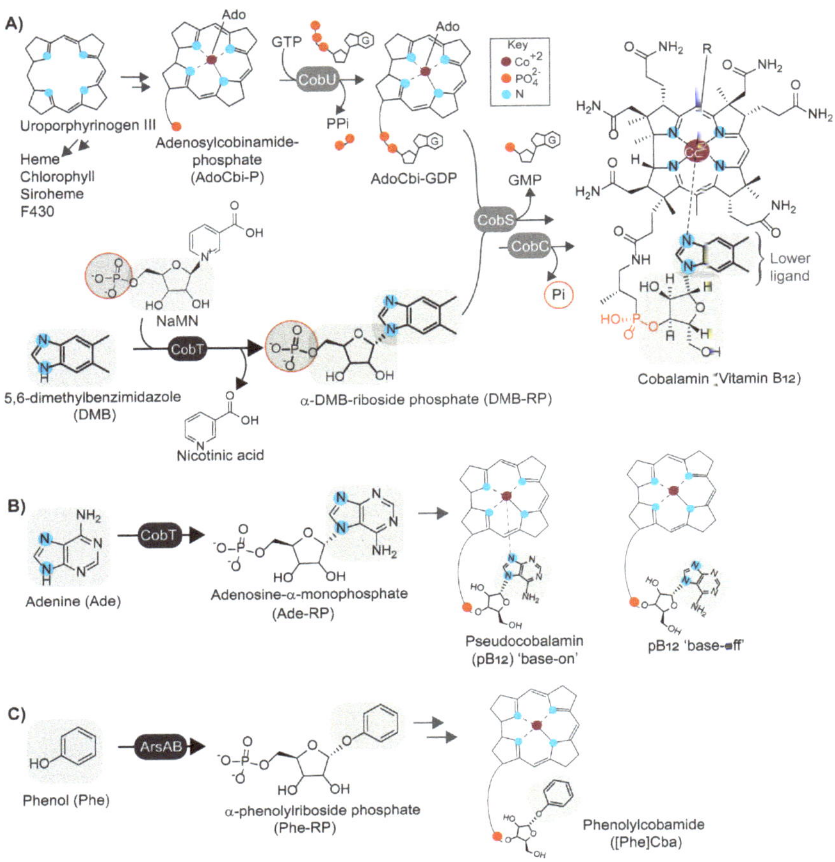

Fig. 1 Overview of the biosynthesis of Vitamin B_{12} and other cobamides and the role of CobT: (A) Vitamin B_{12} biosynthesis is modular—the corrin ring is synthesized, the upper ligand adenosine is attached, then the nucleotide loop is built and the lower ligand is phosphoribosylated. The activated cobinamide (AdoCbi-GDP) is coupled with activated lower ligand (DMB-RP) to finally form cobalamin. This general scheme is conserved across all cobamide biosynthesis where only the lower ligand varies. (B) The formation of pseudocobalamin (pB_{12}) involves CobT-catalyzed activation of adenine (Ade) to form the adenosine-α-monophosphate (α-AMP). The other nitrogen of the Ade ring may coordinate to the corrin cobalt ion to form a 'base-on' pB_{12}, and if it is not coordinated, it forms a 'base-off' pB_{12}. This holds true for all benzimidazolyl and purinyl lower ligands and for any other lower ligand substrate that contains two ring heteroatoms. (C) ArsAB, the class of heterodimeric phosphoribosyltransferases, catalyzes the activation and attachment of phenol and other phenolic/benzyl derivatives as lower ligands to produce phenolyl/benzylcobamides.

Even though the cobamide lower ligand does not directly engage in catalysis, it has been shown to modulate the cofactor's interaction with the enzyme, as well as influence microbial community dynamics of the organisms thriving on cobamide-dependent exchanges (Ma, Beld, & Brahamsa, 2017; Sokolovskaya et al., 2019).

1.1 Lower ligand diversity arises from the late steps of cobamide biosynthesis

Only a small fraction of microorganisms contain the complete set of genes required for *de novo* cobamide biosynthesis (Rodionov et al., 2003; Shelton et al., 2018). Several organisms either import the entire cobamide or conduct partial cobamide biosynthesis through a salvage strategy by acquiring the corrin ring and/or the lower ligand from the environment (Seth & Taga, 2014). Another route involves remodeling an imported cobamide by replacing the lower ligand by another one to make a cofactor that may be favorable for the organism's metabolism (Seth & Taga, 2014). The three routes—*de novo* biosynthesis, salvaging, and remodeling—converge at the nucleotide loop assembly pathway which employs the gene products of the *cobUTSC* operon (Fig. 1A) (Mattes et al., 2017). Specifically, CobU activates adenosylcobinamide-phosphate (AdoCbi-P) by transferring a GMP moiety to the aminopropanol tail to form adenosylcobinamide-GDP (AdoCbi-GDP) (Blanche et al., 1991; Thomas, Thompson, Rayment, & Escalante-Semerena, 2000). In another reaction, CobT independently activates the lower ligands as α-riboside phosphate (α-RP) derivatives. The two activated intermediates, AdoCbi-GDP and the α-RP are combined by the cobalamin-5′-phosphate synthase, CobS (also annotated as *cobV*) (Cameron et al., 1991). Finally, the phosphatase, CobC cleaves the terminal phosphate of the cobalamin-5′-phosphate yielding cobalamin (Zayas & Escalante-Semerena, 2007). Using this pathway, several novel cobamides with natural and unnatural lower ligands have been produced in organisms by providing the corresponding molecule to the growth medium, a process termed as guided biosynthesis (Table 1).

1.2 A historical account of CobT discovery and characterization

The significance of the lower ligand activation step in producing diverse cobamides was apparent with the discovery of the α-phosphoribosyltransferase

Table 1 Diversity of CobT substrates tested *via in vitro* characterization and guided biosynthesis of cobamides.

Class of substrates	R1	R2	R3	R4	Substrate	Guided biosynthesis (G)	*in vitro* characterization (IV)	Primary references
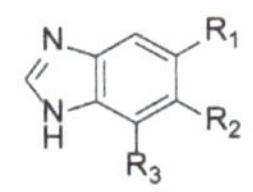 Benzimidazoles	CH_3	CH_3	H		DMB	G	IV	(i) (ii)
	CH_3	H	H		5-MeBza	G	IV	(i) (iii)
	H	H	H		Bza	G	IV	(i) (ii)
	OCH_3	H	H		5-OMeBza	G	IV	(viii) (xii)
	OH	H	H		5-OHBza	G	IV	(viii) (xiv)
	NH_2	H	H		5-NH_2Bza	G	IV	(i) (iii)
	NO_2	H	H		5-NO_2Bza	G	IV	(i) (iv)
	H	H	NH_2		7-NH_2Bza	G		(xvii)
	H	H	CH_3		7-MeBza	G		(xvii)
	Cl	Cl	H		5,6-DCB	G	IV	(i) (iv)
	CF_3	H	H		5-CF_3Bza		IV	(iii)
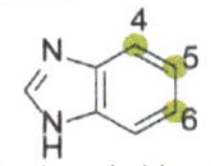 Azabenzimidazoles	**4**	**5**	**6**					
	N	C	C		4-AzaBza	G		(xvi)
	C	N	C		5-AzaBza	G		(xvi)
	C	C	N		6-AzaBza	G		(xvi)
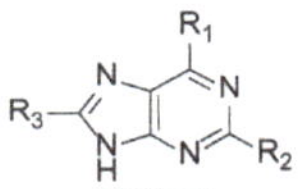 Purines	**R1**	**R2**	**R3**					
	NH_2	H	H		Adenine	G	IV	(i) (iii)
	NH_2	CH_3	H		2-MeAde	G		(vii)
	NH_2	Cl	H		2-Cl-Ade	G		(v)
	NH_2	Cl	Cl		2,8-DCA	G		(i)
	NH_2	SCH_3	H		2-SMeAde	G		(x)
	NH_2	SO_2CH_3	H		2-SO_2MeAde	G		(x)
	NH_2	$SOCH_3$	H		2-SOMeAde	G		(x)
	H	H	H		Purine	G		(v)
	SCH_3	H	H		6-Me-thiopurine	G		(v)
	Cl	H	H		6-Cl-purine	G		(v)
	NH_2	NH_2	H		2,6-DAP	G		(v)
	NH_2	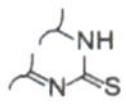	H		2-thioAde	G		(vi)
	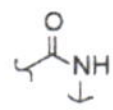	NH_2	H		Guanine	G	IV	(ix)
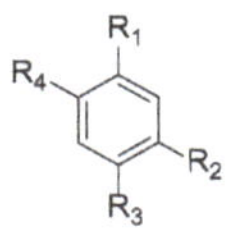 Benzene derivatives	**R1**	**R2**	**R3**	**R4**				
	OH	CH_3	CH_3	H	3,4-dimethylphenol	G		(xv)
	NH_2	H	CH_3	H	p-toluidine	G		(xv)
	OH	H	H	H	Phenol	G	IV	(xv) (xiv)
	OH	H	CH_3	H	p-Cresol	G	IV	(xv) (xiii)
	CH_3	NH_2	NH_2	CH_3	4,5-diMe-1,2-PDA	G	IV	(i) (ix)
	Cl	NH_2	NH_2	Cl	4,5-diCl-1,2-PDA	G		(i)
Others	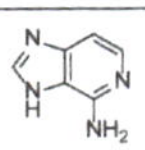 3-deazaAde G (xvii)	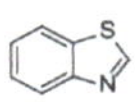 Benzthiazole G (i)	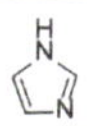 Imidazole G IV (ix)	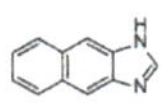 Naphthaimidazole G (viii)	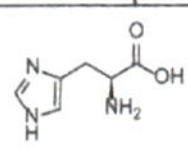 Histidine G IV (ix)	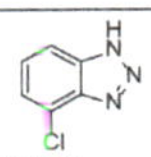 4-ClBtza G (i)		

Table 1 Diversity of CobT substrates tested *via in vitro* characterization and guided biosynthesis of cobamides.—cont'd

Abbreviations: DMB, 5,6-dimethylbenzimidazole; 5-MeBza, 5-methylbenzimidazole; Bza, benzimidazole; 5-OMeBza, 5-methoxybenzimidazole; 5-OHBza, 5-hydroxybenzimidazole; 5-NH_2Bza, 5-aminobenzimidazole; 5-NO_2Bza, 5-nitrobenzimidazole; 7-NH_2Bza, 7-aminobenzimidazole; 7-MeBza, 7-methylbenzimidazole; 5,6-DCB 5,6-dichlorobenzimidazole; 5-CF_3Bza, 5-trifluoromethylbenzimidazole; 4-AzaBza, 4-azabenzimidazole; 5-AzaBza, 5-azabenzimidazole; 6-AzaBza, 6-azabenzimidazole; 2-MeAde, 2-methyladenine; 2-ClAde, 2-chloroadenine; 2,8-DCA, 2,8-dichloroadenine; 2-SMeAde, 2-methylmercaptoadenine, 2-SO_2MeAde, 2-methylsulfonyladenine; 2-SOMeAde, 2-methylsulfinyladenine; 6-Cl-purine, 6-chloropurine; 6-Me-thiopurine, 6-methylthiopurine; 2,6-DAP, 2,6-diaminopurine; 2-thioAde, 2-thioadenine; 4,5-diMe-1,2-PDA, 4,5-dimethyl-1,2-phenylenediamine; 4,5-diCl-1,2-PDA, 4,5-dichloro-1,2-phenylenediamine; 3-deazaAde, 3-deazaadenine; 4-ClBtza, 4-chloro-1-H-benzotriazole

References: (i) Ford, Holdsworth, and Kon (1954), (ii) Friedmann and Harris (1965), (iii) Friedmann (1965), (iv) Fyfe and Friedmann (1969), (v) Kamikubo and Matsuura (1969), (vi) Hayashi and Kamikubo (1971), (vii) Perlman (1971), (viii) Stupperich, Steiner, and Ruhlemann (1986), (ix) Trzebiatowski and Escalante-semerena (1997), (x) Renz, Blickle, and Friedrich (1987), (xi) Krautler, Kohler, and Stupperich (1988), (xii) Cheong et al. (2001), (xiii) Newmister et al. (2012), (xiv) Hazra et al. (2013), (xv) Mok and Taga (2013), (xvi) Schubert et al. (2019), (xvii) Sokolovskaya et al. (2019).

CobT in 1965 (Friedmann & Harris, 1965). The reaction catalyzed by CobT involves the formation of a unique α-glycosidic bond between C1′ of the ribose-phosphate and a heteroatom of the lower ligand substrate (one of the two nitrogens in the case of benzimidazoles and purines). The biochemical validation of the *cobT* gene in this role followed in 1997 (Trzebiatowski & Escalante-semerena, 1997) and since then, CobT homologs from over a dozen organisms have been studied using biochemical and structural approaches and microbiology tools to understand its mechanism of action and how it contributes to cobamide diversity. In 2011 and 2019, two new classes of CobT enzymes were discovered - (i) a CobT-like phosphoribosyltransferase called ArsAB (Aromatic ribotide synthesis) that catalyzes the activation of phenolic molecules by forming an α-glycosidic bond between the C1′ of ribose and the oxygen of phenol to form α-phenolylriboside phosphate (Phe-RP) and (ii) CobT homologs from archaea and cyanobacteria that differ distinctly in sequence and structure from previously characterized CobTs. (Fig. 1C) (Chan & Escalante-Semerena, 2011; Mok & Taga, 2013; Jeter, Mattes, Beattie and Escalante-Semerena, 2019). Once the lower ligand is attached to the cobamide cofactor, benzimidazolyl and purinyl lower ligands may coordinate with the other nitrogen atom in their ring to the corrin cobalt ion to form a 'base-on' cofactor. However, if they do not coordinate, or are unable to coordinate, as in the case for phenolyl lower ligands that do not contain another heteroatom, a 'base-off' cofactor is formed (Fig. 1B, C) (Banerjee & Ragsdale, 2003).

1.3 Functional and molecular features of CobT

Two characteristic features of the CobT family of enzymes that influence cobamide diversity are (i) their inherent substrate preference and (ii) their regioselectivity of activation, which are encoded within sequence and structural features of the homologs (Crofts et al., 2014; Crofts, Seth, Hazra, & Taga, 2013; Hazra, Tran, Crofts, & Taga, 2013). Most CobTs are found to activate multiple substrates, however, the range and relative preference vary across the homologs. For example, *Salmonella enterica* produces pB_{12} as the native cobamide, but produces B_{12} when DMB is available in the growth media (Anderson et al., 2008). However, the *S. enterica* CobT (*Se*CobT) shows a large preference for DMB over adenine when examined by *in vitro* assays as well as heterologous expression (Hazra et al., 2013). Similar studies with CobT homologs from a wide range of microbes reveal DMB as a widely preferred substrate of choice in comparison to most other lower ligand substrates. CobT homologs that prefer DMB also accommodate various benzimidazole derivatives and adenine but interestingly show little or no activity with phenols. In contrast, the ArsAB homologs activate phenols preferentially but also show activity with benzimidazolyl lower ligands (Chan & Escalante-Semerena, 2011; Hazra et al., 2013). Other than these, a surprisingly large range of substrates are also accommodated by CobT homologs as evidenced by *in vitro* analyses and guided biosynthesis efforts (Table 1).

Comparative sequence and structural analyses of *Se*CobT bound to DMB, NaMN, and DMB-riboside phosphate (DMB-RP, also known as α-ribazole) have provided molecular insights into the reaction (Cheong, Escalante-Semerena, & Rayment, 1999, 2002). As per the proposed mechanism, the catalytic residue Glu^{317} abstracts a hydrogen from the protonated nitrogen of DMB and the anion shifts *via* a tautomerization to the other nitrogen, which in turn attacks the C1′ of the NaMN cofactor, thus replacing the nicotinate base and yielding an α-riboside phosphate product (Fig. 1A, 2A). In the case of benzimidazoles and purines, either of the two N atoms present in the five-membered rings can participate in the reaction mechanism, resulting in two distinct product regioisomers if the molecule is asymmetric. That is, DMB is a symmetric molecule hence DMB-RP is the sole product, however, asymmetric substrates such as 5-OMeBza yield two products, namely 5-methoxybenzimidazole-riboside phosphate (5-OMeBza-RP) and 6-OMeBza-RP (Crofts et al., 2014). Interestingly, the two regioisomers are not always synthesized in equal quantities, and their ratio can shed light on the molecular features of the CobT active site that explain the functional nuances of different CobT homologs.

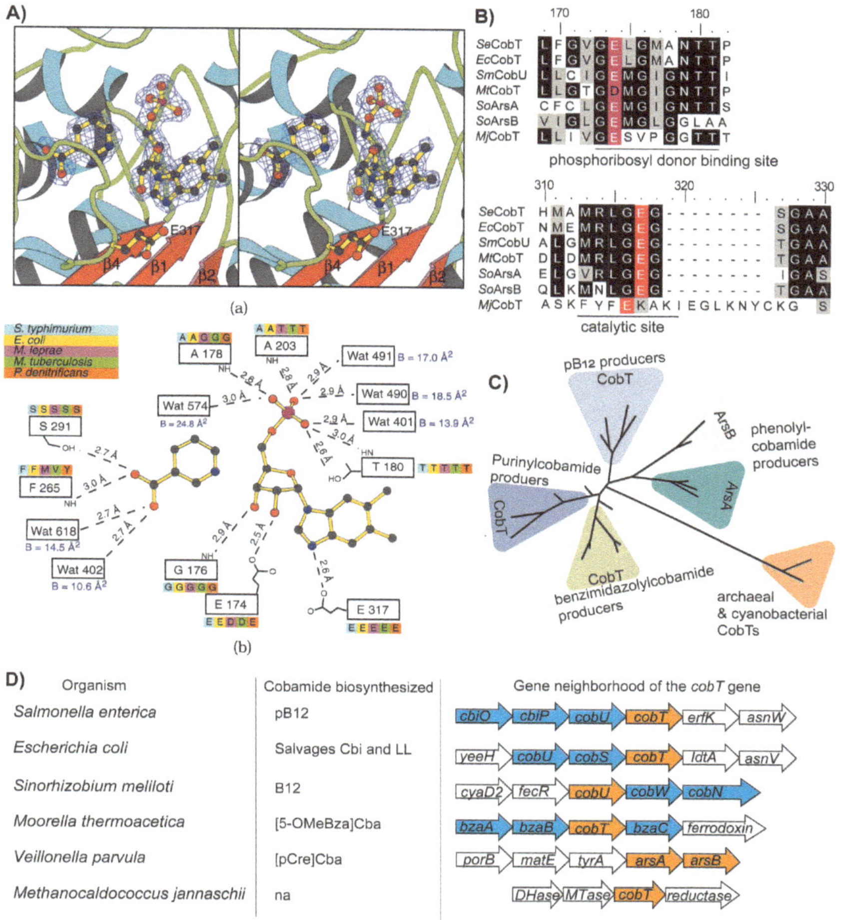

Fig. 2 Bioinformatic traits of CobT homologs (A) A snapshot of the reaction products DMB-RP and nicotinic acid bound in the active site of *Se*CobT (PDB ID 14LE). *(a)* The catalytic residue E^{317} lies in the proximity of the nitrogen of the imidazole ring of the substrate; *(b)* interaction map of the active site residues with the products and the variations in the active site residues across a few CobT homologs are shown (Cheong et al., 1999). (B) The sequence alignment of a subset of CobTs characterized in literature. The conservation of residues in the phosphoribosyl donor site (top panel) and in the catalytic site are shown, and these two regions are characteristic of all CobT homologs we have analyzed so far. The E^{174} and the catalytic residue E^{317} are conserved across all homologs, however, the ArsB sequence show truncation in the cofactor binding site. The sequence of *Mj*CobT show notable differences and is distant from all known bacterial CobT sequences. This method should be used for the sequence analysis of a new CobT for predicting its active site residues. (C) It was found that the distribution of CobT sequences within the clades of a phylogenetic tree is correlated with the nature of native cobamides produced by the organisms. Details of this method are noted in Yan

(Continued)

Fig. 2—Cont'd et al. (2018). A similar tree can be created from various bacteria along with the new CobT being studied. This method will provide insights into the range of substrates to test for activity of a new CobT homolog. (D) The gene neighborhood of *cobT* homologs in several bacterial genomes shows presence of other B_{12} biosynthesis genes and often the *cobUTSC* genes required for nucleotide loop assembly. In anaerobes producing the lower ligand, *cobT* gene is often found localized with the *bza* operon. The aromatic lower ligand activating homologs, *arsA* and *arsB* are always found co-localized and are not necessarily clustered with other cobamide biosynthesis genes. The conservation of *cobT* localization in archaeal genomes in less studied—in this case, we see *MjcobT* encoded in non-cobamide related neighborhood. Such a gene neighborhood analysis will allow for better classification of new CobT homologs. *Panel (A) obtained from Cheong, C. G., Escalante-Semerena, J. C., Rayment, I (1999). The three-dimensional structures of nicotinate mononucleotide: 5,6-Dimethylbenzimidazole phosphoribosyltransferase (CobT) from* Salmonella typhimurium *complexed with 5,6-dimethybenzimidazole and its reaction products determined to 1.9 Å resolution.* Biochemistry, 38*(49), 16125–16135, fig. 7.*

2. Finding new CobT homologs

2.1 Why should we study new CobT homologs?

Biochemical and structural studies with CobT homologs derived from phylogenetically diverse organisms can elucidate the molecular factors that govern choice of substrates. Further, examining candidate *cobTs* from metagenomes of samples that contain cobamides with unique or new lower ligands can identify microbes responsible for producing them. Additionally, experiments are always ongoing to improve the production efficiency of specific cobamides, especially Vitamin B_{12}, using low-cost fermentation conditions for commercial purposes (Balabanova, Averianova, Marchenok, Son, & Tekutyeva, 2021). Introducing appropriate *cobT* homologs in laboratory strains of cobamide producers can aid in the large-scale production of cobamides with natural or synthesized lower ligands. Lastly, CobT catalyzes formation of unique α-riboside derivatives such as α-AMP, distinct from its β-riboside isomer AMP that we are familiar from biochemistry textbooks (Fig. 7). Exploring the CobT reaction mechanism and active site presents us with a potential tool for chemi-enzymatic synthesis of α-ribosides.

2.2 Do all organisms that produce or utilize cobamides encode a *cobT*?

No, many cobamide-utilizing organisms are missing the cobamide biosynthesis genes entirely or partially, and depend on their environment for obtaining the whole cofactor or salvaging the parts that they do not have the genes to produce (Croft, Warren, & Smith, 2006; Rodionov et al., 2003; Shelton et al., 2018). That being said, the *cobT* gene is commonly spread across bacterial and archaeal species and is found in organisms that remodel or salvage cobamides using the *cobUTSC* operon (Ma, Tyrell, & Beld, 2019). The heterodimeric ArsAB enzyme is limited to bacteria from the class Negativicutes of the phylum Firmicutes (Chan & Escalante-Semerena, 2011; Shelton et al., 2018). Finally, a small group of microbes have adopted a CobT-independent pathway to produce cobamides by obtaining activated lower ligands from their environment. For example, *Listeria innocua* imports α-DMB-riboside using a transporter CblT which is subsequently phosphorylated by CblS to form DMB-RP (Gray & Escalante-Semerena, 2010).

2.3 What are ways to find new CobT sequences?

2.3.1 What's in a name?

Searching the databases with the gene name *cobT* results in false hits owing to discrepancies in nomenclature such as *cobT* gene of the *cobNST* operon that encodes the cobaltochelatase. Also, the NaMN:DMB PRT is annotated as *cobU* in certain aerobic organisms which may result in the omission of some sequences from a '*cobT*' search.

2.3.2 Protein databases:

The CobT and ArsAB enzymes are recognized under the same family of proteins as dimethylbenzimidazole phosphoribosyltransferase (NaMN: DMB PRT) and are collectively represented by PF02277 in PFAM, IPR003200 in the InterPro database and EC 2.4.2.21 in the enzyme commission (EC) classification. The archaeal and cyanobacterial CobT sequences differ remarkably from bacterial sequences and are distinctly classified as TIGR00303 in the TIGRFAM database and in InterPro as IPR002805 superfamily. As per the present records, 88% sequences in the PFAM02277 database are single domain DMB-PRT, 8% are fusion proteins with a BluB (annotated nitroreductase involved in DMB biosynthesis) and 2% harbor additional domain(s) for CobS, CobU and/or CobC enzymes. Additionally, there are architectures with repeats of DMB-PRT or fusion enzymes with amidotransferase and alcohol dehydrogenase domains, characterization of which may reveal the occurrence of novel α-glycosidic bond-containing biomolecules. For this chapter, all methods pertain to the single domain DMB-PRT homologs which typically range from 300 to 400 amino acid residues in length.

For using BLASTp (https://blast.ncbi.nlm.nih.gov/Blast.cgi, Altschul, Gish, Miller, Myers, & Lipman, 1990) against bacterial genomes, we recommend the sequences of *Se*CobT, *Sinorhizobium meliloti* (*Sm*CobU) or *Moorella thermoacetica (Mt*CobT) as queries. The archaeal and cyanobacterial sequences can be searched using CobT sequences from *Methanocaldococcus jannaschii* (*Mj*CobT) and *Synechococcus* sp. (Table 3).

2.3.3 Homology based models:

To enable a high-throughput search, we recommend the HMMER web server hosted by EBI (https://www.ebi.ac.uk/Tools/hmmer/) (Potter et al., 2018). The HMMsearch employs hidden Markov models (HMM) that are published on protein databases or can be user-curated for searching the proteomes. The HMMsearch can find distant homologs that otherwise remain undetected when using a single query in a traditional BLASTp search. Additionally, we observe fewer false positive hits with the HMMsearch. The results on HMMER on EBI are accompanied with array of cross-references to multiple databases and literature which are helpful for further analysis.

2.4 What are the hallmarks of CobT sequences?

The detailed structural and biochemical studies of the *Se*CobT, *Sporomusa ovata* (*So*ArsAB), and *Mj*CobT provided the conserved signatures of CobT as well as the features that set the three apart (Cheong et al., 1999; Crofts et al., 2014; Jeter et al., 2019; Newmister et al., 2012).

(i) The *Se*CobT shares 36% and 21% identity with *So*ArsAB and *Mj*CobT respectively. Despite several major differences in the sequences, the active site residues are conserved across archaeal and bacterial CobT sequences (Fig. 2B).

(ii) *Se*CobT and *Mj*CobT exist as homodimers whereas ArsAB exists as a heterodimer. A structure of *Se*CobT with the products DMB-RP and nicotinate shows two equivalent active sites at the interfaces of N-terminal of one subunit and the C-terminal of another subunit. The DMB binds in a partially exposed site lined with hydrophobic residues Leu^{313}/Met^{313}-Arg^{314}-Leu^{315}-Gly^{316}-Glu^{317}-Gly^{318} (L/MRLGEG) which are conserved across bacterial CobTs (Fig. 2A (b) reproduced from Cheong et al., 1999, 2B). The Glu^{174} and Glu^{317} found in proximity to the DMB nitrogens are necessary for *in vivo* activity and Glu^{317} is the catalytic residue (Fig. 2A(a)). The phosphoribosyl donor binds in a cavity defined by a GMANTT consensus sequence (Fig. 2B). In contrast, the ArsAB heterodimer has only one active site that is present at the interface of C-terminal ArsA and the N-terminal ArsB. The ArsB shows a deletion at the NaMN binding site, and a protruding N-terminal loop from ArsA blocks the substrate binding site of ArsB. Hence, despite the presence of the catalytic Glu^{303} in ArsB, its activity is obliterated (Chan et al., 2014; Newmister et al., 2012). In archaeal CobTs, the lack of ligand bound structures means that the placement of the active site in the dimer and residues involved in substrate recognition are currently unknown. Very recently, *Synechococcus* sp. *cobT* was validated as a functional DMB-PRT through complementation in *S. enterica ΔcobT* strain, however the characterization of cyanobacterial CobTs are still in nascent stages. The exploration of the various biochemical features of the lower ligand activation step in archaea and cyanobacteria through the study of CobT will lead to a clearer understanding of cobamide metabolism and diversity in these organisms.

2.5 Is the gene neighborhood of *cobT* important?

In organisms such as *Escherichia coli*, *cobT* is a part of the *cobUTSC* operon whereas other organisms such as *S. enterica, and Sinorhizobium meliloti* have *cobT* adjacent to other cobamide biosynthesis genes (Fig. 2D). The *arsA* and *arsB* genes are typically always found adjacent to one another and contains an overlap of about eight base pairs. However, the gene neighborhood of *M. jannaschii* does not contain any B_{12}-related genes (Fig. 2D). In anaerobic bacteria, *cobT* is found as a part of the *bza* operon which encodes the anaerobic biosynthesis of lower ligands DMB, 5-OHBza, and 5-OMeBza (Hazra et al., 2015). Several organisms also contain multiple *cobT* genes encoded in different gene neighborhoods. The molecular and functional differences and the significance of multiple *cobT* genes encoded within different gene neighborhood in an organism are currently being investigated in our lab.

Recently, we characterized the *cobT* from the *bza* operon of the anaerobe *M. thermoacetica* (*Mt*CobT) (Mathur et al., 2020). The *Mt*CobT exhibits unprecedented regiospecificity and substrate preference for 5-OHBza, contrary to all other CobT homologs characterized previously, providing new light on the previously unexplained regio-specific attachment of DMB in anaerobes. Interestingly, *M. thermoacetica* genome predicts another *cobT* gene in the *cob* operon; however, on closer inspection we find that the gene has been truncated to omit the catalytic residues, thus reducing it to a vestigial gene (data unpublished). Hence, inspecting the gene neighborhood and sequence of the CobT homologs can provide interesting insights into function of this versatile enzyme.

To summarize, the availability of lower ligand substrates in an organism's environment coupled with the substrate promiscuity, preference, and regioselectivity of CobT appear to be the determinants of the cobamide(s) produced by an organism. The cumulative inferences from *in vitro* characterization of CobT homologs mostly concur with the cobamides produced by the native organisms. Hence, the *in vitro* characterization of CobT acts as a reasonable screen for predicting the cobamides produced by an organism. As different organisms tend to demonstrate specificity in their choice of cobamide for cellular metabolism, the cobamides produced and shared within a microbial community ultimately influence the community composition and metabolism (Sokolovskaya, Shelton, & Taga, 2020). Thus, a detailed inquisition into the molecular features of CobT structure and function is an important pursuit. This chapter focuses on methods required to systematically characterize novel CobT homologs.

3. Cloning, overexpression, and enzyme purification of CobT homologs

The optimized method for cloning, heterologous expression, and purification of the CobT enzyme has been detailed previously (Mathur et al., 2020). Some highlights are described below.

3.1 Construction of plasmids encoding *cobT* homologs

The *cobT* homologs are amenable toward cloning and overexpression in *E. coli* strains, which has allowed characterization of over 20 CobTs from bacteria and archaea (Table 3). The majority of *cobT* homologs were cloned in an IPTG-inducible pET28a vector with a N-terminal 6 × histidine (6 × His) tag using either traditional digestion-ligation method or a two-step restriction-free (RF) cloning method (Van Den Ent & Löwe, 2006). The final reaction mix is transformed into *E.coli* DH5α strain and selected on LB agar plates containing 25 μg/mL kanamycin and 0.2%

glucose. The transformants are screened through colony-PCR and gene sequences are confirmed through DNA sequencing.

3.2 Heterologous expression of CobT in *E. coli* BL21(DE3)

Briefly, a 1% overnight culture of the *E. coli* BL21(DE3) pET28a *cobT* strain is inoculated into 1 L LB-Kan broth at 37 °C with shaking at 180 rpm until its $O.D._{600nm}$ reaches 0.6–0.8. Protein expression is induced with 250 μM IPTG and the culture is incubated at 25 °C for 12 h with shaking at 180 rpm. Cells are harvested by centrifugation and stored in −80 °C until further use.

3.3 Purification of CobT using Ni-NTA affinity chromatography and estimating its concentration

3.3.1 Reagents required for protein purification

All buffers contain 50 mM potassium phosphate and 300 mM NaCl, pH 8.0. In addition, lysis buffer contains 5 mM imidazole and 100 μM of the protease inhibitor PMSF, and binding, wash and elution buffers contain 10, 50, and 250 mM imidazole, respectively. The desalting buffer consists of 50 mM potassium phosphate buffer, 150 mM NaCl, 7 μM β-mercaptoethanol at pH 8.0. All reagents and cell lysates should be stored on ice during benchtop purification.

3.3.2 Equipment required

Peristaltic pump (ElectroLab), 5 mL Ni-NTA affinity columns (GE Healthcare Pvt. Ltd.) recharged with 100 mM $NiSO_4{\cdot}6H_2O$, 10 kDa cut-off desalting columns (BioRad laboratories Econo-Pac DG10).

3.3.3 Protein purification

The lysis, loading, wash, and elution was conducted using standard Ni-NTA chromatography and has been described previously (Mathur et al., 2020). To identify fractions with the highest concentration of the eluted protein, we used an inexpensive method involving a paraffin film, Bradford reagent, and SDS-PAGE. On the paraffin layer, spot multiple drops of 10 μL of Bradford reagent and add 5 μL of the eluent fractions. Analyze the fractions that turn dark blue on SDS-PAGE gel and pool the pure fractions (typically, 6 × His-CobT constructs range from 32 to 42 kDa). Finally, the protein is desalted using manufacturer's instructions and stored with 12% glycerol as 200 μL aliquots at −80 °C. The purified CobT shows long shelf life with activity retained even after a year of storage.

3.3.4 Estimating protein concentration

Most CobT homologs studied so far contain tryptophan and/ or tyrosine residues and hence have an absorbance at 280 nm. Thus, to calculate the concentration of the purified protein, record its absorbance (A) at 280 nm as described previously using the Beer-Lambert equation [$A = \varepsilon cl$; A = Absorbance, ε = molar extinction coefficient, c = concentration, l = path length of the cuvette] (Aitken & Learmonth, 2009). The theoretical molar extinction coefficient (ε) of the protein can be estimated using the ExPaSY Protparam tool (Gasteiger et al., 2005). Additionally, the concentration can be measured *via* the Bradford assay (Ernst & Zor, 2010) or also in the case that the CobT homolog lacks tryptophan and tyrosine residues.

4. *In vitro* reconstitution of the phosphoribosyltransferase activity of a CobT homolog

4.1 Obtaining substrates

Materials: The substrates 5,6-Dimethylbenzimidazole (DMB), 5-Methoxybenzimidazole (5-OMeBza), 5-Methylbenzimidazole (5-MeBza), Benzimidazole (Bza), adenine (Ade), Nicotinamide mononucleotide (NMN), Nicotinic acid mononucleotide (NaMN), and Nicotinamide adenine dinucleotide (NAD) were purchased from Merck-Sigma Aldrich. 5-Hydroxybenzimidazole (5-OHBza) was synthesized in-house as described previously (Crofts et al., 2013; Mathur et al., 2020). The modification that we suggest to the previous protocol is that NaBr is a by-product of the neutralization step and poorly soluble in acetonitrile. Thus, we separate the 5-OHBza product from the salt *via* multiple extractions in acetonitrile, evaporate the solvent and proceed with normal phase silica column chromatography to obtain the pure compound. The net yield typically ranges from 30% to 40% and the method can be scaled up to 2–3 times of starting material for higher product yields.

4.2 Choice of the phosphoribosyl donor

The physiological source of the phosphoribosyl group is still unknown and under *in vitro* conditions, CobT utilizes pyridyl biomolecules β-NMN, β-NaMN, β-NAD and β-NaAD as phosphoribosyl donors (Fig. 3, reproduced from Hazra et al., 2013). Several studies summarized here have tested these four molecules. The rate of activity for

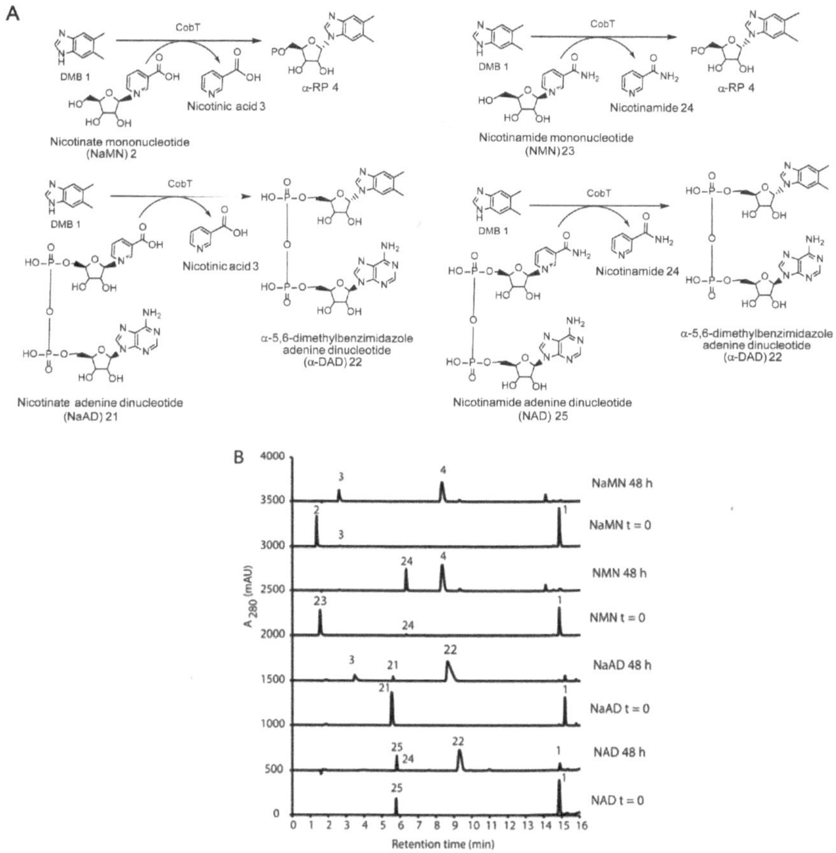

Fig. 3 The diversity of phosphoribosyl donors used for the CobT reaction. (A) NaMN, NMN, NaAD, and NAD are the four phosphoribosyl donors that can be used by CobT along with the lower ligand substrate to produce the corresponding phosphoribosylated products. (B) HPLC analysis of *Sm*CobU with DMB and each of the four phosphoribosyl donors at time points 0 min and 48 h shows that they are utilized to different extents. *Obtained from Hazra, A. B., Tran, J. L. A., Crofts, T. S., Taga, M. E. (2013). Analysis of substrate specificity in CobT homologs reveals widespread preference for DMB, the lower axial ligand of vitamin B_{12}.* Chemistry and Biology, 20*(10), 1275–1285, supplementary fig. S1.*

Propionibacterium freudenreichii CobT (*Pf*CobT) with NaMN is 7-fold higher than its activity with NMN (Friedmann, 1965). The activity of *Clostridium sticklandii* CobT (*Cs*CobT) to activate benzimidazole versus adenine is 860-fold higher with NaMN and 90-fold higher with NMN (Fyfe &

Friedmann, 1969). In the same study, NaAD was shown to support *Cs*CobT-catalyzed activation of benzimidazole 100-fold better than NAD. Similar cofactor preference patterns were recorded with *Se*CobT where the apparent $K_{m(SeCobT:NMN)}$ was 44-fold higher than $K_{m(SeCobT:NaMN)}$, however the net catalysis efficiency (k_{cat}) remains comparable between NMN and NaMN (Trzebiatowski & Escalante-Semerena, 1997). The structure of *Se*CobT bound to the products DMB-RP and nicotinate shows no apparent structural reason for why CobT shows higher affinity for NaMN over NMN (Cheong et al., 1999). Also, *Se*CobT was validated to transfer ADP-ribosyl group from NaAD and NAD to DMB to produce 5,6-dimethylbenzimidazole-α-dinucleotide (α-DAD) which fluxes into the cobamide biosynthesis pathway (Fig. 3). However, the K_m value for *Se*CobT with NaMN is 17 times higher and the catalytic efficiency is ~400-fold higher than that of NAD (Maggio-Hall & Escalante-Semerena, 2003). Later, the *S. meliloti* CobT (*Sm*CobU) was also shown to transfer phosphoribosyl from NaMN, NMN and ADP-ribosyl from NaAD and NAD resulting in DMB-RP and α-DAD respectively (Fig. 3). In general, CobT homologs tested with other benzimidazolyl substrates and adenine showed that only NaMN and NMN gave products for all the bases (Hazra et al., 2013).

For primary characterization of a new CobT homolog, we recommend testing the activity with co-substrates NaMN and NMN. Further biochemical analysis can be conducted with either as the phosphoribosyl donor based on commercial availability and cost-effectiveness of the reagents.

4.3 Reconstitution of the *in vitro* activity of a CobT homolog

The phosphoribosyltransferase activity of a novel CobT homolog is first tested with DMB which is the preferred substrate for all CobT homologs reported so far. Typically, a 250 μL reaction is set up in buffer constituting 50 mM Tris-Cl pH 8.0 with 10 mM $MgCl_2$, 500 μM of DMB, 2 mM NMN or NaMN and 5–20 μM of the purified enzyme. To test the ADP-ribosyltransferase activity, reactions with 5 mM NAD or NaAD are set up in place of NMN or NaMN. In addition to the reaction, three controls are set up without the enzyme, phosphoribosyl donor, and DMB, respectively. The assay mixtures are incubated for 24–48 h at 25 °C. Subsequently, the reactions are quenched with 1% formic acid to precipitate the enzyme which is removed by centrifugation at 14,000 RPM for 20 min at 4 °C. The supernatant is collected, neutralized by

NaOH (final concentration 25 mM), and preliminarily analyzed by thin layer chromatography (TLC solvent system chloroform: methanol:triethylamine in a ration of 18:2:0.1, retention factor of DMB > DMB-RP) followed by reverse phase-HPLC and LC-MS analysis (Fig. 4A, D).

4.4 HPLC, UV and LC-MS analysis of the reconstituted enzyme reaction

The reconstitution of CobT activity is monitored on an Agilent 1260 infinity II UHPLC system coupled with UV–vis and fluorescence spectroscopy. The mass spectrometry experiments are conducted on ScieX X500R-QTOF mass spectrometer system attached to Exion-LC series UHPLC. To enable both HPLC and LC-MS analyses, a reverse phase C-18 column is used with 10 mM ammonium acetate, pH 6.5 (solvent A) and methanol (solvent B) as the mobile phases and a 0.5 mL/min flowrate at 25 °C column temperature. The gradient applied is as follows: 0–2 min 100% A; 2–15 min from 100% A to 70% B; 15–20 min hold at 70% B; 20.5–25 min to 100% B; 25.5–35 min 100% solvent A. The chromatograms at 250, 260, and 280 nm and UV–vis spectra for all peaks are recorded from 190 to 390 nm at a step size of 2 nm. The fluorescence emission of benzimidazoles substrates are recorded at 312 nm upon excitation at 280 nm. The HPLC chromatograms are analyzed on the Agilent OpenLab CDS software.

To monitor the mass of the reactants and products, the LC method is coupled with a quadrupole-time of flight (Q-TOF) mass spectrometer. The MS analysis is conducted in +ESI mode with +4500 V ion spray voltage for benzimidazoles and purines and −ESI mode with −5500 V for phenols with the declustering potential of 80 V and a collision energy of 5 V at 400 °C for both modes of analysis. A two-part MS experiment was optimized that performs a TOF-MS for a mass range of 100 to 1000 Da followed by a targeted Multiple Reaction Monitoring-High Resolution (MRM-HR) analysis using the precursor ion mass and fragment ion mass (Fig. 4E). All LC-MS data was subsequently processed on the ScieX-OS Explorer to obtain extracted ion chromatogram (EIC) and the m/z of the compounds of interest.

The phosphoribosylated product is more polar than the substrate, and hence DMB-RP elutes at 19.9 min and DMB at 21.2 min, whereas the byproduct nicotinamide is observed at 16.8 min (Fig. 4A, top trace). The product is further confirmed by its characteristic UV–vis spectrum

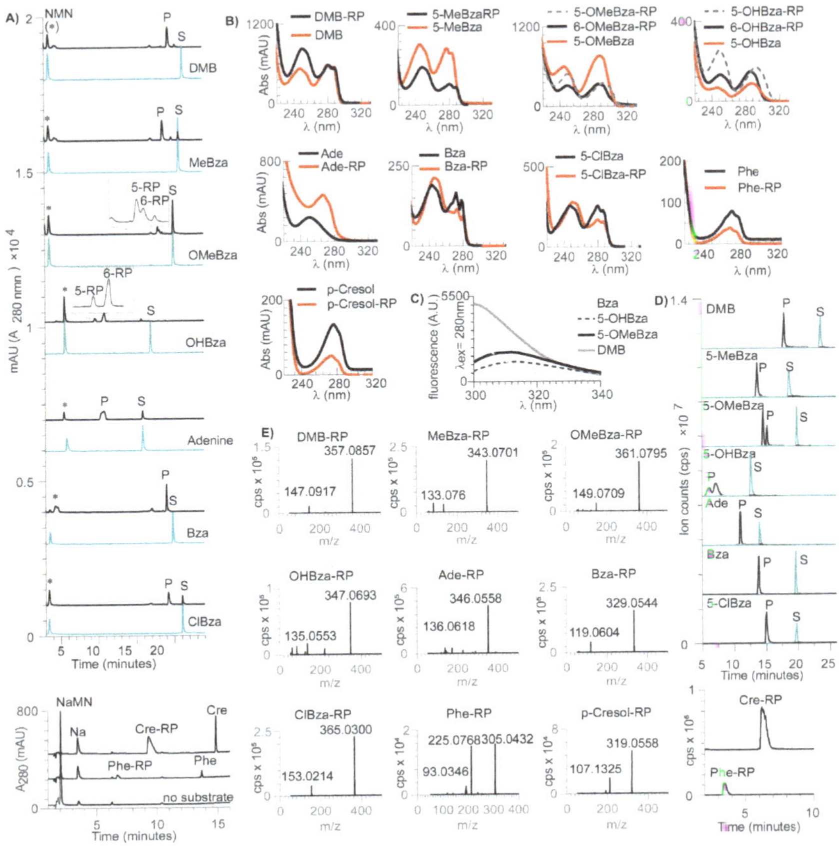

Fig. 4 Characterization of the *in vitro* reconstitution of CobT. (A) The HPLC chromatograms recorded at 280 nm for CobT reactions (black traces) along with a no enzyme control (blue traces). The peaks labeled with 'S' indicates the substrates DMB, 5-MeBza, 5-OMeBza, 5-OHBza, Adenine, Benzimidazole (Bza), and 5-ClBza, respectively and the peaks labeled with 'P' indicates the corresponding product(s) for each CobT reaction. '*' stands for NMN and a peak for the byproduct nicotinamide is visible at ~17 min. The two product peaks from the asymmetric substrates 5-OHBza and 5-OMeBza are the regioisomers, namely 5-OHBza-RP and 6-OHBza-RP, and 5-OMeBza-RP and 6-OMeBza-RP which are indicated as 5-RP and 6-RP in the insets of respective chromatograms. The bottom panel shows the ArsAB reaction of phenol and cresol substrates with NaMN and demonstrates the formation of the corresponding riboside phosphate products and the byproduct nicotinic acid (Na). (B) The characteristic UV–vis spectra of the substrates (red traces) and products (black and gray traces) from each of nine substrates discussed in the chapter are shown here. The differences in the absorbance spectrum helps in identification of product and regiospecificity of the activation. The 5-OHBza-RP and 5-OMeBza-RP show $\varepsilon 250 > \varepsilon 280$ (dashed gray traces), whereas the 6-OHBza-RP and 6-OMeBza-RP show $\varepsilon 250 < \varepsilon 280$ (solid black traces). (C) The benzimidazoles also show fluorescence emission when excited in the range of 275–295 nm. We show emission spectra of Bza, 5-OHBza, 5-OMeBza, and DMB recorded with excitation at 280 nm. Hence, the reaction can also be monitored with HPLC coupled to a fluorescence detector. (D) The extracted ion chromatograms (EIC) for substrates (blue trace) and the product (black trace) from individual CobT reactions analyzed on a LC-MS experiment are shown here, all recorded in ESI+ (positive) mode. The bottom panel shows the EIC of the Phe-RP and Cre-RP products and is in the ESI− (negative) mode. (E) The *m/z* values for the nine CobT reaction products are shown here.

(Fig. 4B, first trace), fluorescence (Fig. 4C), LC-MS (Fig. 4D, top trace), and *m/z* values (Fig. 4E, first trace).

5. Assessing the substrate promiscuity of CobT

The CobT family of enzymes can collectively activate a wide range of substrates, however, individual homologs can activate only a subset of them. Most homologs that activate DMB can also accommodate other benzimidazole derivatives such as 5-methylbenzimidazole (5-MeBza), 5-OHBza, 5-OMeBza, and 5-cholorbenzimidazole (5-ClBza) (Cheong, Escalante-Semerena, & Rayment, 2001; Hazra et al., 2013). Among the characterized CobT enzymes, adenine remains a poorly favored substrate and assays with commonly found purine lower ligands 2-MeAde, 2-SMeAde are challenging due to limitations in synthesizing/producing these molecules (Fyfe & Friedmann, 1969; Hazra et al., 2013). Among substrates for ArsAB enzymes, *in vitro* assays have been conducted with phenol, cresol, and DMB (Chan & Escalante-Semerena, 2011; Hazra et al., 2013). An extensive study to explore the phenol derivatives that can be accommodated as substrates remains to be done. Also, as archaeal CobTs were the last to be discovered, our understanding of substrate promiscuity in these enzymes is still an open field.

To determine the substrate promiscuity, reactions are usually tested with 500 μM substrate, 2 mM NMN, 5–20 μM purified CobT in buffer containing 50 mM Tris-Cl, and 10 mM $MgCl_2$ at 25 °C for 48 h. For each substrate, we set up controls lacking the NMN, enzyme, and the substrate under same conditions as described for full reactions. The reactions are monitored by HPLC and LC-MS as described above, and all characteristic parameters of the products are determined (Fig. 4).

One can also determine the molar extinction coefficient ε of the CobT product using the Beer-Lambert's law equation $A = \varepsilon cl$. To do so, one requires the absorbance (A) and the concentration (c) of the molecule being studied. To determine A, the product is purified by HPLC and analyzed in a UV–vis spectrophotometer and c is determined by subjecting the same sample to ERETIC NMR using benzoic acid as the standard (Akoka, Barantin, & Trierweiler, 1999; Hazra et al., 2013). Using these values, the molar extinction coefficient of each ribosylated lower ligand may be determined. The ε values for some of the riboside derivatives of CobT products are provided (Table 2).

Table 2 Molar extinction coefficients for some of the phosphoribosylated CobT products measured in H_2O at 280 nm.

No.	Phosphoribosylated products	Extinction coefficient in H_2O at 280 nm ($M^{-1}cm^{-1}$)
1.	DMB-RP	6438
2.	5-MeBza-RP	4941
3.	5-OHBza-RP	4443
4.	6-OHBza-RP	7582
5.	5-OMe-RP	4875
6.	5-OMe-RP	9595
7.	Bza-RP	4787
8.	Phe-RP	879
9.	p-Cre-RP	730

6. Analysis of regioselectivity of lower ligand activation using asymmetric lower ligands 5-OMeBza and 5-OHBza as examples

The reactions of CobT with asymmetric lower ligands 5-OHBza and 5-OMeBza have revealed methylation-sensitive regioselectivity while forming the ribosylated product(s) (Crofts et al., 2014; Mathur et al., 2020). For example, *E. coli* CobT activates 5-OMeBza in a regiospecific manner and forms only 5-OMeBza-RP product (Mathur et al., 2020). However, no regioselectivity is seen with 5-OHBza, and both 5-hydroxybenzimidazolylriboside-phoshate (5-OHBza-RP) and 6-OHBza-RP have been observed. Similarly, *Mt*CobT shows regiospecificity with 5-OHBza but not with 5-OMeBza (Mathur et al., 2020). The molecular factors that determine these differences in regioselectivity with the two substrates require extensive comparative structural and biochemical studies.

The regioselectivity of CobT homologs can be determined from reconstitution of phosphoribosyltransferase activity with 5-OHBza or 5-OMeBza or any other asymmetric substrate. We illustrate the experiments with a literature example of 5-OMeBza reacted with *Se*CobT, *Vp*CobT, and *Sm*CobU (Fig. 5, obtained from Crofts et al., 2014). The two products,

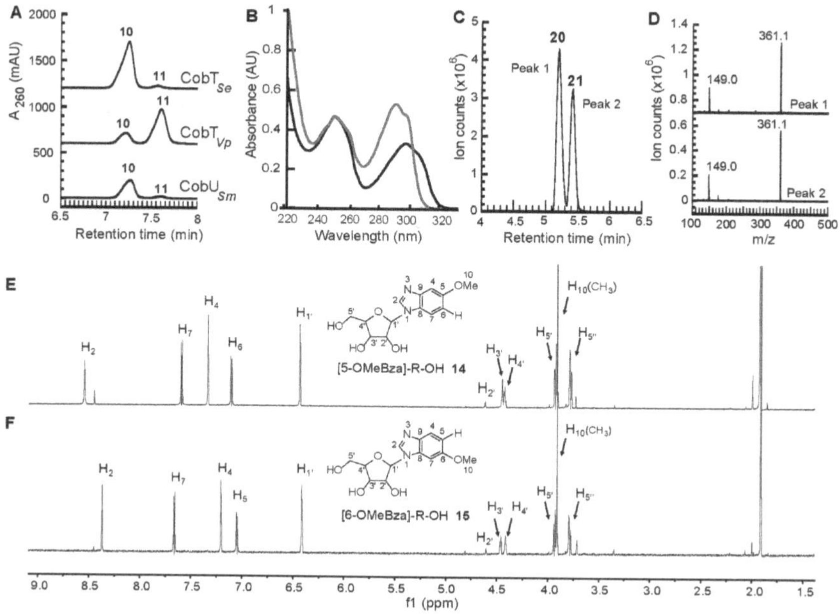

Fig. 5 Characterization of the two regioisomeric products formed by CobT with 5-OMeBza as the substrate. (A) HPLC chromatogram of the reaction of 5-OMeBza with CobT homologs from *S. meliloti, Veilonella parvula* and *S. enterica* show that different enzymes produce different amounts of the two isomers. (B) The UV–vis profile of the two isomeric products show two peaks that are distinctly different from one another—5-OMeBza-RP (black) shows an absorbance maximum at 253 nm and 286 nM, whereas 6-OMeBza-RP shows absorbance maxima at 253 and 300 nm. (C) LC-MS extracted-ion chromatogram for the mass 361.10 shows two distinct peaks, Peak 1 and Peak 2, for the two isomers. (D) *m/z* for the Peak 1 and Peak 2 are identical (E) 1D ^{1}H NMR for the dephosphorylated product 5-OMeBza-R-OH. (F) 1D ^{1}H NMR for the dephosphorylated product 6-OMeBza-R-OH. *Obtained from Crofts, T. S., Hazra, A. B., Tran, J. LA, Sokolovskaya, O. M., Osadchiy, V., Ad, O., et al. (2014). Regiospecific formation of cobamide isomers is directed by CobT.* Biochemistry, *53(49), 7805–7815, fig. 2.*

5-OMeBza-RP and 6-OMeBza-RP elute at 7 min and 7.5 min respectively (Fig. 5A). Additionally, each product has a distinctive non-overlapping UV–vis spectrum: For 5-OMeBza-RP, the absorbance (A) at 250 nm $A_{250} > A_{280}$, whereas 6-OMeBza-RP has $A_{280} > A_{250}$ and identical mass spectrum (Fig. 5B, C, D). The 1D ^{1}H NMR spectrum of the two products show differences in the chemical shifts of peaks from the ribose and the benzimidazole rings (Fig. 5E, F). Further, 2D NMR spectra of the two products from 5-OMeBza confirmed the identity of the 5-OMeBza-RP and 6-OMeBza-RP isomers (Crofts et al., 2014).

7. A pairwise substrate competition assay to determine substrate preference

Enzymes are typically known to be specific for the substrate they use and the reaction they catalyze. However, many CobT homologs can accept a range of substrates to form the corresponding phosphoribosylated product (Table 1). Multi-substrate enzymes not only accommodate several substrates but usually exhibit an innate preference for one substrate over another. This feature is beneficial in efficiently producing a range of similar natural products using the same enzyme machinery. When an enzyme encounters a choice of substrates, the preference depends on the relative affinity of the enzyme for the various substrates, the relative substrate concentrations, and the rate at which they are turned over. Traditionally, determining substrate preference of a multi-substrate enzyme relies on comparing kinetic parameters using individual substrates (Deng, Mao, Wang, Zou, & Ye, 2017; Kuo, Henry, & Andrews, 2016). Apart from being tedious, this method fails to inform on substrate preference when encountering a choice similar to what would be present under physiological conditions. Ideally, one would like to conduct a competition among all the substrates added in one reaction mixture and monitor the rates of substrate turnover to determine relative preference. However, such a method would be limited by the ability to resolve the all substrates and products using reliable analytical techniques (in the case of CobT, reverse-phase HPLC). Here, we describe a pairwise substrate competition assay that allows us to probe relative substrate preferences of CobT and can be generalized for any other multi-substrate enzyme. The pairwise competition involves adding two substrates and monitoring the percent turnover at a time-point when the enzyme just starts to show activity. This assay can be extended to multiple substrates subject to the ability to quantify the products in a reliable analytical manner. A series of reactions is set up with all combinations of substrate pairs and comparison of activity in each case informs on the order of substrate preference.

7.1 Experimental method for pairwise competition assay

In the pairwise competition assays with CobTs, we illustrate here the detailed process and calculations using a literature example (Hazra et al., 2013). Reactions are conducted with all combinations (C) of DMB, 5-MeBza, 5-OHBza, 5-MeBza, Bza, Ade lower ligands, i.e., for comparing preference among six substrates, we set up a series of $^{6}C_{2} = 15$ reactions. The stepwise method for competition between each pair of substrates is described as follows:

1. In a fresh microcentrifuge tube, prepare equimolar mixtures of the substrates, A and B; this mix will be used as a stock solution for substrates
2. Add 2 μL of 90% formic acid each in three fresh tubes for acid-quenching the reactions
3. Prepare the reaction mixture containing 50 mM Tris-Cl pH 7.2, 10 mM $MgCl_2$, 250 μM each of substrate-A and substrate-B, 2 mM NMN and lastly, 1 μM purified enzyme in a volume of 250 μL on ice using pre-chilled reagents
4. Promptly aliquot 50 μL of the reaction mix into one of the acid-treated tubes as the 'zero time point' sample
5. Incubate the reaction in at 25 °C
6. Aliquot 50 μ L from the reaction mix after 0 min, 1 min and an endpoint at 48h or 72h of incubation. In case when no activity is observed at 1 min, the sample can be collected at a later point, say 5 or 10 min, when the enzyme starts to show turnover with at least one of the substrates.
7. Centrifuge the acid-quenched samples at 14,000 rpm at 4 °C for 20 min and collect the supernatant for HPLC analysis
8. Repeat the experiment with all other combinations of the substrates. In reactions where the α-riboside phosphate product was not detectable at 1 min, the concentration of the substrate(s) was increased to 5 mM

7.2 Sample analysis and quantitation

1. To quantify the results from the competition experiments, we generate standard curves of DMB-RP, 5-MeBza-RP, 5-OHBza-RP, 5-MeBza-RP, Bza-RP, and Ade-RP in the range of 1000 pmol to 10 nmol. We determined the concentration of the pure standard of the phosphorylated lower ligand using ERETIC NMR against a standard of 50 mM nicotinic acid and then use the same sample to generate a standard curve on HPLC by plotting the area under the curve at 280 nm (Akoka et al., 1999; Hazra et al., 2013)
2. The samples for each competition are analyzed using the HPLC method described earlier. The concentration of products is derived by interpolating the area under curve for the products peaks from the corresponding standard curves. In the example shown in Fig. 6, a pairwise competition is conducted with DMB and 5-MeBza. In the HPLC chromatogram for 0 min, we see the substrates DMB (peak labeled 1) and 5-MeBza (peak labeled 7) at 14.5 and 15 min, respectively, and the products for either of them are not observed yet (Fig. 6A, obtained from Hazra et al., 2013). At 1 min, we start to see the

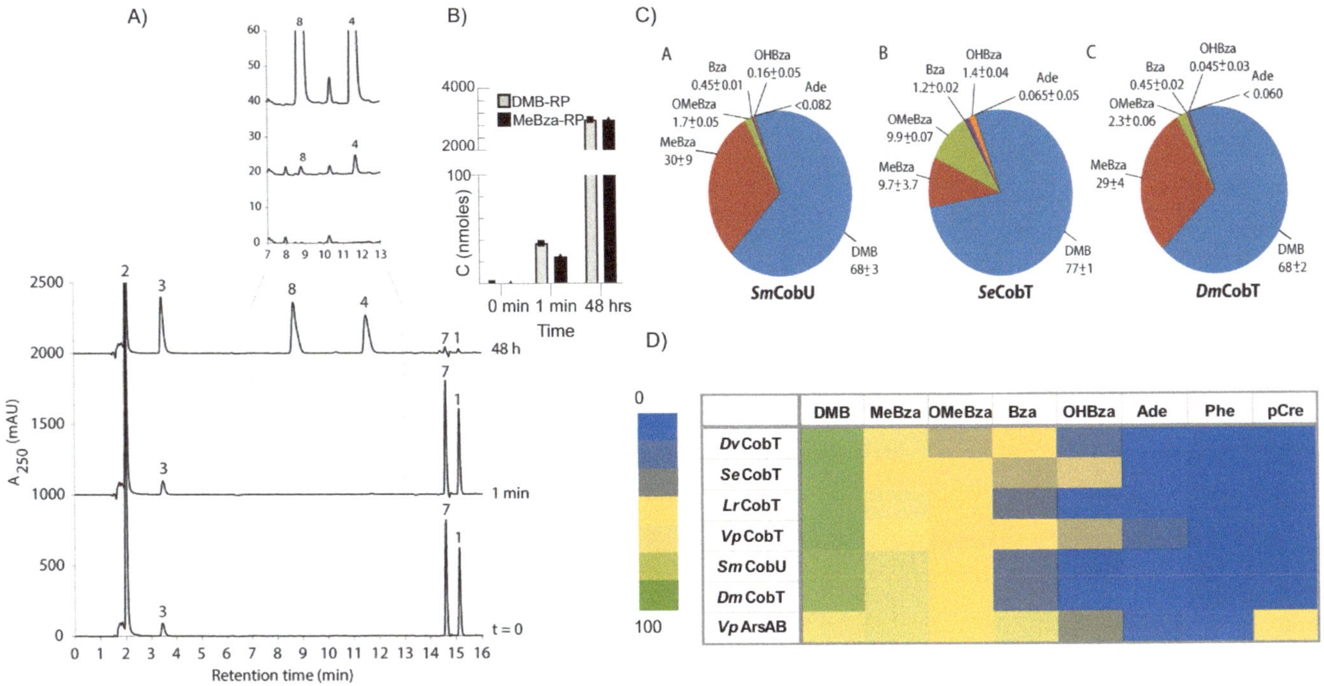

Fig. 6 Pairwise substrate competition assay as a tool to probe relative substrate preference of CobT homologs. (A) HPLC analysis of competition reaction between DMB (peak 1) and 5-MeBza (peak 7) with *Sm*CobU with the amplified inset showing products of the reaction at indicated time points. (B) Bar graph indicating concentration of products DMB-RP and 5-MeBza-RP at indicated time points. This is a step toward calculating the final relative substrate preference of *Sm*CobU. (C) Representative pie-charts of three enzymes to visualize relative substrate percent preference for the individual enzymes tested. (D) Heat map derived from relative percent preferences of the seven CobT enzymes tested to compare overall relative substrate preferences between the different homologs. *Some parts of this figure are obtained from Hazra, A. B., Tran, J. L. A., Crofts, T. S., Taga, M. E. (2013). Analysis of substrate specificity in CobT homologs reveals widespread preference for DMB, the lower axial ligand of vitamin* B_{12}. Chemistry and Biology, *20(10), 1275–1285, figs. 4 and 5.*

build-up of the products of both DMB and 5-MeBza (inset, peaks labeled 4 and 8 represent DMB-RP and 5-MeBza-RP respectively), which are quantified using the corresponding standard curves. On quantitating the reaction progress at 1 min, we find that turnover with DMB is higher than that with 5-MeBza. After 48 h, both substrates are completely consumed (Fig. 6B).

7.3 Data analysis

To calculate the relative preference for each substrate against the other substrates, concentrations of product formed at 1 min of each reaction are used for calculations as follows:

In a pairwise competition between substrate *A vs* substrate *B* in presence of enzyme *E*,

$$A + B \xrightarrow{E} A' + B'$$

The molar ratio (fold-preference) for *A* w.r.t. to *B* is calculated as, $R_{AB} = \frac{[A']}{[B']}$; where $[A']$ and $[B']$ are the concentrations of products at 1 min. In the example shown in Fig. 6, DMB is substrate *A* and 5-MeBza is substrate *B*, and the corresponding products DMB-RP is A' and 5-MeBza-RP is B'. The molar ratio (fold-preference) was found to be $R_{AB} = \frac{[DMB-RP]}{[5-MeBza-RP]} = 8.5$, and informs that DMB is preferred 8.5-fold over 5-MeBza as a substrate.

The next parameter is the molar ratio percent preference (R^{100}) which is the reciprocal of the molar ratio, R_{BA}, × 100, i.e., for a *A vs B* competition, $R^{100}_{BA} = \frac{1}{R_{AB}} \times 100 = \frac{[B']}{[A']} \times 100$ and indicates the moles of B' formed per 100 mol of A'.

In the example data, we normalize molar ratios of product to per 100 mol of DMB-RP, hence when DMB is competed against itself we get, $R^{100}_{AA} = \frac{[DMB-RP]}{[DMB-RP]} \times 100 = 100 \text{ mol of DMB} - \text{RP}$.

Next, when 5-MeBza is competed against DMB, we get:

$$\begin{aligned} R^{100}_{BA} &= \frac{[5 - MeBza - RP]}{[DMB - RP]} \times 100 \\ &= 11.76 \text{ mol of } 5 - \text{MeBza} - \text{RP isproduced per 100 mol of DMB} \\ &\quad - \text{RP produced.} \end{aligned}$$

We then calculate the molar ratio percent preference for the remaining combinations of the competed substrates. Hence, in a scenario where a competition assay is conducted with four substrates, we calculate the preference of substrate A w.r.t. substrates B, C, D, that is, we additionally calculate the $R^{100}_{CA} = \frac{[C']}{[A']} \times 100$, $R^{100}_{DA} = \frac{[D']}{[A']} \times 100$ values.

Next, we derive the relative percent preference for substrate A w.r.t. to the rest of substrates collectively, which reflects how the enzyme would use substrate A when a mixture of all four substrates would be present. Relative% preference for A $(P^A) = \frac{R^{100}_{AA}}{\left(R^{100}_{AA} + R^{100}_{BA} + R^{100}_{CA} + R^{100}_{DA}\right)} \times 100$. Since, $R^{100}_{AA} = 100$; the above equation can be simplified as, $(P^A) = \frac{100}{\left(100 + 100\left(\frac{[B']}{[A']} + \frac{[C']}{[A']} + \frac{[D']}{[A']}\right)\right)} \times 100$ which equals $(P^A) = \frac{1}{\left(1 + \frac{[B']}{[A']} + \frac{[C']}{[A']} + \frac{[D']}{[A']}\right)} \times 100$; where $[B']$, $[C']$, $[D']$ are the concentrations of products for B, C, and D in individual competitions with A. As it is apparent, the P^A can be derived without converting the preference to percent preference in previous steps. However, the precent preference values allow us a more tangible interpretation, that is, one can think that for every 100 mol of A' being made 11.7 mol of B' are made by the enzyme.

Similarly, to calculate the preference of B w.r.t. A, C, D substrates, we determine $R^{100}_{AB} = \frac{[A']}{[B']} \times 100$; $R^{100}_{CB} = \frac{[C']}{[B']} \times 100$; and $R^{100}_{DB} = \frac{[D']}{[B']} \times 100$, where $[A']$, $[B']$, $[C']$, and $[D']$ are the concentrations of products of A, B, C, and D, respectively from the competition reactions between A and B, B and C, and B and D.

Thus, the relative percent preference for substrate B w.r.t. A, C, D would be:

$$
\begin{aligned}
(P^B) &= \frac{R^{100}_{BB}}{\left(R^{100}_{AB} + R^{100}_{BB} + R^{100}_{CB} + R^{100}_{DB}\right)} \times 100 \\
&= \frac{1}{\left(\frac{[A']}{[B']} + 1 + \frac{[C']}{[B']} + \frac{[D']}{[B']}\right)} \times 100
\end{aligned}
$$

Based on these formulae, the molar ratios, molar ratio percent preference, and the relative percent preference values can be calculated for substrates C and D, respectively. The P^A, P^B, P^C, P^D values add up to a 100 and can be analyzed as parts of whole values.

In this example, pairwise competition assays were conducted with DMB, 5-MeBza, 5-OMeBza, 5-OHBza, Bza, and Ade as substrates to compare the intrinsic substrate preference for seven CobT homologs. The molar ratios were calculated for each pair of substrates as described above (Hazra et al., 2013). Finally, the relative percent preference for each substrate, i.e., P^{DMB}, P^{Ade}, P^{MeBza}, P^{OHBza}, P^{OMeBza}, P^{Bza} were plotted on a pie-chart (Fig. 6C) (Hazra et al., 2013). For a comparative visualization of the variations in the substrate preference of the different homologs tested in this way, a heat map can ultimately be created (Fig. 6D). The information we get from this analysis is the following: (i) DMB is widely the most preferred substrate for seven CobT homologs (ii) There exist subtle nuances in preferences for 5-MeBza, 5-OMeBza, and 5-OHBza which hints toward variations in methylation sensitivity of each CobT homolog (iii) Notably, adenine remains a poorly favored substrate by a large margin, which explains why pB_{12} biosynthesis is suppressed in most organisms when any other benzimidazolyl lower ligand is available.

8. Conclusions

In this chapter, we have discussed in detail the steps involved in characterizing CobT, a multi-substrate enzyme that is primarily responsible for shaping cobamide diversity. This being said, we recognize that the other enzymes involved in nucleotide loop assembly (CobS, CobU), lower ligand salvage (CblT, CblS), and remodeling cobamides (CbiZ, CbiR) are also possible contributors to cobamide diversity.

A characteristic property of CobT is its ability to produce unique α-riboside derivatives, which vary from their more commonly occurring β-counterparts (Fig. 7) (Friedmann & Harris, 1965; Ni et al., 2019). The presence of an α-linkage in vitamin B_{12} was revealed by the inability of β-nucleosidases to cleave the lower ligand from the corrin, confirmed in later years by X-ray crystallography (Friedmann & Harris, 1965; Lenhert & Hodgkin, 1968). Owing to the higher stability, and proposed antitumor and antiretroviral properties of α-ribosides, they are being currently investigated for broader applications (Ni et al., 2019). The ability of CobT homologs to form α-ribosidic linkages with a varied range of substrates can contribute efficient chemi-enzymatic methods to synthesize interesting α-riboside derivatives.

We conclude this chapter with a comprehensive list of all CobT homologs studied using *in vitro* methods (Table 3). Each CobT homolog demonstrates certain unique properties which may have been analyzed by techniques different from what we describe here, and we urge our readers to refer to these excellent accounts. We also believe that the bioinformatics and the pairwise competition assay described here can be a starting point for the identification and characterization of other novel multi-substrate enzymes.

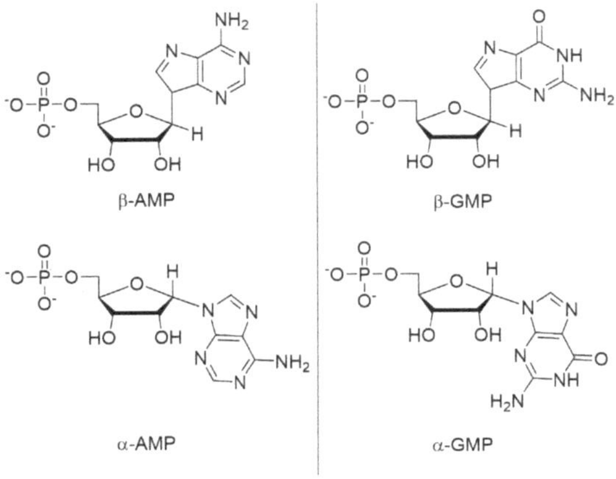

Fig. 7 α- and β-anomeric isomers of AMP and GMP: β-AMP and β-GMP are commonly found in biological systems, and the corresponding α-isomers are synthesized by CobT.

Table 3 A chronological list of the CobTs characterized through *in vitro* characterization, biochemical analysis, and/or structural analysis.

No.	Organism	References
1	*Propionibacterium freudenreichii subsp. shermanii;* D7GJ95 (a BluB/CobT2 fusion protein)	Friedmann and Harris (1965) and Deptula et al. (2015)
2	*Clostridium sticklandii* (*Acetoanaerobium sticklandii*); E3PUT2	Fyfe and Friedmann (1969)
3	*Pseudomonas denitrificans (nom. rej.);* A0A5J6R3E4	Cameron et al. (1991)
4	*Salmonella enterica serovar typhimurium;* Q05603 (PDB: 1L4E)	Trzebiatowski and Escalante-semerena (1997) and Cheong et al. (1999)
5	*Thermus thermophilus HB8;* Q7SIC7 (PDB: 1WX1)	Asada and Kunishima (2005)
6	*Sporomusa ovata (ArsAB);* F6MZ55, F6MZ55 (PDB: 4HDN)	Newmister, Chan, Escalante-Semerena, and Rayment (2012)
8	*Pyrococcus horikoshii;* (PDB: 3U4G)	Cuff et al. (2011)

Table 3 A chronological list of the CobTs characterized through *in vitro* characterization, biochemical analysis, and/or structural analysis.—cont'd

No.	Organism	References
9	*Sinorhizobium meliloti;* P29935	Hazra et al. (2013)
10	*Lactobacillus reuteri;* Q0GL53	
11	*Desulfovibrio vulgaris;* Q725Z4	
12	*Dehalococcoides mccartyi;* A5FRG9	
13	*Veillonella parvula* CobT; D1BPH1	
14	*Veillonella parvula* ArsAB; D1BP26, D1BP27	
15	*Yersinia enterocolitica;* A1JTP8 (PDB: 6B5F)	Stogios et al. (2017)
16	*Desulfitobacterium hafniense strain Y51;* Q24VN9	Yan et al. (2018)
17	*Methanocaldococcus jannaschii;* Q58993 (PDB: 6PTF)	Jeter, Mattes, Beattie, and Escalante-Semerena (2019)
18	*Synechococcus sp.;* A4CRQ3	
19	*Esherichia coli;* P36562	Mathur, Sreyas, Datar, Sathian, and Hazra (2020)
20	*Moorella thermoacetica;* Q2RHR6	
21	*Eubaterium limosum*	*Unpublished data*
22	*Eubacterium barkeri*	
23	*Desulfofundulus thermosubsterraneum*	
24	*Desulfobulbous mediterraneus*	

The links to the UniProt entries for previously reported CobTs and PDB identifiers for the CobT with solved crystal structures are provided next to the name of organisms.

Acknowledgment

A.B.H. acknowledges the Ramalingaswami Re-entry fellowship received from the Ministry of Science and Technology, Government of India Department of Biotechnology (DBT)—BT/RLF/Re-entry/12/2014, Early Career Research Award ECR/2016/000466 received from Department of Science and Technology (DST), Ministry of Science and Technology, India - Science and Engineering Research Board, and IISER Pune for support. Y.M. is supported by a fellowship from IISER Pune and a Council for Scientific and Industrial Research (CSIR)-Senior Research Fellowship, and A.R.V. is supported by an IISER Pune fellowship. We would also like to thank Michiko E. Taga for use of data generated in her laboratory, Sheryl Sreyas for permission to use her data, and Vinayak Mathur for help with formatting equations.

References

Aitken, A., & Learmonth, M. P. (2009). Protein determination by UV absorption. In J. M. Walker (Ed.), *Springer protocols handbooks. The protein protocols handbook* (pp. 3–6). Totowa, NJ: Humana Press.

Akoka, S., Barantin, L., & Trierweiler, M. (1999). Concentration measurement by proton NMR using the ERETIC method. *Analytical Chemistry, 71*(13), 2554–2557.

Allen, R. H., & Stabler, S. P. (2008). Identification and quantitation of cobalamin and cobalamin analogues in human feces. *American Journal of Clinical Nutrition, 87*(5), 1324–1335.

Altschul, S. F., Gish, W., Miller, W., Myers, E. W., & Lipman, D. J. (1990). Basic local alignment search tool. *Journal of Molecular Biology, 215*(3), 403–410.

Anderson, P. J., Lango, J., Carkeet, C., Britten, A., Kra, B., Hammock, B. D., et al. (2008). One pathway can incorporate either adenine or dimethylbenzimidazole as an α-axial ligand of B_{12} cofactors in *salmonella enterica*. *Journal of Bacteriology, 190*(4), 1160–1171.

Asada, Y., & Kunishima, N. (2005). *RCSB PDB—1WX1: Crystal structure of nicotinate-nucleotide-dimethylbenzimidazole phosphoribosyltransferase from Thermus thermophilus HB8.* Retrieved from https://www.rcsb.org/structure/1WX1.

Balabanova, L., Averianova, L., Marchenok, M., Son, O., & Tekutyeva, L. (2021). Microbial and genetic resources for cobalamin (vitamin B_{12}) biosynthesis: From ecosystems to industrial biotechnology. *International Journal of Molecular Sciences, 22*(9), 4522.

Banerjee, R. (1997). The Yin-Yang of cobalamin biochemistry. *Chemistry and Biology, 4*(3), 175–186.

Banerjee, R., & Ragsdale, S. W. (2003). The many faces of vitamin B_{12}: Catalysis by cobalamin-dependent enzymes. *Annual Review of Biochemistry, 72*, 209–247.

Blanche, F., Debussche, L., Famechon, A., Thibaut, D., Cameron, B., & Crouzet, J. (1991). A bifunctional protein from *Pseudomonas denitrificans* carries cobinamide kinase and cobinamide phosphate guanylyltransferase activities. *Journal of Bacteriology, 173*(19), 6052–6057.

Brown, K. L. (2005). Chemistry and enzymology of vitamin B_{12}. *Chemical Reviews, 105*(6), 2075–2149.

Cameron, B., Blanche, F., Rouyez, M., Bisch, D., Famechon, A., Couder, M., et al. (1991). Genetic analysis, nucleotide sequence, and products of two *Pseudomonas denitrificans cob* genes encoding nicotinate- nucleotide: Dimethylbenzimidazole Phosphoribosyltransferase and cobalamin (5'-phosphate) synthase. *Journal of Bacteriology, 173*(19), 6066–6073.

Chan, C. H., & Escalante-Semerena, J. C. (2011). ArsAB, a novel enzyme from *Sporomusa ovata* activates phenolic bases for adenosylcobamide biosynthesis. *Molecular Microbiology, 81*(4), 952–967.

Chan, C. H., Newmister, S. A., Talyor, K., Claas, K. R., Rayment, I., & Escalante-Semerena, J. C. (2014). Dissecting cobamide diversity through structural and functional analyses of the base-activating CobT enzyme of Salmonella enterica. *Biochimica et Biophysica Acta, 1840*(1), 464–475.

Cheong, C. G., Escalante-Semerena, J. C., & Rayment, I. (1999). The three-dimensional structures of nicotinate mononucleotide: 5,6-Dimethylbenzimidazole phosphoribosyltransferase (CobT) from *Salmonella typhimurium* complexed with 5,6-dimethybenzimidazole and its reaction products determined to 1.9 Å resolution. *Biochemistry, 38*(49), 16125–16135.

Cheong, C. G., Escalante-Semerena, J. C., & Rayment, I. (2001). Structural investigation of the biosynthesis of alternative lower ligands for cobamides by nicotinate mononucleotide: 5,6-Dimethylbenzimidazole phosphoribosyltransferase from *Salmonella enterica*. *Journal of Biological Chemistry, 276*(40), 37612–37620.

Cheong, C. G., Escalante-Semerena, J. C., & Rayment, I. (2002). Capture of a labile substrate by expulsion of water molecules from the active site of nicotinate mononucleotide:5,6-Dimethylbenzimidazole phosphoribosyltransferase (CobT) from *Salmonella enterica*. *Journal of Biological Chemistry, 277*(43), 41120–41127.

Croft, M. T., Warren, M. J., & Smith, A. G. (2006). Algae need their vitamins. *Eukaryotic Cell*, *5*(8), 1175–1183.
Crofts, T. S., Hazra, A. B., Tran, J. L. A., Sokolovskaya, O. M., Osadchiy, V., Ad, O., et al. (2014). Regiospecific formation of cobamide isomers is directed by CobT. *Biochemistry*, *53*(49), 7805–7815.
Crofts, T. S., Seth, E. C., Hazra, A. B., & Taga, M. E. (2013). Cobamide structure depends on both lower ligand availability and CobT substrate specificity. *Chemistry and Biology*, *20*(10), 1265–1274.
Cuff, M. E., Evdokimova, E., Mursleen, A., Savchenko, A., Edwards, A., & Joachimiak, A. (2011). *RCSB PDB—3U4G: The structure of CobT from Pyrococcus horikoshii*. Retrieved from https://www.rcsb.org/structure/3U4G.
Deng, Z., Mao, J., Wang, Y., Zou, H., & Ye, M. (2017). Enzyme kinetics for complex system enables accurate determination of specificity constants of numerous substrates in a mixture by proteomics platform. *Molecular and Cellular Proteomics*, *16*(1), 135–145.
Deptula, P., Kylli, P., Chamlagain, B., Holm, L., Kostiainen, R., Piironen, V., et al. (2015). BluB/CobT2 fusion enzyme activity reveals mechanisms responsible for production of active form of vitamin B_{12} by *Propionibacterium freudenreichii*. *Microbial Cell Factories*, *14*(1), 1–12.
Ernst, O., & Zor, T. (2010). Linearization of the Bradford protein assay. *Journal of Visualized Experiments*, *38*, e1918.
Ford, J. E., Holdsworth, E. S., & Kon, S. K. (1954). The biosynthesis of vitamin B_{12}-like compounds. *The Biochemical Journal*, *59*(1), 86–93.
Friedmann, H. C. (1965). Partial purification and properties of a single displacement trans-N-glycosidase. *Journal of Biological Chemistry*, *240*(1), 413–418.
Friedmann, H. C., & Harris, D. L. (1965). The formation of α-Glycosidic 5'-nucleotides by a single displacement trans-N-glycosidase. *Journal of Biological Chemistry*, *240*(1), 406–412.
Fyfe, J. A., & Friedmann, H. C. (1969). Vitamin B_{12} biosynthesis: Enzyme studies on the formation of the α-glycosidic nucleotide precursor. *Journal of Biological Chemistry*, *244*(7), 1659–1666.
Gasteiger, E., Hoogland, C., Gattiker, A., Duvaud, S., Wilkins, M. R., Appel, R. D, et al. (2005). Protein identification and analysis tools on the ExPASy server. In John M. Walker (Ed.), *The Proteomics Protocols Handbook* (pp. 571–607). Humana Press.
Giedyk, M., Goliszewska, K., & Gryko, D. (2015). Vitamin B_{12} catalysed reactions. *Chemical Society Reviews*, *44*, 3391–3404.
Girard, C. L., Santschi, D. E., Stabler, S. P., & Allen, R. H. (2009). Apparent ruminal synthesis and intestinal disappearance of vitamin B_{12} and its analogs in dairy cows. *Journal of Dairy Science*, *92*(9), 4524–4529.
Gray, M. J., & Escalante-Semerena, J. C. (2010). A new pathway for the synthesis of α-ribazole-phosphate in *Listeria innocua*. *Molecular Microbiology*, 77(6), 1429–1438.
Hayashi, M., & Kamikubo, T. (1971). Biosynthesis and properties of 2-thioadenyl cobamide and its coenzyme form. *FEBS Letters*, *15*(3), 213–216.
Hazra, A. B., Han, A. W., Mehta, A. P., Mok, K. C., Osadchiy, V., Begley, T. P., et al. (2015). Anaerobic biosynthesis of the lower ligand of vitamin B_{12}. *Proceedings of the National Academy of Sciences of the United States of America*, *112*(34), 10792–10797.
Hazra, A. B., Tran, J. L. A., Crofts, T. S., & Taga, M. E. (2013). Analysis of substrate specificity in CobT homologs reveals widespread preference for DMB, the lower axial ligand of vitamin B_{12}. *Chemistry and Biology*, *20*(10), 1275–1285.
Jeter, V. L., Mattes, T. A., Beattie, N. R., & Escalante-Semerena, J. C. (2019). A new class of phosphoribosyltransferases involved in cobamide biosynthesis is found in methanogenic archaea and cyanobacteria. *Biochemistry*, *58*(7), 951–964.

Kamikubo, T., & Matsuura, T. (1969). Incorporation of purine bases into vitamin B_{12} molecule. *Agricultural and Biological Chemistry*, *33*(8), 1207–1209.
Kräutler, B., Fieber, W., Ostermann, S., Fasching, M., Ongania, K. H., Gruber, K., et al. (2003). The cofactor of tetrachloroethene reductive dehalogenase of *Dehalospirillum multivorans* is Norpseudo-B_{12}, a new type of a natural corrinoid. *Helvetica Chimica Acta*, *86*(11), 3698–3716.
Krautler, B., Kohler, H. E., & Stupperich, E. (1988). 5′-Methylbenzimidazolyl-cobamides are the corrinoids from some sulfate-reducing and sulfur-metabolizing bacteria. *European Journal of Biochemistry*, *176*(2), 461–469.
Kuo, Y. M., Henry, R. A., & Andrews, A. J. (2016). Measuring specificity in multi-substrate/product systems as a tool to investigate selectivity *in vivo*. *Biochimica et Biophysica Acta - Proteins and Proteomics*, *1864*(1), 70–76.
Lenhert, P., & Hodgkin, D. C. (1968). The structure of vitamin B12 (VII). The X-ray analysis of the vitamin B_{12} cofactor. *Proceedings of the Royal Society*, *84*(August 1961), 45–84.
Ma, A. T., Beld, J., & Brahamsa, B. (2017). An amoebal grazer of cyanobacteria requires cobalamin produced by heterotrophic bacteria. *Applied and Environmental Microbiology*, *83*(10), 1–13.
Ma, A. T., Tyrell, B., & Beld, J. (2019). Specificity of cobamide remodeling, uptake and utilization in *Vibrio cholerae*. *Molecular Microbiology*, *113*(1), 89–102.
Maggio-Hall, L. A., & Escalante-Semerena, J. C. (1999). *In vitro* synthesis of the nucleotide loop of cobalamin by *Salmonella typhimurium* enzymes. *Proceedings of the National Academy of Sciences of the United States of America*, *96*(21), 11798–11803.
Maggio-Hall, L. A., & Escalante-Semerena, J. C. (2003). Alpha-5,6-dimethylbenzimidazole adenine dinucleotide (alpha-DAD), a putative new intermediate of coenzyme B12 biosynthesis in *Salmonella typhimurium*. *Microbiology*, *194*(4), 984–990.
Mathur, Y., Sreyas, S., Datar, P. M., Sathian, M. B., & Hazra, A. B. (2020). CobT and BzaC catalyze the regiospecific activation and methylation of the 5-hydroxybenzimidazole lower ligand in anaerobic cobamide biosynthesis. *Journal of Biological Chemistry*, *295*(31), 10522–10534.
Mattes, T. A., Escalante-Semerena, J. C., Deery, E., & Warren, M. J. (2017). Cobalamin biosynthesis and insertion. In R. A. Scott (Ed.), *Encyclopedia of inorganic and bioinorganic chemistry* (pp. 1–24). John Wiley & Sons, Ltd.
Men, Y., Seth, E. C., Yi, S., Crofts, T. S., Allen, R. H., Taga, M. E., et al. (2015). Identification of specific corrinoids reveals corrinoid modification in dechlorinating microbial communities. *Environmental Microbiology*, *17*(12), 4873–4884.
Mok, K. C., & Taga, M. E. (2013). Growth inhibition of *Sporomusa ovata* by incorporation of benzimidazole bases into cobamides. *Journal of Bacteriology*, *195*(9), 1902–1911.
Moore, S. J., & Warren, M. J. (2012). The anaerobic biosynthesis of vitamin B_{12}. *Biochemical Society Transactions*, *40*(3), 581–586.
Newmister, S. A., Chan, C. H., Escalante-Semerena, J. C., & Rayment, I. (2012). Structural insights into the function of the nicotinate mononucleotide:phenol/p-cresol phosphoribosyltransferase (ArsAB) enzyme from *Sporomusa ovata*. *Biochemistry*, *51*(43), 8571–8582.
Ni, G., Du, Y., Tang, F., Liu, J., Zhao, H., & Chen, Q. (2019). Review of α-nucleosides: From discovery, synthesis to properties and potential applications. *RSC Advances*, *9*(25), 14302–14320.
Perlman, D. (1971). [211], Processes for the biosynthesis of cobamides. *Methods in Enzymology*, *18*(*Pt C*), 75–82. Academic Press.
Potter, S. C., Aurelien, L., Eddy, S. R., Park, Y., Lopez, R., & Finn, R. D. (2018). HMMER web server: 2018 update. *Nucleic Acids Research*, *46*(June), W200–W204.

Renz, P., Blickle, S., & Friedrich, W. (1987). Two new vitamin B_{12} factors from sewage sludge containing 2-methylsulfinyladenine or 2-methylsulfonyladenine as base component. *European Journal of Biochemistry*, *179*, 175–179.
Rodionov, D. A., Vitreschak, A. G., Mironov, A. A., & Gelfand, M. S. (2003). Comparative genomics of the vitamin B_{12} metabolism and regulation in prokaryotes. *Journal of Biological Chemistry*, *278*(42), 41148–41159.
Roth, J. R., Lawrence, J. G., & Bobik, T. A. (1996). Cobalamin (coenzyme B_{12}): Synthesis and biological significance. *Annual Review of Microbiology*, *50*, 137–181.
Schubert, T., von Reuß, S. H., Kunze, C., Paetz, C., Kruse, S., Brand-Schön, P., et al. (2019). Guided cobamide biosynthesis for heterologous production of reductive dehalogenases. *Microbial Biotechnology*, *12*(2), 346–359.
Seth, E. C., & Taga, M. E. (2014). Nutrient cross-feeding in the microbial world. *Frontiers in Microbiology*, *5*, 350.
Shelton, A. N., Seth, E. C., Mok, K. C., Han, A. W., Jackson, S. N., Haft, D. R., et al. (2018). Uneven distribution of cobamide biosynthesis and dependence in bacteria predicted by comparative genomics. *ISME Journal*, *13*(3), 789–804.
Sokolovskaya, O. M., Mok, K. C., Park, J. D., Tran, J. L. A., Quanstrom, K. A., & Taga, M. E. (2019). Cofactor selectivity in methylmalonyl coenzyme a mutase, a model cobamide-dependent enzyme. *MBio*, *10*(5), 27–32.
Sokolovskaya, O. M., Shelton, A. N., & Taga, M. E. (2020). Sharing vitamins: Cobamides unveil microbial interactions. *Science*, *369*(6499).
Stogios, P. J., Skarina, T., McChesney, C., Grimshaw, T., Savchenko, A., & Anderson, W. F. (2017). *RCSB PDB—6B5F: Crystal structure of nicotinate mononucleotide-5,6-dimethylbenzimidazole phosphoribosyltransferase CobT from Yersinia enterocolitica*. Retrieved from https://www.rcsb.org/structure/6B5F.
Stupperich, E., Steiner, I., & Ruhlemann, M. (1986). Isolation and analysis of bacterial cobamides liquid chromatography. *Analytical Biochemistry*, *155*(2), 365–370.
Thomas, M. G., Thompson, T. B., Rayment, I., & Escalante-Semerena, J. C. (2000). Analysis of the adenosylcobinamide kinase/adenosylcobinamide-phosphate guanylyltransferase (CobU) enzyme of *Salmonella typhimurium LT2*. *Journal of Biological Chemistry*, *275*(36), 27576–27586.
Trzebiatowski, J. R., & Escalante-semerena, J. C. (1997). Purification and characterization of CobT, the nicotinate-mononucleotide: 5,6-Dimethylbenzimidazole phosphoribosyltransferase enzyme from *salmonella typhimurium*. *Journal Biological Chemistry*, *272*(28), 17662–17667.
Van Den Ent, F., & Löwe, J. (2006). RF cloning: A restriction-free method for inserting target genes into plasmids. *Journal of Biochemical and Biophysical Methods*, *67*(1), 67–74.
Yan, J., Bi, M., Bourdon, A. K., Farmer, A. T., Wang, P. H., Molenda, O., et al. (2018). Purinyl-cobamide is a native prosthetic group of reductive dehalogenases. *Nature Chemical Biology*, *14*(1), 8–14.
Zayas, C. L., & Escalante-Semerena, J. C. (2007). Reassessment of the late steps of coenzyme B_{12} synthesis in *Salmonella enterica*: Evidence that dephosphorylation of adenosylcobalamin- 5′-phosphate by the CobC phosphatase is the last step of the pathway. *Journal of Bacteriology*, *189*(6), 2210–2218.
Zhang, Y., Rodionov, D. A., Gelfand, M. S., & Gladyshev, V. N. (2009). Comparative genomic analyses of nickel, cobalt and vitamin B_{12} utilization. *BMC Genomics*, *10*(78), 1471–2164.

CHAPTER THREE

Purification and detection of vitamin B_{12} analogs

Kenny C. Mok, Zachary F. Hallberg, and Michiko E. Taga*
Department of Plant & Microbial Biology, University of California, Berkeley, CA, United States
*Corresponding author: e-mail address: taga@berkeley.edu

Contents

Methods in Enzymology, Volume 668
ISSN 0076-6879
https://doi.org/10.1016/bs.mie.2021.11.023

Abstract

Cobamides are a family of enzyme cofactors that are required by organisms in all domains of life. Over a dozen cobamides exist in nature although only cobalamin (vitamin B_{12}), the cobamide required by humans, has been studied extensively. Cobamides are exclusively produced by a subset of prokaryotes. Importantly, the bacteria and archaea that synthesize cobamides *de novo* typically produce a single type of cobamide, and furthermore, organisms that use cobamides are selective for certain cobamides. Therefore, a detailed understanding of the cobamide-dependent metabolism of an organism or microbial community of interest requires experiments performed with a variety of cobamides. A notable challenge is that cobalamin is the only cobamide that is commercially available at present. In this chapter, we describe methods to extract, purify, and quantify various cobamides from bacteria for use in laboratory experiments.

1. Introduction

The class of vitamin cofactors known as cobamides is comprised of over a dozen structurally-diverse members. It is primarily known for its founding member, cobalamin, or vitamin B_{12}, due to its role in human health. The cobalamin deficiency disease pernicious anemia was first described in 1849 (Addison, 1849), and since that time the role of cobalamin as an enzyme cofactor in humans and many other organisms has been studied extensively (Banerjee & Ragsdale, 2003). Interestingly, cobamides are the only vitamins produced exclusively by prokaryotic organisms (Roth, Lawrence, & Bobik, 1996).

The major structural component of cobalamin is a distinctive modified tetrapyrrole known as a corrin ring (Fig. 1A). All molecules containing a corrin ring are collectively termed corrinoids. Cobamides are a subset of corrinoids that contain a central cobalt ion, an array of peripheral amide groups decorating the corrin ring, and a nucleotide loop consisting of an aminopropanol linker covalently tethered to a ribose 3-phosphate moiety (The nomenclature of corrinoids (1973 recommendations), 1974). One related family of cofactors, norcobamides, substitutes the aminopropanol linker with an ethanolamine linker, and has been found in *Sulfurospirillum multivorans* (Kräutler et al., 2003). In the case of "complete" corrinoids—cobamides that are utilized as enzyme cofactors—this nucleotide loop contains a heterocyclic base with an N- or O-glycosyl linkage to the ribose moiety, termed the "lower ligand" because it sometimes associates with the cobalt axially as an α-ligand. The numerous cobamide forms differ

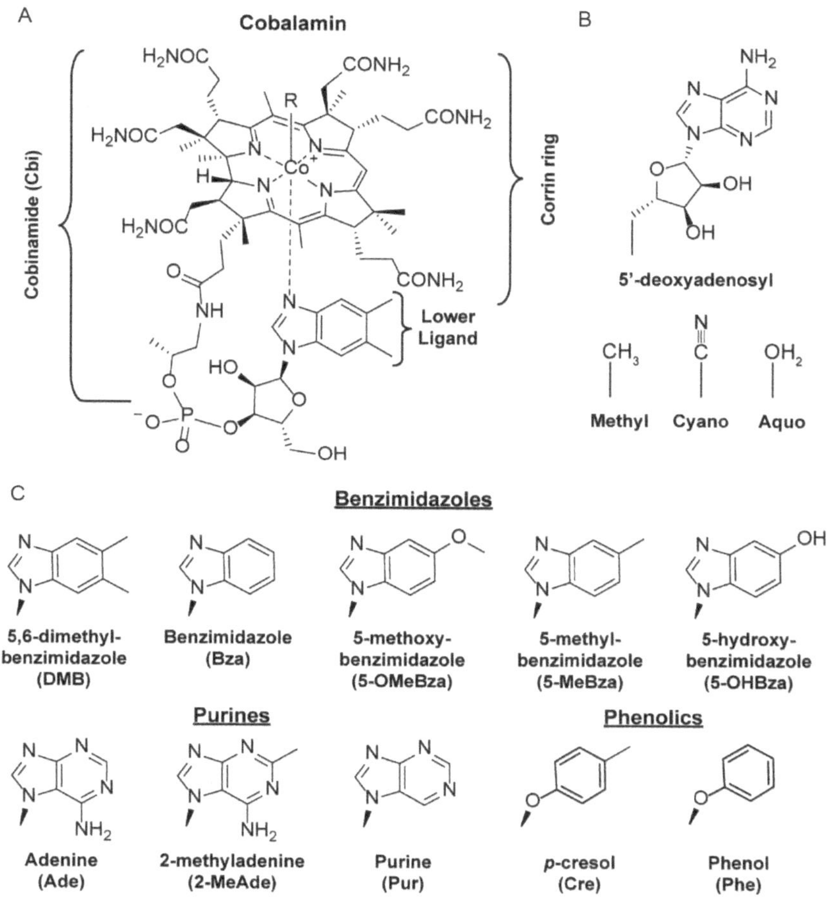

Fig. 1 Cobamide structure. (A) Cobamides are composed of a corrin ring containing a central cobalt ion and an upper (R) and lower ligand. In the cobamide shown, cobalamin, the lower ligand is 5,6-dimethylbenzimidazole. The precursor cobinamide is also denoted. (B) Upper ligands (R) commonly found in cobamides. (C) Examples of lower ligands found in naturally occurring cobamide cofactors. The three structural classes are indicated. Abbreviations for the lower ligand bases are shown in parentheses.

primarily in the structure of the lower ligand. For example, while 5,6-dimethylbenzimidazole (DMB) is the lower ligand of cobalamin, a variety of other benzimidazoles, purines, and phenolics have been found as lower ligands in cobamides isolated from natural environments or laboratory cultures (Fig. 1C) (Renz, 1999; Yan et al., 2018). Often when a cobamide is bound by an enzyme, the lower ligand does not associate with the cobalt ion and so is not directly involved in catalysis (Bommer et al., 2014; Drennan, Huang, Drummond, Matthews, & Ludwig, 1994; Mancia et al., 1996), yet its structure can still influence the catalytic capability of the cobamide as a cofactor (Lengyel, Mazumder, & Ochoa, 1960; Poppe, Stupperich, Hull, Buckel, & Retey, 1997; Sokolovskaya et al., 2021).

Key to discovering the structural diversity of cobamide cofactors was the development of a microbiological bioassay for cobalamin based on growth measurements of the bacterium *Lactobacillus lactis* (Shorb, 1947a, 1947b, 1948). Researchers found that compounds similar to cobalamin, with adenine and 2-methyladenine as lower ligands rather than DMB [pseudovitamin B_{12} (pseudoB_{12}) and 2-methyladeninylcobamide ([2-MeAde]Cba), respectively], had activity in the microbiological assay but not in experiments with animals because animals are more restrictive in the cobamide forms they could use (Coates, Ford, Harrison, Kon, & Porter, 1952; Dion, Calkins, & Pfiffner, 1952, 1954; Ford & Porter, 1952; Pfiffner et al., 1951).

PseudoB_{12} and [2-MeAde]Cba were the first two "alternate" cobamide cofactors identified, but in total, 15 other natural vitamin B_{12} analogs have been described (Fig. 1) (Kräutler et al., 2003; Renz, 1999; Yan et al., 2018). These various cobamides are produced *de novo* by only a subset of bacteria and archaea. However, the majority of prokaryotic genomes have been found to encode at least one cobamide-dependent enzyme (Shelton et al., 2019) and thus, "sharing" of this nutrient is considered a critical functional interaction in microbial communities (Sokolovskaya, Shelton, & Taga, 2020). Importantly though, it appears that most microbes have specific preferences for only a few cobamide forms (Helliwell et al., 2016; Mok & Taga, 2013; Shelton, Lyu, & Taga, 2020; Yan et al., 2018; Yi et al., 2012). Therefore, deciphering how microbiomes are shaped by cobamides requires not only knowledge about which microbes produce and require them, but also an understanding of the relative cobamide preferences of each organism.

Though the importance of cobamide structural diversity has become increasingly clear, the majority of research on cobamides remains focused on cobalamin. This is primarily due to its connection to human health, as cobalamin is the form specifically required by humans. In addition, cobalamin is the only form that is commercially available, and therefore research on alternate cobamides requires additional effort to purify them from microbial cultures. Alternate cobamides have been purified from a number of bacteria and archaea that are capable of producing cobamides at high levels under laboratory conditions (Hoffmann et al., 2000; Stupperich, Eisinger, & Kräutler, 1988; Yan et al., 2018; Yi et al., 2012). Furthermore, using a process known as "guided biosynthesis," where an alternate lower ligand base provided in the growth medium is incorporated as the cobamide lower ligand, some bacteria can also be made to synthesize a variety of cobamides of interest (Allen & Stabler, 2008; Mok & Taga, 2013; Perlman & Barrett, 1958; Sokolovskaya et al., 2019). Guided biosynthesis can be used to produce the range of cobamides that are found in nature as well as unnatural analogs (Mok & Taga, 2013; Sokolovskaya et al., 2019). In this chapter, we describe procedures for performing guided biosynthesis of cobamides, extracting corrinoids from bacterial cultures, and quantifying corrinoids (Fig. 2).

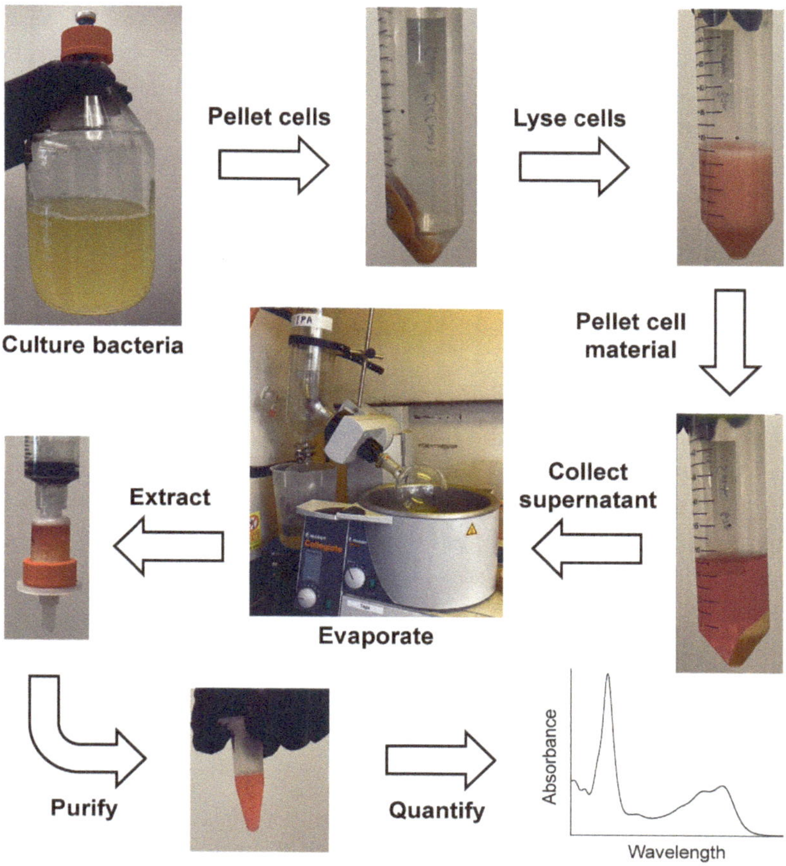

Fig. 2 Flowchart for corrinoid production, extraction, purification, and quantification.

2. Cobamide production in bacteria

In this section, we describe procedures for cobamide production in *Sporomusa ovata* and *Propionibacterium acidi-propionici*, two bacteria that we prefer due to their high production levels and versatility in guided biosynthesis (Fig. 2). Alternatively, specific cobamides of interest can be extracted from the microbes that naturally produce them (Hoffmann et al., 2000; Irion & Ljungdahl, 1965; Kräutler, Kohler, & Stupperich, 1988; Stupperich et al., 1988; Stupperich, Eisinger, & Kräutler, 1989; Yan et al., 2018; Yi et al., 2012), or from commercial microbial products such as *Spirulina* tablets (Heal et al., 2017), but these methods will not be discussed here.

Sporomusa ovata DSM 2662 natively synthesizes *p*-cresolylcobamide ([Cre]Cba) (Stupperich et al., 1988), and can produce other phenolyl as well as benzimidazolyl cobamides via guided biosynthesis (Mok & Taga, 2013). *Propionibacterium acidi-propionici* DSM 20273 natively produces pseudoB_{12} (Hoffmann et al., 2000) and can additionally produce other purinyl and benzimidazolyl cobamides by guided biosynthesis (Sokolovskaya et al., 2019). Growth of these two strains requires anaerobic conditions. The reader should refer to (Balch, Fox, Magrum, Woese, & Wolfe, 1979; Hungate, 1969; Strobel, 2009; Wolfe, 2011) for details regarding anaerobic culturing. The procedures below describe culturing of anaerobes in sealed bottles. They may be adapted for other culturing methods, such as growth in an anaerobic chamber or fermenter.

For laboratories in which anaerobic growth of bacteria is not feasible, *Salmonella enterica* serovar Typhimurium can alternatively be used for synthesizing benzimidazolyl and purinyl cobamides under aerobic conditions (Gray & Escalante-Semerena, 2009). The cobamide yield for *S. enterica* is lower than for *S. ovata* and *P. acidi-propionici* (Table 1), but the culturing steps are simpler.

Table 1 Expected cobamide yields.

Organism	Wet cell weight (per L)	Cobamide yield (mg per L)
S. ovata	2–3 g	Benzimidazolyl cobamides: 0.2–0.6 Phenolyl cobamides: 0.5–1
P. acidi-propionici	30–35 g	Purinyl cobamides: 1.5–2.5 Benzimidazolyl cobamides: up to 0.7
S. enterica	2–3 g	Benzimidazolyl cobamides 0.1–0.2 pseudoB_{12}: 0.04–0.07 [2-MeAde]Cba: 0.004–0.02

2.1 *S. ovata* DSM 2662 culturing and guided biosynthesis

S. ovata is one of the few bacteria known to be capable of synthesizing phenolyl cobamides. It can also produce benzimidazolyl cobamides by guided biosynthesis, but does not appear to incorporate purines effectively into cobamides. *S. ovata* can use a variety of substrates as carbon and energy sources, such as methanol, H_2+CO_2, and fructose (Möller, Oßmer, Howard, Gottschalk, & Hippe, 1984), though *S. ovata* generally produces higher levels of cobamides when grown on betaine. Similar amounts of phenolyl cobamides can be produced when methanol is used as the growth substrate, but growth of *S. ovata* is inhibited by the presence of benzimidazole bases under this condition (Mok & Taga, 2013).

In addition to preparing 1 L of the growth medium for cobamide production by *S. ovata*, Hungate tubes containing 10 mL of the medium should also be made. These are used for reviving *S. ovata* from a frozen glycerol stock. 10 mL of culture is needed for inoculation of each liter of growth medium.

2.2 Protocol

1. Prepare medium for cobamide production and reviving of *S. ovata* from a frozen glycerol stock as described in Section 5.
2. Inoculate 10 mL medium with 0.5 mL thawed *S. ovata* glycerol stock. Incubate at 30 °C for 60–72 h. If growth does not reach an OD_{600} of 0.7, transfer 0.5 ml of the culture into 10 mL of fresh medium and incubate for 60–72 h.
3. Inoculate 1 L growth medium containing the desired lower ligand base with a 10 mL culture of *S. ovata*. Incubate at 30 °C for 72 h.

2.3 *P. acidi-propionici* DSM 20273 culturing and guided biosynthesis

P. acidi-propionici can be used to produce high levels of purinyl cobamides (Table 1). In addition, depending on the particular benzimidazole, *P. acidi-propionici* may produce similar or higher levels of the benzimidazolyl cobamide compared to *S. ovata*, though unlike *S. ovata*, *P. acidi-propionici* cultures require daily maintenance during growth in batch culture (see Safety considerations). A frozen glycerol stock of *P. acidi-propionici* is revived in DSM medium 91 and then used to inoculate the Guided Biosynthesis medium (see Section 5).

2.4 Safety considerations

P. acidi-propionici produces substantial amounts of gas during growth in the Guided Biosynthesis medium. When growing in sealed bottles, do not fill more than halfway with medium. Vent gas every 12 h to prevent pressure buildup (see Section 2.5), which could cause an explosion.

2.5 Protocol

1. Prepare DSM medium 91, Guided Biosynthesis medium, and bicarbonate solution as described in Section 5, as well as 100 mL of sterile anoxic 50% glucose (w/v).
2. Inoculate a total of 20 mL of DSM medium 91 with 1 mL of a thawed *P. acidi-propionici* glycerol stock and incubate at 37 °C until the OD_{600} exceeds 1.0, about 48–72 h.
3. Add the 20 mL *P. acidi-propionici* culture to 1 L Guided Biosynthesis medium containing the desired lower ligand base. Incubate at 37 °C for 5 days. Approximately every 12 h, add 7.5 mL of 50% glucose solution and adjust the culture to pH 6–7 using the bicarbonate solution. To avoid overpressurization of the sealed bottle, release the gas produced during culturing by inserting a sterile 22 gauge needle into the rubber stopper before and after adding these solutions.

2.6 *S. enterica* serovar Typhimurium strain LT2 culturing and guided biosynthesis

S. enterica is able to produce cobamides *de novo*, though synthesis of the corrin ring occurs only under anaerobic conditions (Roth et al., 1996). However, if the cobamide precursor cobinamide is included in the growth medium during aerobic growth, *S. enterica* can be used for the production of purinyl and benzimidazolyl cobamides by guided biosynthesis. Cobamide yields are generally lower for *S. enterica* compared to *S. ovata* and *P. acidi-propionici*, particularly for purinyl cobamides, with *S. enterica* producing 20- to 60-fold less pseudoB_{12} and 100- to 600-fold less [2-MeAde]Cba than *P. acidi-propionici* (Table 1).

2.7 Safety considerations

S. enterica is a RG2 organism; ensure that your laboratory and any shared equipment are approved for BSL2 work.

2.8 Protocol

1. Prepare 1 L of NCE medium as described in Section 5.
2. Grow a 5 mL culture of *S. enterica* in Luria-Bertani (LB) or NCE broth overnight at 37 °C.
3. Inoculate 1 L of NCE medium containing the desired lower ligand base with the 5 mL culture and incubate with aeration at 37 °C for 3 days.

3. Extraction and purification of corrinoids from bacterial cultures

3.1 Corrinoid extraction and purification from cell pellets

After culturing the bacteria as described in Section 2, their corrinoids will need to be extracted from the cells and purified. The protocol described below is adapted from (Gray & Escalante-Semerena, 2009). It can be tailored for smaller or larger cultures and for other applications, such as characterizing the corrinoids in a sample of interest.

Cells are lysed by incubating in methanol (MeOH) at 60 °C, and the upper ligand of the corrinoids is replaced with a cyano group following the addition of potassium cyanide (KCN). Corrinoids are light-stable in this form, as the Co-CN bond is not sensitive to photolysis in the presence of white light, unlike corrinoids with a 5′-deoxyadenosine or methyl group as the upper ligand. The conversion to the cyano form also increases the extraction efficiency of corrinoids from cells.

Due to its acute toxicity, treatment with KCN is not permitted in all laboratories. If the use of KCN is not allowed, corrinoids can be extracted in the aquo form by exposing the samples to white light following cell lysis. Alternatively, extractions can be performed in the dark or under red light to extract the adenosyl, methyl, and aquo forms of the corrinoids that are present in the cell (Fig. 1B).

Corrinoids are extracted using C18 solid phase extraction cartridges or columns (e.g., Waters Corp., WAT020515). Choose the appropriately sized cartridge for your application. Typically, 100–400 mg of C18 sorbent is sufficient for binding the corrinoids (and other hydrophobic compounds) present in the cellular fraction of a 1 L culture; C18 cartridges and columns are generally reusable following elution of bound material if the sorbent becomes saturated (corrinoids are pigmented molecules, typically reddish or purple in color, which makes them visible when bound to the C18 sorbent (Fig. 2)). Corrinoids can be further purified following extraction, typically with the use of a high-performance liquid chromatography instrument (HPLC).

3.2 Safety considerations

KCN is acutely toxic and a lethal poison, and thus transferring of solid KCN should always be performed in a fume hood. This will also prevent exposure to HCN gas, which is released from solid KCN upon exposure to moisture in the air. It is recommended that a stock solution of KCN (100 mg/mL in H_2O) be prepared and stored in the fume hood. Subsequent steps in which the extractions are not in sealed containers should also be performed in a fume hood, as KCN is converted to HCN upon exposure to H_2O. Cyanide liquid and solid waste must be collected for hazardous waste disposal; contact your Environmental Health and Safety department to determine whether work with KCN is allowed at your institution and for proper waste collection and disposal procedures.

3.3 Equipment

Centrifuge
Conical/centrifuge tubes
Water bath
Rotary evaporator or vacuum concentrator
Reverse-phase C18 solid phase extraction cartridge/column
HPLC

3.4 Protocol

1. Pellet cells by centrifugation. Discard supernatant and transfer the cell pellet to 50 mL conical tubes or centrifuge tubes (up to 5 g per tube).
2. For each 5 g of wet cell weight, resuspend the pellet in 2 mL H_2O and then add 25 mL MeOH. If the estimated corrinoid yield is known, add a 100-fold molar excess of KCN to the tube (e.g., 2 mg of KCN per 5 g *P. acidi-propionici* for pseudoB_{12}; 5 mg of KCN per 3 g *S. ovata* for [Cre]Cba); if unknown, add 2 mg KCN per gram of wet cell weight. Ensure that the caps are securely sealed and vortex for 1 min.
3. Incubate at 60 °C for 1.5 h with vortexing for 30 s every 20 min.
4. Centrifuge extractions at $\geq 7000 \times g$ for 20 min to remove cellular debris. Transfer the supernatant to a new container and use a rotary evaporator or vacuum concentrator to remove the solvent. Use these instruments only in a fume hood to prevent HCN exposure. Filtration (0.22 µm) of samples after centrifugation will improve concentration under vacuum.

5. Resuspend the dried sample in 10–20 mL H_2O. If the sample could not be completely dried, dilute the remaining volume at least 10-fold with H_2O to ensure that the corrinoids present will be retained by the C18 sorbent (corrinoids elute from C18 sorbent in MeOH). Prepare the C18 solid phase extraction cartridge according to the manufacturer's instructions. Apply the cell extract to the C18 cartridge.
6. Wash the cartridge with 10–15 column bed volumes of H_2O. Elute corrinoids with 2.5–5 bed volumes of 100% MeOH. It may be possible to remove some impurities by including a second wash step with low concentrations of MeOH (20–40%) prior to elution; in addition, impurities that are more hydrophobic than corrinoids may be excluded from the eluate by eluting with reduced concentrations of MeOH (cyanated corrinoids generally elute from C18 sorbents in 70% MeOH). The second wash step is not recommended when aquo-corrinoids are being extracted, as they are less hydrophobic than cyanated corrinoids. Dispose of the flowthrough and wash as cyanide waste.
7. Remove MeOH from the eluted sample by evaporation and resuspend in H_2O for subsequent purification by HPLC or another form of chromatography (Baum, Menezes, Svatoš, & Schubert, 2020). Skip the subsequent steps if your application does not require HPLC purification.
8. Run the sample on HPLC and collect the corrinoid peak(s). The HPLC method used will depend on the instrument and columns available to the lab (see (Sokolovskaya et al., 2019) for examples). Reverse-phase C18-based columns are preferable for separation and purification of corrinoids. Note that cyanated phenolyl cobamides and the precursor cobinamide commonly elute in two peaks due to mono-cyanation at either of the two axial positions (Mok et al., 2020; Stupperich et al., 1989).
9. Remove solvent by evaporation and resuspend the corrinoid in H_2O. Apply to a C18 cartridge and wash with 5–10 column bed volumes of H_2O to remove acid and residual salts from the HPLC purification. Elute corrinoid with 100% MeOH. Remove MeOH by evaporation and store the corrinoid dry or resuspended in H_2O (pH 7) at −20 °C.

3.4.1 Corrinoid extraction from bacterial culture supernatants

In some microbial cultures, a considerable fraction of the corrinoids will be present in the culture supernatant. For example, when *P. acidi-propionici* is cultured as described above, up to half of the corrinoids produced can be found in the supernatant. These corrinoids can be extracted using the above procedure if they are first concentrated with a C18 solid phase extraction cartridge prior to the addition of KCN in order to decrease the volume of cyanide-containing material that is being handled. Elute corrinoids from the C18 cartridge with 100% MeOH and add KCN at a final concentration of 0.01% (w/v) and continue with step 3 of the protocol above (periodic vortexing is not necessary). Concentrating the corrinoids in supernatants can be time consuming if large volumes are involved. Increased amounts of C18 sorbent are also required due to the abundance of other hydrophobic compounds present in culture supernatants.

4. Methods for quantification and detection of corrinoids

Several methods have been developed to measure the concentration of corrinoids in a sample. If a corrinoid is pure, UV–Vis absorbance can be used to calculate its concentration. If a sample is not pure, the concentration of the corrinoid(s) can be determined by HPLC, with UV–Vis, or MS detection, by comparison to an authentic standard(s) of known concentration. An *Escherichia coli* bioassay can also be used to measure the total corrinoids in a sample without the need for purification.

4.1 Quantification of corrinoids by absorbance

Cobalt-containing corrinoids have distinct UV–Vis spectra (Fig. 3), which enables the use of absorbance measurements to quantify them. To determine the concentration of a purified cyanated corrinoid, measure its absorbance at the appropriate wavelength (Table 2) using a spectrophotometer. Calculate the concentration c using the Beer-Lambert law, $c = \frac{A}{\varepsilon l}$, where A is the absorbance at wavelength λ, ε is the extinction coefficient at wavelength λ, and l is the optical pathlength.

The concentration of any cobalt-containing corrinoid can also be determined by cyanating the Co ion at both axial positions and using an extinction coefficient for dicyanocobinamide ($(CN)_2Cbi$) (Table 2). Dicyanation is performed by incubating the corrinoid (10–50 μM) with 0.1 M KCN and exposing it to white light for 30–60 min. A color change to purple is indicative of conversion to the dicyano form of the corrinoid.

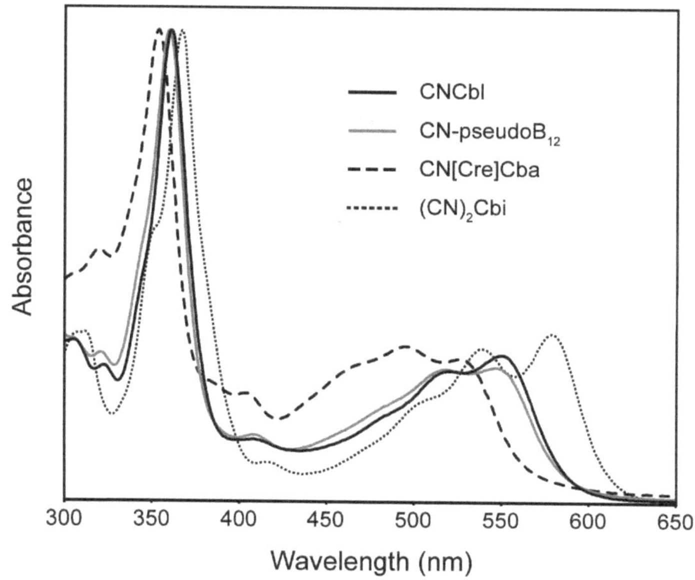

Fig. 3 UV–Vis spectra of select corrinoids.

Table 2 Extinction coefficients of select corrinoids.

Corrinoid	**Extinction coefficient at specified wavelength (mM^{-1} cm^{-1})**	**Source**
Benzimidazolyl cobamides	7.4 (518 nm)	Hogenkamp (1975)
Purinyl cobamides	7.9 (548 nm)	Hoffmann et al. (2000)
Phenolyl cobamides	9.5 (495 nm)	Stupperich et al. (1988)
Dicyanated corrinoids	9.9 (583 nm)	Schneider (1987)

4.2 Identification and quantification of corrinoids by HPLC and LCMS

The concentration of a corrinoid can be determined by HPLC with UV–Vis detection or by (LC)MS by comparison to a standard curve generated from a purified sample of that corrinoid. Table 3 lists the $M+H^+$ and $M+2H^+$ m/z values for the cyano form of common corrinoids (Fig. 1), which can be used as references for MS analysis. The relative abundances of the H^+ vs $2H^+$ species will depend on the particular MS instrument used.

Table 3 Mass values of cyanated corrinoids.

Corrinoid name or abbreviation[a]	Exact mass	m/z ($M+H^+$)	m/z ($M+2H^+$)
Cyanocobalamin	1354.5674	1355.5747	678.2910
CN[Bza]Cba	1326.5361	1327.5434	664.2753
CN[5-MeBza]Cba	1340.5518	1341.5591	671.2832
CN[5-OMeBza]Cba	1356.5467	1357.5540	679.2806
CN[5-OHBza]Cba	1342.5310	1343.5383	672.2728
CN-pseudoB_{12}	1343.5375	1344.5448	672.7760
CN[2-MeAde]Cba	1357.5532	1358.5605	679.7839
CN[Pur]Cba	1328.5266	1329.5339	665.2706
CN[Cre]Cba	1316.5405	1317.5478	659.2775
CN[Phe]Cba	1302.5249	1303.5322	652.2697
$(CN)_2$Cbi	1041.4953	1042.5026	521.7549
$CNCbi^+$	1015.4923	1015.4923	508.2500

[a]Structures are shown in Fig. 1.

4.3 *E. coli* bioassay for detecting and estimating concentration of corrinoids

The *E. coli* bioassay offers a simple method for determining whether a sample contains corrinoids (cobamides or precursors such as cobinamide) and estimating the total corrinoid concentration without the need to perform a solid-phase extraction (Kirmiz et al., 2020). In the absence of methionine, growth of an *E. coli* Δ*metE* strain is dependent on the cobamide-dependent methionine synthase MetH. This strain can either acquire cobamides from the environment to use directly with MetH or complete cobamide biosynthesis from certain cobamide precursors (Hazra et al., 2015). Therefore, this strain can be used to measure a variety of corrinoids in a sample. Furthermore, the growth yield of the Δ*metE* strain with a sample of interest is a proxy for corrinoid concentration, which can be determined if a standard curve with a purified corrinoid is generated (Fig. 4). This bioassay is capable of detecting concentrations of cobalamin as low as 50 pM, which can be up to an order of magnitude more sensitive than MS, five orders more sensitive than HPLC, and seven orders more sensitive than UV–Vis absorbance. Thus, this bioassay is ideal for samples that may contain low concentrations of corrinoids. However, because it relies only on *E. coli* growth measurements, this bioassay does not provide structural information about the corrinoid(s) present in a

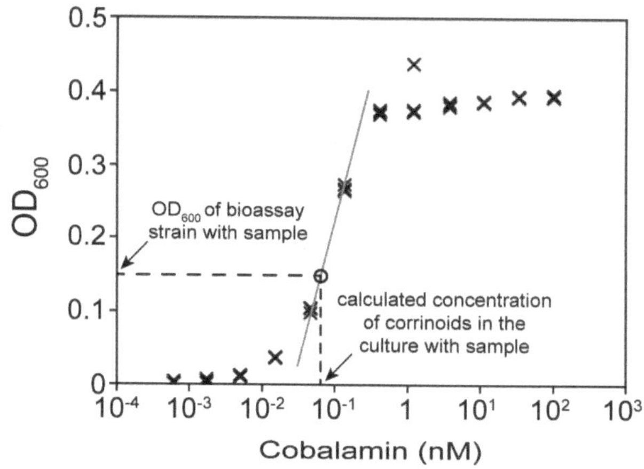

Fig. 4 Quantification of corrinoids in a sample using the *E. coli* bioassay strain. Growth yield after 12h is shown for the *E. coli* Δ*metE* strain incubated with serial dilutions of cobalamin. The concentration of corrinoids in a sample can be estimated if the OD_{600} of the culture containing the sample falls on the sloped portion of the standard curve, as marked by the gray line; if the OD_{600} of the culture is too high, use dilutions of the sample. Convert OD_{600} to corrinoid concentration based on the standard curve, as shown in the example (circle with dashed lines indicating the measured OD_{600} and the corresponding calculated corrinoid concentration).

sample. Additionally, because of differences in the growth response of *E. coli* to different corrinoids, if the identity of the corrinoid(s) in a sample is not known, its concentration can only be estimated.

Three *E. coli* strains are required for the bioassay: the cobamide-dependent bioassay strain (Δ*metE*) and two control strains. The wild type control strain MG1655 tests for any growth-inhibiting compounds present in the sample, and the Δ*metE*Δ*metH* control strain tests whether any methionine in the sample contributes to growth of the bioassay strain.

A traditional spectrophotometer is sufficient for measuring the OD_{600} of cultures grown in an incubator, though growth in a 96-well plate and OD_{600} measurements in a plate reader requires less sample volume and enables analysis of samples in higher throughput. The protocol is described for 2 mL cultures of the bioassay strains grown in individual tubes; culture volumes can be scaled down to 200 μL for growth in 96-well plates.

4.4 Protocol

1. Sterilize sample to be tested (e.g., culture supernatant, cell extract or environmental sample) by filtration (0.22 μm). All subsequent steps should also be performed aseptically.

2. For estimating the corrinoid concentration in a sample, prepare dilutions of the sample being tested as well as a cobalamin standard of known concentration for the standard curve (Fig. 4). If the identity of the cobamide in the sample is known, a purified standard of that cobamide should be used for the standard curve, if available.
3. Grow 3 mL cultures of the *E. coli* bioassay strains in LB or M9 glucose minimal medium (see Section 5) overnight at 37 °C. Supplement M9 glucose medium with L-Methionine (L-Met) (1 mM).
4. Remove 1 mL of each culture and pellet cells by centrifugation (2000–6000 × *g*). Discard supernatant.
5. Resuspend each cell pellet in 1 mL M9 glucose medium (without L-Met), centrifuge, and remove supernatant. Repeat this wash step twice. Resuspend the washed pellet in 1 mL M9 medium.
6. Set up 2 mL cultures with each *E. coli* strain as follows, with three replicates of each sample being tested:
   1.8 mL M9 glucose medium
   0.2 mL sample for testing
   0.02 mL washed *E. coli* cells
7. For each *E. coli* strain, include controls with L-Met (1 mM), cobalamin (10 nM), and no sample added. Additionally, samples "spiked" with cobalamin (10 nM) can be included to test for any effects the samples may have on corrinoid detection.
8. Incubate cultures at 37 °C. Measure OD_{600} after 12 h of growth.

5. Appendix

5.1 Equipment

Incubator

<u>For anaerobic culturing:</u>

Gassing setup

Sealed/serum bottles

Hungate tubes

Rubber stoppers

Crimp seals

Crimper

Needles and syringes

5.2 *S. ovata* DSM 2662 medium

Prepare 1 L Base medium (Table 4), excluding L-cysteine, Na_2S, and $NaHCO_3$ in a 2 L bottle to be used for anaerobic culturing or in a container for subsequent transfer to Hungate tubes. Sparge with 80% N_2 and 20% CO_2 for 30–45 min to make anoxic and then add L-cysteine, Na_2S, and $NaHCO_3$. If preparing Hungate culture tubes, dispense 10 mL into each tube under 80% N_2 and 20% CO_2. Seal containers and autoclave to sterilize.

Add remaining reagents to autoclaved medium after cooling to room temperature. For the 1 L medium for cobamide production, first mix betaine, K_2HPO_4, KH_2PO_4, Vitamin solution, and the desired lower ligand in a final volume of 30 mL H_2O and sparge with N_2 for 10 min to make anoxic. Sterilize by filtration (0.22 μm) and add to 1 L Base medium. For 10 mL culture tubes, first make a 20× stock solution in a serum bottle containing betaine, K_2HPO_4, KH_2PO_4, and Vitamin solution; sparge with N_2 to make anoxic and filter sterilize. Add 0.5 mL stock solution per 10 mL culture tube prior to inoculation with *S. ovata*. Store 20× stock solution at 4 °C in dark conditions. Adjust the final pH of the medium to 7.0–7.2, if necessary. Note that some benzimidazoles may not be completely soluble in H_2O at mM concentrations and should first be dissolved in 0.5–1 mL DMSO before diluting in H_2O.

5.3 *P. acidi-propionici* DSM 20273 media

Media should be prepared anoxically. However, since *P. acidi-propionici* is a facultative anaerobe, the addition of reducing agents is not strictly necessary, though the reader may wish to add them (see *S. ovata* medium for amounts (Table 4)).

Mix reagents for DSM medium 91 (Table 7) and make anoxic by sparging with N_2 for 30–45 min. Dispense 10 or 20 mL aliquots under N_2 into Hungate tubes or serum bottles, respectively. Seal containers and autoclave to sterilize.

Mix reagents for base of Guided Biosynthesis medium (Table 8) in 2 L bottle and sparge with N_2 for 30–45 min. Seal and autoclave to sterilize. To add remaining reagents, first mix in 30 mL H_2O and sparge with N_2 for 10 min to make anoxic. Sterilize by filtration and add to autoclaved medium after it has cooled to room temperature. Dicyanocobinamide ($(CN)_2Cbi$) is not required for cobamide production, but addition of 5 mg $(CN)_2Cbi$ (5 μM) increases the yield in the cellular fraction by two- to fourfold. While addition of higher concentrations of $(CN)_2Cbi$ will

Table 4 *S. ovata* Medium for Guided Biosynthesis.
Base medium

Reagent	Amount
NH_4Cl	0.5 g
$MgSO_4 \cdot 7H_2O$	0.5 g
$CaCl_2 \cdot 2H_2O$	0.025 g
NaCl	2.25 g
$FeSO_4 \cdot 7H_2O$	0.002 g
Trace elements solution SL-10 (Table 5)	3 mL
Na_2SeO_3	0.017 mg
Yeast Extract	2 g
Casitone/Pancreatic digest of casein	2 g
Na-resazurin solution (0.1% w/v)	0.25 mL
L-cysteine-HCl · H_2O	0.3 g
$Na_2S \cdot 9H_2O$	0.3 g
$NaHCO_3$	4 g
H_2O	Up to 970 mL
To add after sterilization of Base medium	
Betaine (anhydrous)	5.86 g
K_2HPO_4	0.35 g
KH_2PO_4	0.23 g
Wolin's Vitamin solution (Table 6)	1 ml
Lower ligand base	*p*-cresol (99%), 1 μL (10 μM final conc.) Phenol (89%), 0.4 mL (4 mM final conc.) Bza, 59.1 mg (0.5 mM final conc.) 5-MeBza, 66.1 mg (0.5 mM final conc.) 5-OMeBza, 148.2 mg (1 mM final conc.) 5-OHBza[a], 268.3 mg (2 mM final conc.)

[a]Not commercially available; can be synthesized from 5-OMeBza as described in Crofts, Seth, Hazra, and Taga (2013).
Adapted from dsmz.de and Möller, B., Oßmer, R., Howard, B. H., Gottschalk, G., & Hippe, H. (1984). Sporomusa, a new genus of gram-negative anaerobic bacteria including *Sporomusa sphaeroides* spec. Nov. and *Sporomusa ovata* spec. Nov. Arch. Microbiol., 139(4), 388–396.

Table 5 Trace elements solution SL-10.[a]

Reagent	**Amount**
Conc. HCl (37%)	6.7 mL
$FeCl_2 \cdot 4H_2O$	1.5 g
$CoCl_2 \cdot 6H_2O$	0.19 g
$MnCl_2 \cdot 4H_2O$	0.1 g
$ZnCl_2$	0.070 g
H_3BO_3	0.006 g
$Na_2MoO_4 \cdot 2H_2O$	0.036 g
$NiCl_2 \cdot 6H_2O$	0.024 g
$CuCl_2 \cdot 2H_2O$	0.002 g
H_2O	Up to 1 L

[a]First dissolve $FeCl_2$ in conc. HCl before diluting with H_2O and adding remaining salts. Store at room temperature in dark conditions.

Table 6 Wolin's vitamin solution (B_{12} omitted).[a]

Reagent	**Amount**
Biotin	0.02 g
Folic acid	0.02 g
Pyridoxine-HCl	0.1 g
Riboflavin	0.05 g
Thiamine-HCl	0.05 g
Nicotinic acid	0.05 g
Ca-D-pantothenate	0.05 g
p-Aminobenzoic acid	0.05 g
Thioctic acid (lipoic acid)	0.05 g
H_2O	1 L

[a]Adjust pH to 7.5 with NaOH to dissolve vitamins. Store in dark conditions at 4 °C (working solution) or –20 °C (long-term storage).

Table 7 DSM medium 91.

Reagent	Amount
Casein peptone, tryptic digest	10 g
Yeast extract	5 g
Na-lactate (60% w/w solution)	16.7 mL
H_2O	Up to 1 L

Adjust pH to 7.0–7.2.
Adapted from dsmz.de.

Table 8 *P. acidi-propionici* Medium for Guided Biosynthesis.
Base medium

Reagent	Amount
$NaH_2PO_4 \cdot H_2O$	1.7 g
K_3PO_4	1.66 g
Casein hydrolysate	10 g
Yeast Extract	5 g
H_2O	970 mL
To add after sterilization of Base medium	
Glucose	7.5 g
$MgCl_2 \cdot 6H_2O$	0.6 g
$FeCl_2 \cdot 4H_2O$	0.009 g
$CoCl_2 \cdot 6H_2O$	0.017 g
Ca-D-pantothenate	0.004 g
D-biotin	0.0003 g
Dicyanocobinamide	0.005 g
Lower ligand base	Adenine hemisulfate, 0.13 g (0.7 mM final conc.) 2-methyladenine, 0.11 g (0.7 mM final conc.) Purine, 0.084 g (0.7 mM final conc.) benzimidazoles, 0.5–1 mM (final conc. in 1 L)

Adapted from Hoffmann, B., Oberhuber, M., Stupperich, E., Bothe, H., Buckel, W., Konrat, R., & Kräutler, B. (2000). Native corrinoids from *Clostridium cochlearium* are adeninylcobamides: Spectroscopic analysis and identification of pseudovitamin B12 and factor A. J Bacteriol, 182(17), 4773–4782.

Table 9 Bicarbonate solution.

Reagent	**Amount**
$KHCO_3$	10 g
$NaHCO_3$	2.5 g
H_2O	100 mL

Table 10 NCE Medium.
50×NCE medium

Reagent	**Amount**
K_2HPO_4	247 g
KH_2PO_4	197 g
$NaNH_4PO_4 \cdot 4H_2O$	175 g
H_2O	Up to 1 L
To add to 1×NCE (per L)	
Trace Elements Solution SL-10 (Table 5)	1 mL
Glycerol	736 μL
1,2-propanediol (99%)	5.9 mL
$MgSO_4$	0.12 g
Dicyanocobinamide	0.001 g
Lower ligand base (optional)	Benzimidazoles—0.075 g

Adapted from Gray, M. J., & Escalante-Semerena, J. C. (2009). In vivo analysis of cobinamide salvaging in *Rhodobacter sphaeroides* strain 2.4.1. J Bacteriol, 191(12), 3842–3851. https://doi.org/10.1128/JB.00230-09.

increase cobamide production overall, the excess cobamide is found primarily in the culture supernatant. Purines have low solubility in H_2O; dissolve first in 3 mL of 1 N HCl or NaOH. Adjust pH of medium to 6–7.

Prepare bicarbonate solution (Table 9) in a serum bottle and sparge with 80% N_2 and 20% CO_2 for 10–15 min to make anoxic. Seal bottle and autoclave to sterilize.

5.4 *S. enterica* serovar Typhimurium strain LT2 medium

A 50×NCE stock solution (Table 10) should be prepared and diluted to 1×and autoclaved as needed. The remaining components should be mixed

Table 11 M9 glucose minimal medium.
10×M9 salts

Reagent	Amount
Na_2HPO_4	60 g
KH_2PO_4	30 g
NaCl	5 g
NH_4Cl	10 g
H_2O	Up to 1 L
Adjust pH to 7.4	
M9 Glucose Minimal medium	
10×M9 Salts	100 mL
$MgSO_4$ (50 mM)	20 mL
$CaCl_2$ (50 mM)	2 mL
Glucose (20% w/v)	10 mL
H_2O	868 mL

Adapted from Sambrook, J. (2001). Molecular cloning: a laboratory manual. Cold Spring Harbor, N.Y.: Cold Spring Harbor Laboratory Press.

in 30 mL H_2O (Table 10), sterilized by filtration, and added to 1×NCE that has been cooled to room temperature to avoid precipitation of the phosphate.

5.5 *E. coli* MG1655 medium

Make 10×M9 Salts solution (Table 11). Autoclave to sterilize. To prepare M9 Glucose Minimal medium, autoclave M9 salts in H_2O and cool to room temperature prior to adding $MgSO_4$, $CaCl_2$, and glucose from sterile stock solutions (Table 11). Alternatively, non-sterile reagents can be added to 1× M9 salts and sterilized by filtration.

Acknowledgments

We thank Sebastian Gude, Gordon Pherribo, Olga Sokolovskaya, Catherine Fields, Matt Weaver, Dennis Suazo for critical reading of the manuscript, and Jared Leadbetter for helpful discussions about anaerobic culturing. This work was funded by National Institutes of Health grant R35GM139633 and Department of Energy grant DE-SC0020155 to M.E.T..

References

Addison, T. (1849). Chronic suprarenal insufficiency, usually due to tuberculosis of suprarenal capsule. *Lond Med Gazette, 43*, 517–518.

Allen, R. H., & Stabler, S. P. (2008). Identification and quantitation of cobalamin and cobalamin analogues in human feces. *The American Journal of Clinical Nutrition, 87*(5), 1324–1335. https://doi.org/10.1093/ajcn/87.5.1324.

Balch, W. E., Fox, G. E., Magrum, L. J., Woese, C. R., & Wolfe, R. S. (1979). Methanogens: Reevaluation of a unique biological group. *Microbiological Reviews, 43*(2), 260–296. https://doi.org/10.1128/mr.43.2.260-296.1979.

Banerjee, R., & Ragsdale, S. W. (2003). The many faces of vitamin B_{12}: Catalysis by cobalamin-dependent enzymes. *Annual Review of Biochemistry, 72*(1), 209–247. https://doi.org/10.1146/annurev.biochem.72.121801.161828.

Baum, C., Menezes, R. C., Svatoš, A., & Schubert, T. (2020). Cobamide remodeling in the freshwater microalga *Chlamydomonas reinhardtii*. *FEMS Microbiology Letters, 367*(20). https://doi.org/10.1093/femsle/fnaa171.

Bommer, M., Kunze, C., Fesseler, J., Schubert, T., Diekert, G., & Dobbek, H. (2014). Structural basis for organohalide respiration. *Science, 346*(6208), 455–458. https://doi.org/10.1126/science.1258118.

Coates, M. E., Ford, J. E., Harrison, G. F., Kon, S. K., & Porter, J. W. (1952). Some properties of vitamin B_{12}-like factors from calf faeces. 2. Biological activities and interrelationships. *The Biochemical Journal, 51*(1), vi.

Crofts, T. S., Seth, E. C., Hazra, A. B., & Taga, M. E. (2013). Cobamide structure depends on both lower ligand availability and CobT substrate specificity. *Chemistry & Biology, 20*(10), 1265–1274. https://doi.org/10.1016/j.chembiol.2013.08.006.

Dion, H. W., Calkins, D. G., & Pfiffner, J. J. (1952). Hydrolysis products of Pseudovitamin B_{12}. *Journal of the American Chemical Society, 74*(4), 1108. https://doi.org/10.1021/ja01124a535.

Dion, H. W., Calkins, D. G., & Pfiffner, J. J. (1954). 2-Methyladenine, an hydrolysis product of pseudovitamin B_{12d}. *Journal of the American Chemical Society, 76*(3), 948–949. https://doi.org/10.1021/ja01632a113.

Drennan, C. L., Huang, S., Drummond, J. T., Matthews, R. G., & Ludwig, M. L. (1994). How a protein binds B_{12}: A 3.0 Å X-ray structure of B_{12}-binding domains of methionine synthase. *Science, 266*(5191), 1669–1674. https://doi.org/10.1126/science.7992050.

Ford, J. E., & Porter, J. W. G. (1952). Some properties of vitamin B_{12}-like factors from calf faeces. 1. Characteristics of different fractions. *The Biochemical Journal, 51*(1), v–vi.

Gray, M. J., & Escalante-Semerena, J. C. (2009). In vivo analysis of cobinamide salvaging in *Rhodobacter sphaeroides* strain 2.4.1. *Journal of Bacteriology, 191*(12), 3842–3851. https://doi.org/10.1128/JB.00230-09.

Hazra, A. B., Han, A. W., Mehta, A. P., Mok, K. C., Osadchiy, V., Begley, T. P., & Taga, M. E. (2015). Anaerobic biosynthesis of the lower ligand of vitamin B_{12}. *Proceedings of the National Academy of Sciences of the United States of America, 112*(34), 10792–10797. https://doi.org/10.1073/pnas.1509132112.

Heal, K. R., Qin, W., Ribalet, F., Bertagnolli, A. D., Coyote-Maestas, W., Hmelo, L. R., et al. (2017). Two distinct pools of B_{12} analogs reveal community interdependencies in the ocean. *Proceedings of the National Academy of Sciences of the United States of America, 114*(2), 364–369. https://doi.org/10.1073/pnas.1608462114.

Helliwell, K. E., Lawrence, A. D., Holzer, A., Kudahl, U. J., Sasso, S., Kräutler, B., et al. (2016). Cyanobacteria and eukaryotic algae use different chemical variants of vitamin B_{12}. *Current Biology, 26*(8), 999–1008. https://doi.org/10.1016/j.cub.2016.02.041.

Hoffmann, B., Oberhuber, M., Stupperich, E., Bothe, H., Buckel, W., Konrat, R., et al. (2000). Native corrinoids from *Clostridium cochlearium* are adeninylcobamides: Spectroscopic analysis and identification of pseudovitamin B_{12} and Factor A. *Journal of Bacteriology, 182*(17), 4773–4782. https://doi.org/10.1128/JB.182.17.4773-4782.2000.

Hogenkamp, H. P. C. (1975). The chemistry of cobalamins and related compounds. In B. M. Babior (Ed.), *Cobalamin biochemistry and pathophysiology* (pp 21–74). John Wiley and Sons, Inc.
Hungate, R. E. (1969). A roll tube method for cultivation of strict anaerobes. In J. R. Norris & D. W. Ribbons (Eds.), *Vol. 3B. Methods in microbiology* (pp. 117–132). New York: Academic Press.
Irion, E., & Ljungdahl, L. (1965). Isolation of factor III_m coenzyme and cobyric acid coenzyme plus other B_{12} factors from *Clostridium thermoaceticum*. *Biochemistry*, *4*(12), 2780–2790. https://doi.org/10.1021/bi00888a031.
Kirmiz, N., Galindo, K., Cross, K. L., Luna, E., Rhoades, N., Podar, M., & Flores, G. E. (2020). Comparative genomics guides elucidation of vitamin B_{12} biosynthesis in novel human-associated *Akkermansia* strains. *Applied and Environmental Microbiology*, *86*(3). https://doi.org/10.1128/AEM.02117-19.
Kräutler, B., Fieber, W., Ostermann, S., Fasching, M., Ongania, K. H., Gruber, K., et al. (2003). The cofactor of tetrachloroethene reductive dehalogenase of *Dehalospirillum multivorans* is norpseudo-B_{12}, a new type of a natural corrinoid. *Helvetica Chimica Acta*, *86*(11), 3698–3716. https://doi.org/10.1002/hlca.200390313.
Kräutler, B., Kohler, H. P., & Stupperich, E. (1988). 5′-Methylbenzimidazolyl-cobamides are the corrinoids from some sulfate-reducing and sulfur-metabolizing bacteria. *European Journal of Biochemistry*, *176*(2), 461–469. https://doi.org/10.1111/j.1432-1033.1988.tb14303.x.
Lengyel, P., Mazumder, R., & Ochoa, S. (1960). Mammalian methylmalonyl isomerase and vitamin B_{12} coenzymes. *Proceedings of the National Academy of Sciences of the United States of America*, *46*(10), 1312–1318. https://doi.org/10.1073/pnas.46.10.1312.
Mancia, F., Keep, N. H., Nakagawa, A., Leadlay, P. F., McSweeney, S., Rasmussen, B., et al. (1996). How coenzyme B_{12} radicals are generated: The crystal structure of methylmalonyl-coenzyme A mutase at 2 Å resolution. *Structure*, *4*(3), 339–350. https://doi.org/10.1016/s0969-2126(96)00037-8.
Mok, K. C., Sokolovskaya, O. M., Nicolas, A. M., Hallberg, Z. F., Deutschbauer, A., Carlson, H. K., et al. (2020). Identification of a novel cobamide remodeling enzyme in the beneficial human gut bacterium *Akkermansia muciniphila*. *mBio*, *11*(6). https://doi.org/10.1128/mBio.02507-20.
Mok, K. C., & Taga, M. E. (2013). Growth inhibition of *Sporomusa ovata* by incorporation of benzimidazole bases into cobamides. *Journal of Bacteriology*, *195*(9), 1902–1911. https://doi.org/10.1128/jb.01282-12.
Möller, B., Oßmer, R., Howard, B. H., Gottschalk, G., & Hippe, H. (1984). *Sporomusa*, a new genus of gram-negative anaerobic bacteria including *Sporomusa sphaeroides* spec. nov. and *Sporomusa ovata* spec. nov. *Archives of Microbiology*, *139*(4), 388–396. https://doi.org/10.1007/BF00408385.
Perlman, D., & Barrett, J. B. (1958). Biosynthesis of cobalamin analogues by *Propionibacterium arabinosum*. *Canadian Journal of Microbiology*, *4*(1), 9–15. https://doi.org/10.1139/m58-002.
Pfiffner, J. J., Calkins, D. G., Peterson, R. G., Bird, O. D., McGlohon, V., & Stipek, R. W. (1951). Amer Chem Soc, 120th meeting, abst. of papers, 22c.
Poppe, L., Stupperich, E., Hull, W. E., Buckel, T., & Retey, J. (1997). A base-off analogue of coenzyme-B_{12} with a modified nucleotide loop; ^{1}H-NMR structure analysis and kinetic studies with *(R)*-methylmalonyl-CoA mutase, glycerol dehydratase, and diol dehydratase. *European Journal of Biochemistry*, *250*(2), 303–307. https://doi.org/10.1111/j.1432-1033.1997.0303a.x.
Renz, P. (1999). Biosynthesis of the 5,6-dimethylbenzimidazole moiety of cobalamin and of the other bases found in natural corrinoids. In R. Banerjee (Ed.), *Chemistry and biochemistry of B_{12}* (pp. 557–575). New York: John Wiley & Sons, Inc.
Roth, J. R., Lawrence, J. G., & Bobik, T. A. (1996). Cobalamin (coenzyme B_{12}): Synthesis and biological significance. *Annual Review of Microbiology*, *50*(1), 137–181. https://doi.org/10.1146/annurev.micro.50.1.137.

Schneider, Z. (1987). Chemistry of cobalamin and related compounds. *Comprehensive B_{12}: Chemistry, Biochemistry, Nutrition, Ecology, Medicine*. De Gruyter.
Shelton, A. N., Lyu, X., & Taga, M. E. (2020). Flexible cobamide metabolism in *Clostridioides* (*Clostridium*) *difficile* 630 Δ*erm*. *Journal of Bacteriology*, *202*(2). https://doi.org/10.1128/JB.00584-19.
Shelton, A. N., Seth, E. C., Mok, K. C., Han, A. W., Jackson, S. N., Haft, D. R., et al. (2019). Uneven distribution of cobamide biosynthesis and dependence in bacteria predicted by comparative genomics. *The ISME Journal*, *13*(3), 789–804. https://doi.org/10.1038/s41396-018-0304-9.
Shorb, M. S. (1947a). Unidentified essential growth factors for *Lactobacillus lactis* found in refined liver extracts and in certain natural materials. *Journal of Bacteriology*, *53*(5), 669. https://doi.org/10.1128/JB.53.5.669-671.1947.
Shorb, M. S. (1947b). Unidentified growth factors for *Lactobacillus lactis* in refined liver extracts. *The Journal of Biological Chemistry*, *169*(2), 455.
Shorb, M. S. (1948). Activity of vitamin B_{12} for the growth of *Lactobacillus lactis*. *Science*, *107*(2781), 397–398. https://doi.org/10.1126/science.107.2781.397.
Sokolovskaya, O. M., Mok, K. C., Park, J. D., Tran, J. L. A., Quanstrom, K. A., & Taga, M. E. (2019). Cofactor selectivity in methylmalonyl coenzyme a mutase, a model cobamide-dependent enzyme. *mBio*, *10*(5), e01303–e01319. https://doi.org/10.1128/mBio.01303-19.
Sokolovskaya, O. M., Plessl, T., Bailey, H., Mackinnon, S., Baumgartner, M. R., Yue, W. W., et al. (2021). Naturally occurring cobalamin (B_{12}) analogs can function as cofactors for human methylmalonyl-CoA mutase. *Biochimie*, *183*, 35–43. https://doi.org/10.1016/j.biochi.2020.06.014.
Sokolovskaya, O. M., Shelton, A. N., & Taga, M. E. (2020). Sharing vitamins: Cobamides unveil microbial interactions. *Science*, *369*(6499). https://doi.org/10.1126/science.aba0165.
Strobel, H. J. (2009). Basic laboratory culture methods for anaerobic bacteria. In J. Mielenz (Ed.), *Vol. 581. Biofuels. Methods in molecular biology (methods and protocols)* (pp. 247–261). Totowa, NJ: Humana Press.
Stupperich, E., Eisinger, H. J., & Kräutler, B. (1988). Diversity of corrinoids in acetogenic bacteria. *p*-cresolylcobamide from *Sporomusa ovata*, 5-methoxy-6-methylbenzimidazolylcobamide from *Clostridium formicoaceticum* and vitamin B_{12} from *Acetobacterium woodii*. *European Journal of Biochemistry*, *172*(2), 459–464.
Stupperich, E., Eisinger, H. J., & Kräutler, B. (1989). Identification of phenolyl cobamide from the homoacetogenic bacterium *Sporomusa ovata*. *European Journal of Biochemistry*, *186*(3), 657–661.
The nomenclature of corrinoids (1973 recommendations). (1974). IUPAC-IUB Commission on biochemical nomenclature. *Biochemistry*, *13*(7), 1555–1560. https://doi.org/10.1021/bi00704a036.
Wolfe, R. S. (2011). Techniques for cultivating methanogens. In A. C. Rosenzweig & S. W. Ragsdale (Eds.), *Vol. 494. Methods in Enzymology* (pp. 1–22). Academic Press.
Yan, J., Bi, M., Bourdon, A. K., Farmer, A. T., Wang, P. H., Molenda, O., et al. (2018). Purinyl-cobamide is a native prosthetic group of reductive dehalogenases. *Nature Chemical Biology*, *14*(1), 8–14. https://doi.org/10.1038/nchembio.2512.
Yi, S., Seth, E. C., Men, Y. J., Stabler, S. P., Allen, R. H., Alvarez-Cohen, L., et al. (2012). Versatility in corrinoid salvaging and remodeling pathways supports corrinoid-dependent metabolism in *Dehalococcoides mccartyi*. *Applied and Environmental Microbiology*, *78*(21), 7745–7752. https://doi.org/10.1128/aem.02150-12.

Further reading

Sambrook, J. (2001). *Molecular cloning: A laboratory manual*. Cold Spring Harbor, N.Y: Cold Spring Harbor Laboratory Press.

CHAPTER FOUR

A method for the efficient adenosylation of corrinoids

Flavia G. Costa[†], **Elizabeth A. Villa, and Jorge C. Escalante-Semerena***
Department of Microbiology, University of Georgia, Athens, GA, United States
*Corresponding author: e-mail address: jcescala@uga.edu

Contents

Abstract

Adenosylcobamides (AdoCbas) are coenzymes required by organisms from all domains of life to perform challenging chemical reactions. AdoCbas are characterized by a cobalt-containing tetrapyrrole ring, where an adenosyl group is covalently attached to the cobalt ion *via* a unique Co–C organometallic bond. During catalysis, this bond is homolytically cleaved by AdoCba-dependent enzymes to form an adenosyl radical that is critical for intra-molecular rearrangements.

The formation of the Co–C bond is catalyzed by a family of enzymes known as ATP: Co(I)rrinoid adenosyltransferases (ACATs). ACATs adenosylate Cbas in two steps: (I) they

[†] Current address: Department of Immunology and Microbiology, University of Colorado – Anschutz Medical Campus, Aurora, CO, United States.

Methods in Enzymology, Volume 668
ISSN 0076-6879
https://doi.org/10.1016/bs.mie.2021.09.013

generate a planar, Co(II) four-coordinate Cba to facilitate the reduction of Co(II) to Co(I), and (II) they transfer the adenosyl group from ATP to the Co(I) ion.

To synthesize adenosylated corrinoids *in vitro*, it is imperative that anoxic conditions are maintained to avoid oxidation of Co(II) or Co(I) ions. Here we describe a method for the enzymatic synthesis and quantification of specific AdoCbas.

1. Introduction to ATP:Co(I)rrinoid adenosyltransferases (ACATs)

Cobamides are coenzymes that are characterized by a cobalt-containing tetrapyrrole ring (Mattes, Deery, Warren, & Escalante-Semerena, 2017). Adenosylcobalamin (AdoCbl), a coenzyme that is essential to humans, is an example of a cobamide (Mattes et al., 2017). While many organisms across all domains of life utilize cobamides, these coenzymes are known to be produced *de novo* in only a handful of prokaryotes. The light-sensitive nature of the Co–C bond in adenosylated cobamides complicates purification from biological sources (Weissbach, Ladd, Volcani, Smyth, & Barker, 1960). Thus, *in vitro* production of adenosylcobamides ensures a substantially higher yield.

ATP:Co(I)rrinoid adenosyltransferases (ACATs) refers to enzymes that adenosylate cobamides, including cobalamin, at the expense of ATP (Mera & Escalante-Semerena, 2010b). This reaction results in the attachment of the adenosyl group from ATP to the corrinoid, forming an organometallic Co–C bond at the β face of the corrin ring. For this reaction to occur, the cobalt ion of the tetrapyrrole ring must be reduced in sequential one-electron transfers from Co(III) → Co(II) → Co(I). The conversion of Co(III) → Co(II) is thermodynamically feasible since the redox potential for the Co(III)/Co(II) couple is high enough to be driven by free dihydroflavins (Fonseca & Escalante-Semerena, 2000; Mera & Escalante-Semerena, 2010a). The Co(II) → Co(I) step, however, is unfavorable, as the redox potential of the reduction (−610 mV) (Lexa & Saveant, 1976) exceeds that of any known physiological reductant in the cell. ACAT enzymes facilitate this step by promoting the formation of a four-coordinate Co(II)cobamide in the active site, which raises the redox potential and permits reduction of Co(II) → Co(I) by dihydroflavins or a flavoprotein (Fonseca & Escalante-Semerena, 2001; Mera & Escalante-Semerena, 2010a; Stich, Yamanishi, Banerjee, & Brunold, 2005).

ACATs are categorized into three known families, named CobA, PduO and EutT (Buan, Suh, & Escalante-Semerena, 2004; Escalante-Semerena,

Suh, & Roth, 1990; Johnson et al., 2001). EutT ACATs are further classified into two classes, metal-containing and metal-less EutTs (Costa & Escalante-Semerena, 2018). While CobA is evolutionarily unrelated to PduO and EutT, PduO and EutT have significant shared sequence characteristics (Costa, Greenhalgh, Brunold, & Escalante-Semerena, 2020).

CobA was the first ACAT identified (Escalante-Semerena et al., 1990). CobA ACATs are characterized by an inverted P-loop motif in the first β-strand that binds the ATP substrate, and conserved phenylalanine and tryptophan residues (FXXW) together displace the base of the corrinoid substrate (Moore, Newmister, Rayment, & Escalante-Semerena, 2012). The CobA from *Salmonella enterica* can adenosylate both incomplete and complete cobamides, and so it is a useful ACAT homolog for adenosylation of incomplete cobamides (Escalante-Semerena et al., 1990; Suh, 1994).

PduO ACATs are found in two forms. One form, such as the homolog from *Salmonella enterica*, has an extended C-terminus that is not required for activity of the ACAT (Johnson, Buszko, & Bobik, 2004). Recent work suggests this region may bind heme, however the physiological role of this region is uncertain (Ortiz de Orue Lucana et al., 2016). The full length PduO from *S. enterica* is difficult to purify, and a complete crystal structure has not been elucidated to date. Another type of PduO lacks this C-terminal extension. The PduO from *Lactobacillus reuteri* and humans, for example, lack this C-terminal extension, and several structures have been elucidated of this form of PduO (Moon, Park, Jang, Kim, & Chi, 2008; Park, Chi, & Moon, 2011; St Maurice et al., 2007). Both forms of PduO are characterized by a conserved phenylalanine residue that is responsible for displacing the lower ligand of the corrinoid substrate (Mera, St Maurice, Rayment, & Escalante-Semerena, 2009). The ATP-binding residues are located throughout the amino acid sequence, forming an inter-subunit ATP-binding site when the protein is folded.

EutT ACATs are similar in sequence to PduO ACATs, however EutT ACATs have an extended N-terminal region of unknown function (Costa et al., 2020). They lack the conserved phenylalanine residue of PduOs, however they share the same ATP-binding site as PduO. EutT ACATs are further subdivided into homologs that contain a metal-binding motif (CXXC) and homologs that lack this motif (Costa & Escalante-Semerena, 2018; Moore, Mera, & Escalante-Semerena, 2014). This metal-binding site is populated with Zn(II) when purified aerobically, and Fe(II) when purified anaerobically (Moore et al., 2014). A recent EutT homolog from *Listeria monocytogenes* was found to have the highest product formation of the ACATs studied thus far, which is useful for *in vitro* production of AdoCbl (Stracey, Costa, Escalante-Semerena, & Brunold, 2018).

This chapter describes the purification of the CobA enzyme of *S. enterica* and the EutT enzyme of *L. monocytogenes* for the rapid adenosylation of cobamides, preparation of reaction substrates, and best practices for adenosylation assays.

2. Expression and purification of CobA from *Salmonella enterica*

For the adenosylation of incomplete corrinoids, such as cobyric acid (Cby) and cobinamide (Cbi), the housekeeping ACAT from *S. enterica* (i.e., the CobA enzyme) is our enzyme of choice to generate AdoCby, AdoCbi, and AdoCbas. This section will outline the expression and purification process that has been optimized for production of this enzyme.

2.1 Cloning of *cobA*

In the method described here, we use plasmid pCOB15 (plasmid laboratory collection) to express untagged CobA protein for purification (Suh & Escalante-Semerena, 1995). Plasmid pCOB15 consists of *cobA* cloned into the *Nde*I-*Bam*HI sites of Novagen's cloning vector pET3a.

Transform 100 ng of the CobA-encoding plasmid into JE2369 (*metE*205 *ara*-9 Δ902(*cobA*-*trp*)/pGP1–2; T7 rpo^{+} T7 RNAP kan^{+}) carrying a plasmid used to produce T7 RNA polymerase (RNAP; e.g., pGP1–2, ATCC 40175) (Table 1). This transformation produces strain JE2884 (Table 1). The resulting strain is grown at 30 °C on lysogeny broth agar (LBA, 10 g/L tryptone, 10 g/L NaCl, 5 g/L yeast extract, 15 g/L agar) supplemented with kanamycin (50 μg/mL) and ampicillin (100 μg/mL).

2.2 Medium, cell growth and *cobA* gene expression

2.2.1 Growth medium

Prepare 6 L of lysogeny broth (LB, 10 g/L tryptone, 10 g/L NaCl, 5 g/L yeast extract) supplemented with glucose (4 g/L glucose) and aliquot 2 L into each of three 4-L flasks equipped with metal springs to improve aeration. Autoclave, then add ampicillin (50 μg/mL) and kanamycin (50 μg/mL) to each flask from filter-sterilized stocks.

Table 1 Strains used in these protocols.

Strain number	Species	Genotype	Reference
JE2369	*S. enterica* LT2	*metE*205 *ara*-9 Δ902(*cobA*-*trp*) / pGP1–2 T7 *rpo*$^{+}$ T7 RNAP *kan*$^{+}$	Suh and Escalante-Semerena (1993)
JE2884	*S. enterica* LT2	*metE*205 *ara*-9 Δ902(*cobA*-*trp*)/ pGP1–2 T7 *rpo*$^{+}$ T7 RNAP *kan*$^{+}$ pCOBA15 *bla*$^{+}$ *cobA*$^{+}$	Suh and Escalante-Semerena (1993)
JE21634	*E. coli* DH5α	F^{-} φ80*lac*ZΔM15 Δ(*lac*ZYA-*arg*F) U169 *rec*A1 *end*A1 *hsd*R17(r_K^-, m_K^+) *pho*A *sup*E44 λ^{-} *thi-1* *gyr*A96 *rel*A1/ pEUT14 *eutT*$^{+}$ *bla*$^{+}$	Costa and Escalante-Semerena (2018)
JE23613	*S. enterica* LT2	metE205 ara-9 Δ*pduO*522 Δ*cobA*1465 Δ*eutT*1141	Costa and Escalante-Semerena (2018)

Strains referenced in the text are listed in this table. Reference is where the strain first appears in the literature.

2.2.2 Cell growth

Add 4 mL of LB to the LBA plate containing strain JE2884, and make a cell suspension. Aliquot 1 mL of this cell suspension into each of the 3 flasks. Move the flasks into a 30 °C incubator shaking at 150 RPM. At an absorbance of ~0.7 at 650 nm wavelength, centrifuge 1 mL of cells from each flask and store at −80 °C to use as a *cobA* expression control for SDS-PAGE analysis.

2.2.3 Production of CobA

Incubate the culture at 42 °C for 30 min to induce production of the T7 polymerase. Then, incubate at 37 °C for 1.5 h to allow expression of the *cobA* gene. Transfer 1 mL from each flask into microcentrifuge tubes. Centrifuge and discard the supernatant as in Section 2.2.2 and store the cell pellets at −80 °C to ascertain protein production prior to purification using SDS-PAGE analysis. CobA is identified at 25 kDa in the induction condition. Pellet the rest of the 6 L of cells by centrifugation, discard supernatant, and store pellets in 50-mL plastic conical tubes at −80 °C.

2.3 Purification of CobA

2.3.1 Preparation of cell-free extracts

Once successful production of CobA is confirmed with SDS-PAGE analysis of the microfuge tube samples, suspend the larger frozen cell pellets in a total of 150 mL of buffer A. Lyse the cells on ice by sonication for 2 s at half-duty on/off at 60% power setting for 1 min. Rest the suspension for 5 min on ice and repeat sonication two more times. Clarify whole cell-free extracts by centrifugation at 40,000 × *g* at 4 °C for 2 h.

2.3.2 Ammonium sulfate precipitation

Calculate the dry weight of protease-free ammonium sulfate required to obtain 25% (w/v) saturation in the cell-free extract at 4 °C and crush the ammonium sulfate into a fine powder. Add the crushed ammonium sulfate to the cell-free extract a beaker in small amounts while stirring slowly, taking care to avoid accumulation of undissolved ammonium sulfate. Protein will begin to precipitate and the cell-free extract will appear opaque.

Tip: When stirring, do not stir so vigorously that bubbles (denatured protein) form, but not so slowly that the ammonium sulfate crystals accumulate on the bottom.

Centrifuge the cell-free extract at 10,000 × *g* for 10 min and move the supernatant into a fresh beaker. Calculate the dry weight of ammonium sulfate required to increase the ammonium sulfate concentration from 25% to 45% (w/v) saturation in the cell-free extract at 4 °C. Repeat the process of adding the ammonium sulfate centrifugation. Decant the supernatant and move forward with the precipitated protein pellet.

2.3.3 Hydrophobic chromatography

Resuspend the precipitated protein pellet in 70 mL of buffer B (buffer A with 20% w/v ammonium sulfate). This solution is loaded onto a HiPrep Phenyl HP column (Cytiva) that is equilibrated in buffer B at a rate of 0.9 mL/min. Wash the column with 3.5 column volumes (CV) of buffer B, and apply a buffer gradient from 100% buffer B to 100% buffer A over 3.5 CV. Apply another 3.5 CV of buffer A. CobA elutes at the end of the gradient. The presence of CobA in the eluted fractions can be checked by SDS-PAGE analysis.

2.3.4 Affinity chromatography

Pool fractions containing CobA, and add potassium chloride (KCl) to a final concentration of 0.2 M. The solution is loaded onto a HiTrap Blue High-Performance column (Cytiva) that is equilibrated with buffer C (buffer A supplemented with 0.2 M KCl). Wash column with 3 CV of buffer C, then follow with a 20 CV gradient from 0.2 to 2.0 M KCl. Check fractions for homogenous CobA using SDS-PAGE analysis. Pool together

CobA containing fractions and concentrate sample to at least 1 mg/mL using a Centriprep 10 device following manufacturer's instructions (Amicon). Dialyze against 1 L of buffer A thrice, at least 2 h each time. Use liquid nitrogen to flash freeze 100 μL samples, collect all pellets and store at −80 °C until used. CobA is stable for months at −80 °C.

3. Production and purification of EutT from *Listeria monocytogenes* in *E. coli*

Several homologs from all three families of ACATs have been successfully expressed in *E. coli* and purified to homogeneity that can perform the adenosylation of cobalamin. However, the *Listeria monocytogenes* EutT product yield is the highest analyzed (~98%) (Stracey et al., 2018). Because it is a metal-less EutT it does not require anaerobic purification. This section will describe the purification approach of this homolog.

3.1 Cloning of *eutT*

eutT from *Listeria monocytogenes* strain 10403S can be rapidly cloned using a high efficiency cloning procedure described elsewhere (VanDrisse & Escalante-Semerena, 2016). We used this method to generate a plasmid that encoded EutT with a N-terminal, rTEV-cleavable hexahistidine (H_6) tag named pEUT140, which is stored in DH5α as strain JE21634 (Costa & Escalante-Semerena, 2018).

3.2 Growth medium and *eutT* gene expression

3.2.1 Growth medium

Prepare 2 L of Terrific Broth [TB; tryptone (24 g/L), yeast extract (24 g/L), glycerol (4 mL/L), KH_2PO_4 (17 mM), K_2HPO_4 (72 mM), $MgSO_4$ (2 mM)] supplemented with ampicillin (100 μg/mL) and chloramphenicol (25 μg/mL) in each of three 4-L flasks with metal springs or in baffled flasks

3.2.2 Production of EutT

Transform 100 ng of the expression vector into chemically competent *E. coli* BL21-CodonPlus cells. Plate on LBA plates supplemented with ampicillin (100 μg/mL) and chloramphenicol (25 μg/mL). Incubate plates at 37 °C overnight. Add 4 mL of fresh LB, and use a sterilized plate spreader to generate a cell slurry. Transfer 1 mL of this cell suspension into each of the three flasks. Move the flasks into an incubator pre-heated to 37 °C and incubate cultures shaking at 150 RPM.

Measure the optical density of the cultures. At an absorbance of ~0.5 at 600 nm, transfer 1 mL from each flask into microcentrifuge tubes. Centrifuge and store the cell pellets at −80 °C to use as a negative control for the expression of the *eutT* gene by SDS-PAGE analysis.

Add IPTG (powder) to each flask to a final concentration of 1 mM in the medium. Incubate at 15 °C for 36 h. Transfer 1 mL from each flask into microcentrifuge tubes. Repeat centrifugation and store cell pellets at −80 °C to check for successful *eutT* expression by SDS-PAGE analysis. His_6-EutT is identified at 33 kDa in the induction condition. Pellet the rest of the cells by centrifugation, ideally at 6000 × *g* for 15 min at 4 °C in a refrigerated floor centrifuge. Discard supernatant, and store pellets in sterile 50-mL plastic tubes at −80 °C.

3.3 Purification of EutT

Once successful EutT production is confirmed, resuspend the large cell pellets in EutT buffer A supplemented with lysozyme (1.0 mg/mL), and DNase (0.1 mg/mL) at a ratio of 5 mL:1 g of cell pellet. Cell slurry is lysed by passing it once through a cell disruptor (Constant Systems) at 1.91 kPa. Protease inhibitor phenylmethanesulfonylfluoride (PMSF, 1 mM) is added after lysis.

Tip: PMSF should be added after cell disruption to avoid excess foaming of the lysate in the cell disruptor.

Centrifuge the cell lysate at 40,000 × *g* for 30 min in a floor refrigerated centrifuge cooled to 4 °C. Decant supernatant into a cold beaker. Using a 0.45 μm filter, filter the supernatant to remove any particulates prior to the purification procedure. Load the filtered supernatant onto a 5 mL HisTrap FF column (Cytiva) equilibrated with EutT buffer A at a rate of 1 mL/min. Wash unbound protein with 10 CV of EutT buffer A. Apply a gradient of EutT buffer B [HEPES (50 mM, pH 7.5 at 4 °C), NaCl (0.5 M), imidazole (500 mM), and TCEP (1 mM)] from 0% to 100% over 10 CV; collect 4-mL fractions. EutT will begin to elute as a large peak at ~26% buffer B and will continue to elute for the rest of the gradient. Five CV of 100% B are applied after the end of the gradient to ensure that the protein is eluted entirely. Check for presence of EutT in fractions where the protein is anticipated to be using SDS-PAGE, and pool fractions together that have >90% EutT present.

Tip: Initial fractions containing EutT are lower in purity than later fractions. For highest purity, it is best to select the fractions with the largest concentration of EutT and the fractions thereafter.

Add rTEV protease to the pooled fractions at a concentration of 1 mg of rTEV per 100 mg of purified EutT. This mixture is incubated at room temperature for 2 h before it is moved into dialysis tubing (10,000 MWCO) and equilibrated against 1 L of buffer A three times. Each dialysis step should equilibrate for a minimum of 1 h.

Load the dialyzed and cleaved EutT protein solution again onto a 5 mL HisTrap FF column equilibrated with EutT buffer A at a rate of 1 mL/min. Capture the flow-through in 4-mL fractions and confirm the purity of EutT using SDS-PAGE analysis. Pool fractions that contain purified, cleaved EutT at ~30 kDa. EutT is then concentrated using an Amicon Ultra 15 centrifugal filter unit (10,000 Da molecular mass cutoff) at the recommended speed in 5-min runs. Between centrifugations, gently resuspend the protein to prevent aggregation.

Dialyze the EutT protein in 1 L of storage buffer [HEPES (50 mM, pH 7.5 at 4 °C), NaCl (0.5 M), and TCEP (1 mM)] three times at 4 °C, minimum 2 h per dialysis step. Store purified and concentrated EutT protein at −80 °C after flash freezing 100 mL samples in liquid nitrogen.

4. Preparation of corrinoid substrates

Some corrinoid substrates, such as hydroxylcobalamin (HOCbl), are readily available commercially. Others, such as hydroxylcobinamide (HOCbi) and hydroxylcobyric acid (HOCby) substrate must be synthesized from commercially available dicyanocobinamide $(CN)_2Cbi$ and prepared dicyanocobyric acid [$(CN)_2Cby$]. Here, we describe how to prepare these substrates for the corrinoid adenosylation assays.

4.1 Synthesis and purification of HOCbi and HOCby

4.1.1 HPLC purification of dicyano-cobyric acid [$(CN)_2Cby$]

$(CN)_2Cby$ is not commercially available and is enzymatically generated from $(CN)_2Cbi$ or CNCbl. The following procedure uses $(CN)_2Cbi$ (Sigma-Aldrich Chemical Co.) as the substrate. $(CN)_2Cbi$ is incubated with the hexahistidine (H_6), N-terminally tagged bifunctional adenosylcobinamide amidohydrolase/adenosylcobalamin 5′-phosphate phophatase (H_6-CbiS) enzyme from *Methanopyrus kandleri* (H_6_*Mk*CbiS), purification is described in Section 4.1.2 (Woodson & Escalante-Semerena, 2006). The following reaction conditions are recommended: KH_2PO_4 buffer (0.5 M, pH 9), 10.6 μM H_6-CbiS enzyme, and 60 μM CN_2Cbi in reaction mixtures ranging from 0.1–5 mL. Reaction mixtures are degassed under a stream of N_2 as depicted in Fig. 1 and incubated for 2 h at 90 °C in a Balch vial. Under these conditions, the conversion of CN_2Cbi to CN_2Cby will result in a color change from pink to orange.

Tip: Protein stored for extended periods at −80 °C may exhibit reduced activity. We have extended reaction times up to overnight to maximize yield.

After incubation, reactions are exposed to oxygen and quickly frozen, then thawed and centrifuged at 16,000 × *g* for 10 min to precipitate protein. The supernatant is filtered using Corning Spin-X centrifuge filters and injected into a Shimadzu HPLC system equipped with a Phenomenex Synergi 4 μ Hydro-RP column equilibrated with 98% buffer A

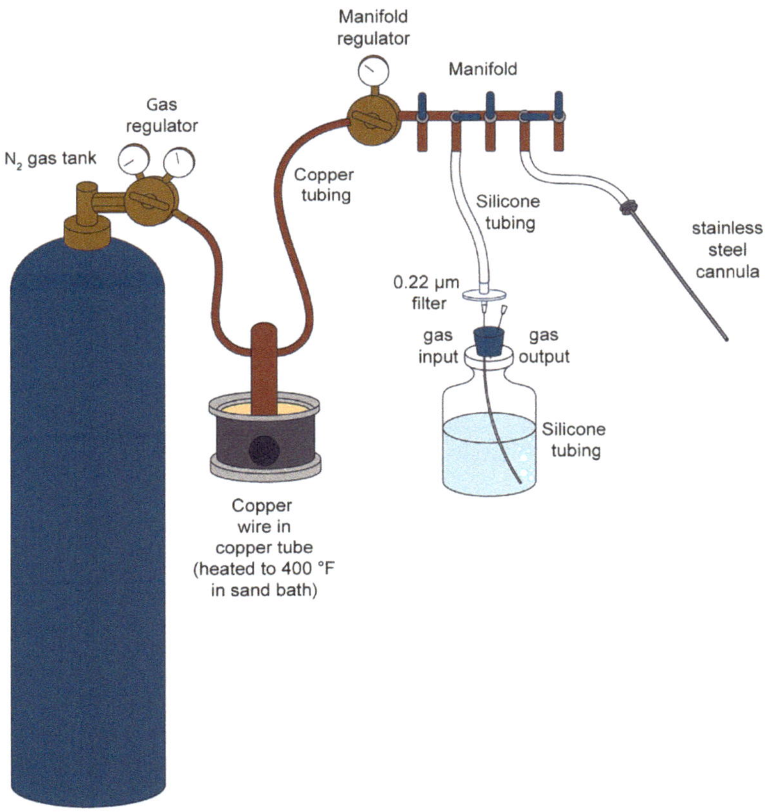

Fig. 1 Setup for N_2 sparging of enzyme assay stocks. An N_2 gas tank is connected to a gas regulator set to an output of 20 psi. The gas is transported to the manifold through copper tubing. A large copper tube containing copper wire is heated at 350 °F to scrub traces of oxygen. At the manifold, a second regulator further controls gas output to 10 psi. N_2 gas is streamed into a stoppered bottle through a 22-gauge needle and a thin polyethylene tube to sparge the reagent. A second needle is included for gas output.

[KH_2PO_4 (0.1 M, pH 6.5) + KCN (10 mM)], 2% buffer B [KH_2PO_4 (0.1 M, pH 8) + KCN (10 mM)]: acetonitrile (100%) at a ratio of 1:1. The column is developed at a flow rate of 1 mL/min. CN_2Cby can be separated from CN_2Cbi by first allowing the column to develop for 1 min after injection, followed by a 5-min linear gradient to 75% buffer A/25% buffer B, then a second linear gradient of 15 min to 65% buffer A/35% buffer B. A clean-up step is then performed on fractions containing CN_2Cby. We recommend loading the material onto a C18 SepPak, eluting with methanol (100%) and subsequently evaporating using a SpeedVac centrifuge. Dried CN_2Cby can then be resuspended in your buffer of choice or H_2O.

4.1.2 Overproduction and purification of the H_6-MkCbiS enzyme

cbiS (locus tag MK1672) from *Methanopyrus kandleri* was cloned into pTEV5 resulting in plasmid p*Mk*CBIS13 (Rocco, Dennison, Klenchin, Rayment, & Escalante-Semerena, 2008), which encoded the wild-type CbiS protein with a hexahistidine tag fused to its N-terminus. The resulting plasmid was electroporated into *E. coli* BL21-λDE3, and the resulting strain is inoculated in 10 mL of nutrient broth (Difco) and incubated overnight (~16 h) at 37 °C with shaking at 200 rpm. Use this overnight culture to inoculate 1.5 L of TB. Grow at 37 °C with shaking at 200 rpm to an absorbance at 600 nm of 0.8, then induce $cbiS^+$ expression by adding isopropyl-β-D-1-thiogalactopyranoside (IPTG) to a final concentration of 1 mM. Incubate with shaking at 150 rpm at 20 °C for an additional 18 h. Harvest cells by centrifugation at 6000 rpm for 15 min at 4 °C. *Enzyme purification.* Resuspend cell pellet in buffer A [Na_2HPO_4 (20 mM), pH 7.4, NaCl (500 mM), imidazole (20 mM)]. Add DNase and lysozyme to a final concentration of 0.025 and 1 mM, respectively. Lyse cells using a Cell Disruptor at 1.7 kPa, or by French Press.

Tip: We have observed that use of sonication to lyse cells may reduce the yield of soluble CbiS.

The lysate is cleared by centrifugation at 43,000 × *g* for 1 h, at 4 °C, then loaded onto a Ni-NTA affinity chromatography column. CbiS purification is achieved by washing with the following buffers (5 column volumes each): buffer A [Na_2HPO_4 (20 mM, pH 7.4) + NaCl (500 mM) + imidazole (20 mM)], buffer B [Na_2HPO_4 (20 mM, pH 7.4) + NaCl (500 mM) + imidazole (50 mM)], buffer C [Na_2HPO_4 (20 mM, pH 7.4) + NaCl (500 mM) + imidazole (200 mM)]. Pool fractions containing His-CbiS and dialyze against 1 L of buffer D [Trizma base, (20 mM, pH 7.4) + NaCl (500 mM) + glycerol (30%)] at least three times. Flash freeze protein in liquid N_2 and store at −80 °C until used.

Tip: Cleavage of the H_6 tag is not necessary for CbiS activity. In fact, removal of this tag reduces activity of the enzyme.

4.1.3 Synthesis of HOCbi and HOCby

Dissolve 3–5 mg of dicyanocobinamide $(CN)_2Cbi$ or cyanocobyric acid (CNCby) in 5 mL of MilliQ water in a small Balch vial. To N_2-sparge cut a piece of polyethylene tubing (Beckton Dickinson, 3 mm (O.D.) × 0.7 mm (I.D.)) at a length equal to the height of the Balch vial. Insert two 22-gauge needles into a stopper and attach the tubing to one needle (Fig. 1). Stopper the Balch vial and attach the N_2 line. Sparge the solution with N_2 gas for 20 min.

Remove both needles from the stopper. Use a stainless-steel cannula to produce a stream of N_2 gas, sparge a Hamilton syringe in a stream of N_2 gas, and transfer this solution to a sealed, N_2 gas sparged vial containing ~27 mg $NaBH_4$ and a stirring bar. Stir overnight under a slow stream of N_2. Then, slowly re-oxidize the solution by exposing it to air over several hours. At this point, the solution should be brownish purple for Cbi, and a very similar color for Cby.

We recommend loading the material onto a C18 SepPak, eluting with methanol (100%) and subsequently evaporating using a SpeedVac centrifuge. Resuspend the dried corrinoid in MilliQ water or buffer of choice. This solution can be quantified by using the molar absorptivity values and wavelengths found in Table 2.

Table 2 Molar absorptivity values for the corresponding wavelengths for measurement of each corrinoid.

Corrinoid	Wavelength (nm)	Molar absorptivity (solution) (M^{-1} cm^{-1})	Source
H_2OCbl	351	22,600	Hogenkamp, Pailes, and Brownson (1971)
HOCbl	357	17,500	
CNCbl and cyano-pseudocobalamin	361	27,500	
$(CN)_2Cbi$ and $(CN)_2Cby$	368	30,400	Unpublished
$(H_2O)_2Cbi$ and $(H_2O)_2Cby$	348	25,200	Ford, Nichols, and Shambee (1991)
$(HO)_2Cbi$ and $(HO)_2Cby$	344	19,300	

All molar absorptivity values were determined in water unless otherwise specified.

5. Measuring adenosyltransferase activity

The primary method to measure adenosyltransferase activity *in vitro* is by monitoring changes in wavelength of the corrinoid as it converts from a Co(I) or Co(II) state to a Co(III) state. Alternatively, growth of selected *S. enterica* strains can be used to determine successful product formation. This section will describe how to measure the formation of adenosylated corrinoids spectrophotometrically in a cuvette or well plate, as well as different biological approaches to confirm product formation.

5.1 *In vitro* enzyme assay

The assay can be set up in a well plate in an anaerobic chamber equipped with a well plate reader, or in a sealed cuvette that is sparged with oxygen-free N_2.

5.1.1 Corrinoid adenosylation assay stocks

5.1.1.1 Reaction buffer

The reaction buffer should be prepared in a Balch vial and sparged with oxygen-free N_2 as described in Section 4.1.3 and shown in Fig. 1. The reaction buffer for the CobA enzyme from *S. enterica* is Tris·HCl buffer (200 mM, pH 8.0 at 37 °C). The reaction buffer for the EutT enzyme from *L. monocytogenes* is a modified Tris·HCl buffer (100 mM, pH 7.5 at 37 °C) + KCl (500 mM).

5.1.1.2 Corrinoid

The corrinoid substrate should be prepared in MilliQ H_2O and sparged with oxygen-free N_2 gas. The recommended concentration for the corrinoid stock is 1–5 mM, depending on the solubility of the corrinoid substrate

5.1.1.3 Nucleoside triphosphate substrate

The nucleoside triphosphate substrate, i.e., ATP, should be prepared fresh in a sealed Balch vial containing reaction buffer previously sparged with oxygen-free N_2 gas. This reaction mixture should be supplemented with divalent cations present at an equimolar concentration to ATP. Use $MgCl_2$ for EutT, and $MnCl_2$ for CobA

5.1.2 Co(II)Corrinoid assay-specific stocks

5.1.2.1 Preparation of NAD(P)H:flavin oxidoreductase (Fre)

For a detailed description of the isolation of Fre see Buan and Escalante-Semerena (2005) and Fonseca and Escalante-Semerena (2000).

An hour prior to assays, an aliquot of Fre enzyme should be thawed on ice, and then moved into the anaerobic chamber or into a Balch vial that has been flushed with oxygen-free N_2 gas

5.1.2.2 Preparation of flavodoxin A (FldA) and FldA reductase (Fpr)

For adenosylation reactions with CobA, FldA and Fpr are necessary to reduce the corrinoid from Co(III) to Co(II). Purification of FldA and Fpr should be done as described elsewhere (Buan & Escalante-Semerena, 2005; Fonseca & Escalante-Semerena, 2001). An hour prior to assays, an aliquot of FldA and Fpr enzymes should be thawed on ice, and then moved into the anaerobic chamber in a sealed Balch vial that has been sparged with oxygen-free N_2 gas

5.1.2.3 Preparation of NADH or NADPH

NADH (100 mM) and NADPH (100 mM) stocks should be prepared fresh in a Balch vial containing reaction buffer that has been sparged with oxygen-free N_2 gas

5.1.2.4 Preparation of FAD^+

FAD^+ (2.5 mM) should be prepared fresh in a Balch vial containing reaction buffer previously sparged with oxygen-free N_2 gas

5.1.3 Co(I)Corrinoid assay-specific stocks

5.1.3.1 Preparation of Ti(III)citrate

To prepare a 200 mM stock of Ti(III)citrate, weigh out 1.9 g of anhydrous citric acid (Sigma, 192.2 g/mol). Suspend in 40 mL and adjust pH to 8.0 with a 1 M NaOH solution. Move this solution into a Balch vial and sparge the solution with oxygen-free N_2 gas as described in Section 4.1.1. Move the Balch vial containing the citric acid solution and a bottle of Ti(III)chloride to an anaerobic chamber. In the chamber, open the vial containing the citric acid solution and transfer 10.3 mL of Ti(III)chloride. Seal the vial again before removing it from the anaerobic chamber. Reduced Ti(III)citrate should be a deep purple/black color. White precipitate indicates that oxidation of Ti(II)citrate occurred and should not be used for Co(I)Cbl assays

5.1.4 Co(II) corrinoid assay

In this assay, the Co(III) corrinoid is reduced to its Co(II). When the corrinoid is reduced to the Co(II) state, the mixture will appear orange-yellow (Fig. 2).

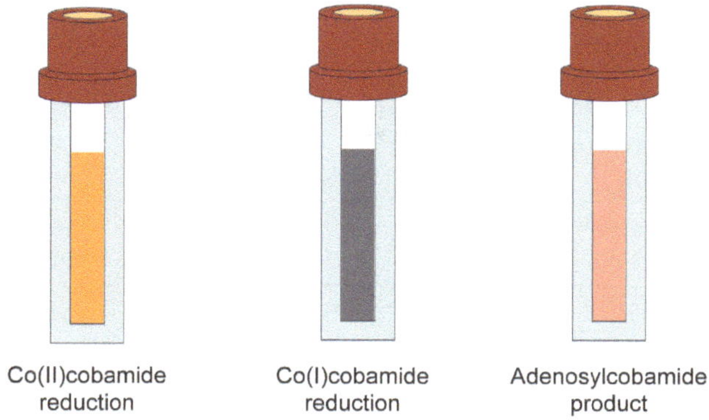

Fig. 2 Appearance of the adenosylcobalamin reaction with two different reduction approaches and successful adenosylation of cobalamin. When the cobalt has been successfully reduced from Co(III) to Co(II), the cobalamin appears orange. When the cobalt has been successfully reduced from Co(III) to Co(I), the cobalamin is colorless, and the reaction appears gray due to the Ti(III)citrate. When the adenosylcobalamin product forms successfully, the reaction appears pink in color.

5.1.4.1 96 well plate setup

For the plate format in both the Co(I) and Co(II) assay, a quartz 96-well plate is ideal. The quartz allows for higher light transmissibility and retains less oxygen. However, a plastic 96-well plate that has been kept under anoxic conditions for at least 3 days is also suitable for this assay. Other plastics needed for the experiment, such as pipette tips and microfuge tubes, should also be kept under anoxic conditions for at least 3 days to minimize reaction exposure to oxygen. *The assay should be set up entirely in the anaerobic chamber.*

For each reaction, the reaction components are divided between two wells. In one well, the corrinoid substrate (1 nM – 1 μM), and the reducing system [Fre (40 nM), NADH (1 mM), and FAD (25 μM)] for EutT, or [FldA (2 μM), Fpr (3 μM), and NADPH (1 mM)] for CobA, are mixed together with the reaction buffer in a total of 150 μL. In the second well, the ACAT enzyme (5–500 nM), and ATP substrate (Mg·ATP for EutT or Mn·ATP for CobA, 1 mM) are mixed in reaction buffer for a total volume of 150 μL. This allows for both pre-incubation of the enzyme with the substrate that binds first, and reduction of Co(III) to Co(II). Incubate the well plate in the well plate reader at 37 °C for 15 min. As mentioned above, the solution containing Co(II) corrinoid should be orange. The reaction is started by adding the 150 μL of Co(II)corrinoid mixture to the well containing the ACAT enzyme and ATP. The reaction is monitored at the wavelength specified for each corrinoid in Table 3. The wells should be monitored every 10 s over the course of 20 min, with pathlength correction.

Table 3 Molar absorptivity values for the corresponding wavelengths for monitoring adenosylcorrinoid formation.

Corrinoid	Wavelength (nm)	Molar absorptivity (solution) (M^{-1} cm^{-1})	Notes	Source
AdoCbl	525	8000	0.01 M phosphate buffer pH 7.1	Hogenkamp et al. (1971)
	525	7877	0.1 M Tris 0.5 M KCl pH 7.5 (37 °C)	Costa and Escalante-Semerena (2018)
AdoCbi and AdoCby	458	9300		Hay and Finke (1987)

All molar absorptivity values were determined in water unless otherwise specified.

Note: Keep in mind that the reagent and enzyme concentrations suggested refer to the final concentration in the total reaction volume. When setting up the reactions, each well will have 2× the concentration of reagents and enzymes prior to mixing.

In the data analysis, the linear range of the reaction should be isolated, and the maximum slope over 10 points should be recorded. To maintain consistency between samples, the frequency of timepoints, and the number of points used to determine the slope, should be consistent across all experiments. This slope, depending on the well plate reader used, will be ΔOD/(unit of time). This can be converted to [moles of adenosylated corrinoid formed]/unit of time using the corresponding molar absorptivity values listed in Table 3, and 1 cm as the pathlength.

5.1.4.2 Cuvette setup

In the absence of an anoxic chamber equipped with a well plate reader, sealed quartz cuvettes with septa in the screwcaps (Starna Cells, Product No. 9-Q-10-GL14-S) are also suitable. The spectrophotometer should accommodate these cuvettes, which are taller than standard cuvettes, and be able to control the temperature of the cuvette. *Additionally, the cuvettes will need to be kept in a dark room, or under red light over the course of the reaction.* To transfer reaction stocks into the sealed quartz cuvette, Hamilton syringes of 10 μL, 50 μL, and 1 mL volume are recommended.

To begin, the quartz cuvettes are sealed with their plastic screwcaps containing rubber septa. Two 18-gauge needles are inserted into the septum, and one needle is connected to the oxygen-free N_2 gas manifold setup illustrated in Fig. 3. The N_2 gas is streamed into the cuvette while the reaction is prepared.

Use a nitrogen-flushed Hamilton syringe to transfer reagents from stopped vials into the septum-sealed cuvette. The order of addition should be: (i) reaction buffer, (ii) corrinoid reductant(s), (iii) corrinoid substrate, and

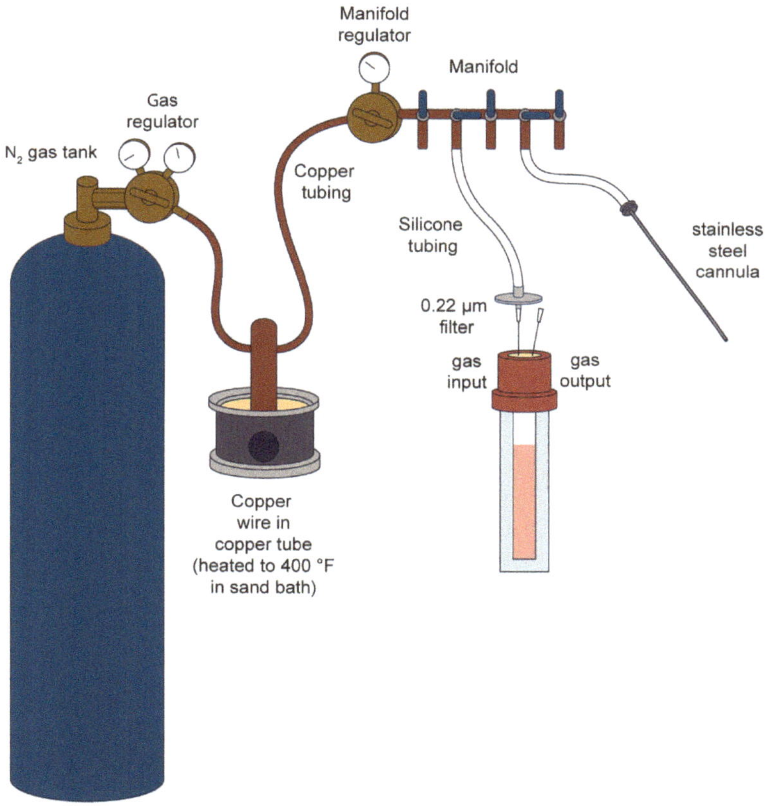

Fig. 3 Set up for preparation of the enzyme reaction in cuvettes. The gas connections as described in Fig. 1 are identical in this setup. The silicone tubing is connected to a 0.22 μM filter and to an 22-gauge needle. This needle and a second needle are inserted into the septum of a sealed 1-cm pathlength cuvette with 1 mL volume. N_2 gas is streamed through the cuvette while the components of the reaction are added. A stainless-steel cannula is also connected to the manifold, and N_2 gas is streamed from the cannula in order to flush the Hamilton syringe used to transfer reagents into the cuvette.

(iv) enzyme. Pull the 18 gauge needles out of the rubber septum, and invert cuvette a few times to mix the solution. Warm the cuvette to temperature in the cuvette reader for 15 min and add the nucleotide substrate with a flushed Hamilton syringe to start the reaction. When the nucleotide substrate is added, invert cuvette a few times to mix. Monitor the reaction at the wavelength specified for each corrinoid in Table 2 for 20–30 min.

The reaction rate should be obtained from the linear range, and converted to moles of adenosylated corrinoid formed/unit of time using the molar absorptivity value in Table 2 (or an experimentally determined molar absorptivity value in your reaction buffer) and a pathlength of 1 cm.

5.1.5 Co(I) corrinoid assay

In this assay, the corrinoid is chemically reduced to the Co(I) state using Ti(III)citrate. This assay just probes the ability of the enzyme to capture the 5′-adenosyl group of ATP to form the Co–C bond

5.1.5.1 Well plate setup

Follow the general setup instructions for the Co(II) assay well plate setup in Section 5.1.4.1. However, the corrinoid reducing agent should be changed to Ti(III)citrate (2.5 mM). Successful reduction of the corrinoid will result in a reaction mixture that appears gray in color (Fig. 2).

5.1.5.2 Cuvette setup

Follow the general setup instructions for the Co(II) assay cuvette setup in Section 5.1.4.2. However, the reducing agent should be changed to Ti(III)citrate (2.5 mM).

5.1.6 Bioassay approaches

In addition to monitoring the adenosylation of corrinoids by changes in absorbance intensity at a particular wavelength, formation of the adenosylated corrinoid can also be confirmed using a bioassay approach

Formation of AdoCbl can be assayed by supplementing the filter-sterilized reaction mixture into an ethanolamine minimal medium with a *S. enterica* Δ*eutT* Δ*pduO* Δ*cobA* (ΔACAT) strain (Table 1) (Costa & Escalante-Semerena, 2018). This strain cannot grow on ethanolamine as a carbon source, and growth with supplemented AdoCbl is dose-dependent (Fig. 4). The strain used in Fig. 4 is JE23613 (Table 1). The media recipe is as follows: No-carbon E phosphate medium [NCE, 1 × (Berkowitz, Hushon, Whitfield,

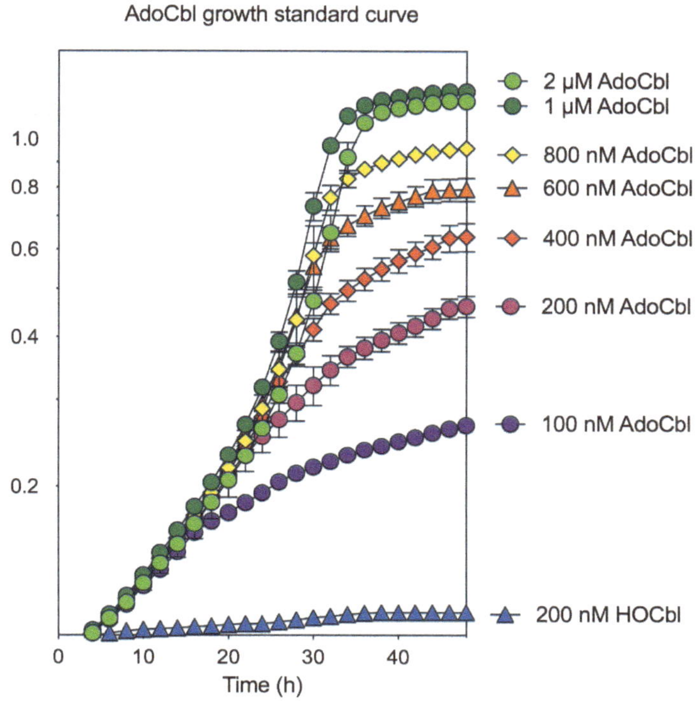

Fig. 4 Bioassay using an ΔACAT strain in ethanolamine minimal medium. An ΔACAT strain is grown on ethanolamine minimal medium supplemented with defined concentrations of adenosylcobalamin (100 nM–2 μM). Hydroxocobalamin (200 nM) was added as a negative control.

Roth, & Ames, 1968)], Wolfe's trace minerals [TM, 1 × (Balch & Wolfe, 1976)], $MgSO_4$ (1 mM), and ethanolamine hydrochloride (90 mM). 190 μL of the medium is added to each well, followed by 5 μL of the reporter strain grown overnight in Difco nutrient broth (NB, 8 g/L) supplemented with NaCl (5 g/L) and 5 μL of the filter-sterilized reaction mixture. The plate should be prepared and kept in a dark room or wrapped in aluminum foil until the plate is inserted into a well plate reader to monitor growth at a wavelength of 600 nm.

Formation of AdoCbi or AdoCby can be assayed by supplementing 0.5–5 μL of the filter-sterilized reaction mixture into glycerol minimal media with *S. enterica* Δ*cobA* strain (Table 1) (Buan, Rehfeld, & Escalante-Semerena, 2006). This assay is much more sensitive, with full growth occurring with supplementation of AdoCbi or AdoCby at nanomolar concentrations (O'Toole & Escalante-Semerena, 1993).

6. Conclusions

Adenosylated cobamides are critical for metabolic function of many organisms across all domains of life. This methodology describes the efficient *in vitro* adenosylation of cobamides for applications in research and biotechnology.

Acknowledgments

This work was supported by NIH grant R35-GM130399 to J.C.E.-S. and F.G.C. was supported in part by NIH grant F31-GM122394.

References

Balch, W. E., & Wolfe, R. S. (1976). New approach to the cultivation of methanogenic bacteria: 2-Mercaptoethanesulfonic acid (HS-CoM)-dependent growth of *Methanobacterium ruminantium* in a pressurized atmosphere. *Applied and Environmental Microbiology, 32*, 781–791. https://doi.org/10.1128/aem.32.6.781-791.1976.

Berkowitz, D., Hushon, J. M., Whitfield, H. J., Jr., Roth, J., & Ames, B. N. (1968). Procedure for identifying nonsense mutations. *Journal of Bacteriology, 96*, 215–220. https://doi.org/10.1128/jb.96.1.215-220.1968.

Buan, N. R., & Escalante-Semerena, J. C. (2005). Computer-assisted docking of flavodoxin with the ATP:Co(I)rrinoid adenosyltransferase (CobA) enzyme reveals residues critical for protein-protein interactions but not for catalysis. *The Journal of Biological Chemistry, 280*, 40948–40956. https://doi.org/10.1074/jbc.M506713200.

Buan, N. R., Rehfeld, K., & Escalante-Semerena, J. C. (2006). Studies of the CobA-type ATP:Co(I)rrinoid adenosyltransferase enzyme of *Methanosarcina mazei* strain Gö1. *Journal of Bacteriology, 188*, 3543–3550. https://doi.org/10.1128/JB.188.10.3543-3550.2006.

Buan, N. R., Suh, S. J., & Escalante-Semerena, J. C. (2004). The *eutT* gene of *Salmonella enterica* encodes an oxygen-labile, metal-containing ATP:Corrinoid adenosyltransferase enzyme. *Journal of Bacteriology, 186*, 5708–5714. https://doi.org/10.1128/JB.186.17.5708-5714.2004.

Costa, F. G., & Escalante-Semerena, J. C. (2018). A new class of EutT ATP:Co(I)rrinoid adenosyltransferases found in *Listeria monocytogenes* and other *Firmicutes* does not require a metal ion for activity. *Biochemistry, 57*(34), 5076–5087. https://doi.org/10.1021/acs.biochem.8b00715.

Costa, F. G., Greenhalgh, E. D., Brunold, T. C., & Escalante-Semerena, J. C. (2020). Mutational and functional analyses of substrate binding and catalysis of the *Listeria monocytogenes* EutT ATP:Co(I)rrinoid adenosyltransferase. *Biochemistry, 59*(10) 1124–1136. https://doi.org/10.1021/acs.biochem.0c00078.

Escalante-Semerena, J. C., Suh, S. J., & Roth, J. R. (1990). *cobA* function is required for both de novo cobalamin biosynthesis and assimilation of exogenous corrinoids in *Salmonella typhimurium*. *Journal of Bacteriology, 172*, 273–280. https://doi.org/10.1128/jb.172.1.273-280.1990.

Fonseca, M. V., & Escalante-Semerena, J. C. (2000). Reduction of cob(III)alamin to cob(II) alamin in *Salmonella enterica* serovar Typhimurium LT2. *Journal of Bacteriology, 182*, 4304–4309. https://doi.org/10.1128/JB.182.15.4304-4309.2000.

Fonseca, M. V., & Escalante-Semerena, J. C. (2001). An in vitro reducing system for the enzymic conversion of cobalamin to adenosylcobalamin. *The Journal of Biological Chemistry*, *276*, 32101–32108. https://doi.org/10.1074/jbc.M102510200.

Ford, S. H., Nichols, A., & Shambee, M. (1991). The preparation and characterization of the diaquo- forms of several incomplete corrinoids:Cobyric acid, cobinamide, and three isomeric cobinic acid pentaamides. *Journal of Inorganic Biochemistry*, *41*, 235–244. https://doi.org/10.1016/0162-0134(91)80015-a.

Hay, B. P., & Finke, R. G. (1987). Thermolysis of the Co-C bond in adenosylcorrins .3. Quantification of the axial base effect in adenosylcobalamin by the synthesis and thermolysis of axial base-free adenosylcobinamide—Insights into the energetics of enzyme-assisted cobalt carbon bond homolysis. *Journal of the American Chemical Society*, *109*, 8012–8018. https://doi.org/10.1021/ja00260a011.

Hogenkamp, H. P. C., Pailes, W. H., & Brownson, C. (1971). Preparation of 5′-deoxyadenosylcobalamin and analogs containing modified nucleosides. In D. B. McCormick, & L. D. Wright (Eds.), *Vol. 18C. Vitamins and coenzymes* (pp. 57–71). Academic Press, Inc. https://doi.org/10.1016/S0076-6879(71)18009-3.

Johnson, C. L., Buszko, M. L., & Bobik, T. A. (2004). Purification and initial characterization of the *Salmonella enterica* PduO ATP:Cob(I)alamin adenosyltransferase. *Journal of Bacteriology*, *186*, 7881–7887. https://doi.org/10.1128/JB.186.23.7881-7887.2004.

Johnson, C. L., Pechonick, E., Park, S. D., Havemann, G. D., Leal, N. A., & Bobik, T. A. (2001). Functional genomic, biochemical, and genetic characterization of the *Salmonella pduO* gene, an ATP:Cob(I)alamin adenosyltransferase gene. *Journal of Bacteriology*, *183*, 1577–1584. https://doi.org/10.1128/JB.183.5.1577-1584.2001.

Lexa, D., & Saveant, J.-M. (1976). Electrochemistry of vitamin B12. 1. Role of the base-on/base-off reaction in the oxidoreduction mechanism of the B12r-B12s system. *Journal of the American Chemical Society*, *98*, 2652–2658.

Mattes, T. A., Deery, E., Warren, M. J., & Escalante-Semerena, J. C. (2017). Cobalamin biosynthesis and insertion. In R. A. Scott (Ed.), *Encyclopedia of inorganic and bioinorganic chemistry* (pp. 1–24). John Wiley & Sons, Ltd. https://doi.org/10.1002/9781119951438.eibc2489.

Mera, P. E., & Escalante-Semerena, J. C. (2010a). Dihydroflavin-driven adenosylation of 4-coordinate Co(II) corrinoids: Are cobalamin reductases enzymes or electron transfer proteins? *The Journal of Biological Chemistry*, *285*, 2911–2917. https://doi.org/10.1074/jbc.m109.059485.

Mera, P. E., & Escalante-Semerena, J. C. (2010b). Multiple roles of ATP:Cob(I)alamin adenosyltransferases in the conversion of B(12) to coenzyme B (12). *Applied Microbiology and Biotechnology*, *88*, 41–48. https://doi.org/10.1007/s00253-010-2773-2.

Mera, P. E., St Maurice, M., Rayment, I., & Escalante-Semerena, J. C. (2009). Residue Phe112 of the human-type corrinoid adenosyltransferase (PduO) enzyme of *Lactobacillus reuteri* is critical to the formation of the four-coordinate Co(II) corrinoid substrate and to the activity of the enzyme. *Biochemistry*, *48*, 3138–3145. https://doi.org/10.1021/bi9000134.

Moon, J. H., Park, A. K., Jang, E. H., Kim, H. S., & Chi, Y. M. (2008). Crystal structure of a PduO-type ATP:Cobalamin adenosyltransferase from *Burkholderia thailandensis*. *Proteins*, *72*, 1066–1070. https://doi.org/10.1002/prot.22084.

Moore, T. C., Mera, P. E., & Escalante-Semerena, J. C. (2014). The EutT enzyme of *Salmonella enterica* is a unique ATP:Cob(I)alamin adenosyltransferase metalloprotein that requires ferrous ions for maximal activity. *Journal of Bacteriology*, *196*, 903–910. https://doi.org/10.1128/JB.01304-13.

Moore, T. C., Newmister, S. A., Rayment, I., & Escalante-Semerena, J. C. (2012). Structural insights into the mechanism of four-coordinate cob(II)alamin formation in the active site of the *Salmonella enterica* ATP:Co(I)rrinoid adenosyltransferase (CobA) enzyme: Critical role of residues Phe91 and Trp93. *Biochemistry*, *51*, 9647–9657. https://doi.org/10.1021/bi301378d.

Ortiz de Orue Lucana, D., Hickey, N., Hensel, M., Klare, J. P., Geremia, S., Tiufiakova, T., et al. (2016). The crystal structure of the C-terminal domain of the *Salmonella enterica* PduO protein: An old fold with a new heme-binding mode. *Frontiers in Microbiology*, 7, 1010. https://doi.org/10.3389/fmicb.2016.01010.

O'Toole, G. A., & Escalante-Semerena, J. C. (1993). *cobU*-dependent assimilation of non-adenosylated cobinamide in *cobA* mutants of *Salmonella typhimurium*. *Journal of Bacteriology*, *175*, 6328–6336. https://doi.org/10.1128/jb.175.19.6328-6336.1993.

Park, A. K., Chi, Y. M., & Moon, J. H. (2011). Crystal structure of PduO-type ATP:Cob(I)alamin adenosyltransferase from *Bacillus cereus* in a complex with ATP. *Biochemical and Biophysical Research Communications*, *408*, 417–421. https://doi.org/10.1016/j.bbrc.2011.04.036.

Rocco, C. J., Dennison, K. L., Klenchin, V. A., Rayment, I., & Escalante-Semerena, J. C. (2008). Construction and use of new cloning vectors for the rapid isolation of recombinant proteins from *Escherichia coli*. *Plasmid*, *59*, 231–237. https://doi.org/10.1016/j.plasmid.2008.01.001.

St Maurice, M., Mera, P. E., Taranto, M. P., Sesma, F., Escalante-Semerena, J. C., & Rayment, I. (2007). Structural characterization of the active site of the PduO-type ATP:Co(I)rrinoid adenosyltransferase from *Lactobacillus reuteri*. *The Journal of Biological Chemistry*, *282*, 2596–2605. https://doi.org/10.1074/jbc.M609557200.

Stich, T. A., Yamanishi, M., Banerjee, R., & Brunold, T. C. (2005). Spectroscopic evidence for the formation of a four-coordinate Co^{2+} cobalamin species upon binding to the human ATP:Cobalamin adenosyltransferase. *Journal of the American Chemical Society*, *127*, 7660–7661. https://doi.org/10.1021/ja050546r.

Stracey, N. G., Costa, F. G., Escalante-Semerena, J. C., & Brunold, T. C. (2018). Spectroscopic study of the EutT adenosyltransferase from *Listeria monocytogenes*: Evidence for the formation of a four-coordinate cob(II)alamin intermediate. *Biochemistry*, *57*(34), 5088–5095. https://doi.org/10.1021/acs.biochem.8b00743.

Suh, S.-J. (1994). *The role of CobA in adenosylation of corrinoids in Salmonella typhimurium*. Ph.D., University of Wisconsin-Madison.

Suh, S. J., & Escalante-Semerena, J. C. (1993). Cloning, sequencing and overexpression of *cobA* which encodes ATP:Corrinoid adenosyltransferase in *Salmonella typhimurium*. *Gene*, *129*, 93–97. https://doi.org/10.1016/0378-1119(93)90701-4.

Suh, S., & Escalante-Semerena, J. C. (1995). Purification and initial characterization of the ATP:Corrinoid adenosyltransferase encoded by the *cobA* gene of *Salmonella typhimurium*. *Journal of Bacteriology*, *177*, 921–925. https://doi.org/10.1128/jb.177.4.921-925.1995.

VanDrisse, C. M., & Escalante-Semerena, J. C. (2016). New high-cloning-efficiency vectors for complementation studies and recombinant protein overproduction in *Escherichia coli* and *Salmonella enterica*. *Plasmid*, *86*, 1–6. https://doi.org/10.1016/j.plasmid.2016.05.001.

Weissbach, H., Ladd, J. N., Volcani, B. E., Smyth, R. D., & Barker, H. A. (1960). Struture of the adenylcobamide coenzyme: Degradation by cyanide, acid, and light. *The Journal of Biological Chemistry*, *235*, 1462–1473.

Woodson, J. D., & Escalante-Semerena, J. C. (2006). The *cbiS* gene of the archaeon *Methanopyrus kandleri* AV19 encodes a bifunctional enzyme with adenosylcobinamide amidohydrolase and alpha-ribazole-phosphate phosphatase activities. *Journal of Bacteriology*, *188*, 4227–4235. https://doi.org/10.1128/jb.00227-06.

CHAPTER FIVE

A method for the production, purification and liposome reconstitution of cobamide synthase

Elizabeth A. Villa and Jorge C. Escalante-Semerena*
Department of Microbiology, University of Georgia, Athens, GA, United States
*Corresponding author: e-mail address: jcescala@uga.edu

Contents

Abstract

Cobamides are essential for the performance of a variety of reactions such methyl transfers, carbon skeleton rearrangements, and eliminations in both prokaryotes and eukaryotes. However, cobamide biosynthesis is limited to a subset of bacteria and archaea. The biosynthesis pathway culminates with the activation and attachment of a lower ligand to the corrin ring; this branch of the pathway is known as nucleotide loop assembly (NLA) pathway. The cobamide synthase (CobS) enzyme is the penultimate step in NLA pathway, and catalyzes the attachment of an α-ribotide to the activated corrin ring.

Methods in Enzymology, Volume 668
ISSN 0076-6879
https://doi.org/10.1016/bs.mie.2021.11.015

While other NLA enzymes have been well-studied, studies of CobS have proven difficult to date. CobS is an integral membrane protein, and limitations have been largely due to difficulties in protein purification. Here we provide a method to purify CobS, reconstitute protein in proteoliposomes, and assay for its activity.

1. Introduction: B_{12} and the nucleotide loop assembly (NLA) pathway

Adenosylcobamides (AdoCbas) are complex biomolecules whose *de novo* synthesis requires approximately 30 enzymes (Costa, Greenhalgh, Brunold, & Escalante-Semerena, 2020). AdoCbas are made up of a central Co ion with four equatorial bonds to the pyrrolic nitrogens of cyclic tetrapyrroles, and two axial ligands. The upper ligand is a 5′-deoxyadenosyl group derived from ATP, and the lower ligand is usually to a nitrogenous nucleobase (i.e., purines, or derivatives of benzimidazole) and in a few instances phenol or *p*-cresol (Anderson et al., 2008; Chan & Escalante-Semerena, 2011; Johnson & Escalante-Semerena, 1992). One commonly-found AdoCba is coenzyme B_{12} (CoB_{12}), a.k.a. adenosylcobalamin (AdoCbl), in which 5,6-dimethylbenzimidazole (DMB) is - the lower ligand. *De novo* CoB_{12} biosynthesis is restricted to a subset of prokaryotes. During *de novo* biosynthesis, the cobalt-containing cyclic tetrapyrrole, or corrin ring, is formed first and subsequently adenosylated (see chapter "A method for the efficient adenosylation of corrinoids" by Costa and Escalante-Semerena). The late steps of CoB_{12} biosynthesis, also known as the nucleotide loop assembly (NLA) pathway, involve the activation of the corrin ring and attachment of the lower ligand (Fig. 1). In *Salmonella enterica*, the NLA pathway comprises the gene products of the *cbiB* gene, the *cobUST* operon and the *cobC* gene. Briefly, the NLA pathway reactions are the following:

(1) ATP + adenosylcobyric acid + (*R*)-1-aminopropan-2-yl phosphate = ADP + phosphate + adenosylcobinamide phosphate (AdoCbi-P)

The reaction is catalyzed by the AdoCbi-P synthase (EC 6.3.1.10), a.k.a. CbiB.

(2) GTP + adenosylcobinamide phosphate = diphosphate + adenosylcobinamide-GDP (AdoCbi-GDP)

The reaction is catalyzed by the AdoCbi kinase/AdoCbi-P guanylyltransferase (EC 2.7.7.62), a.k.a. CobU.

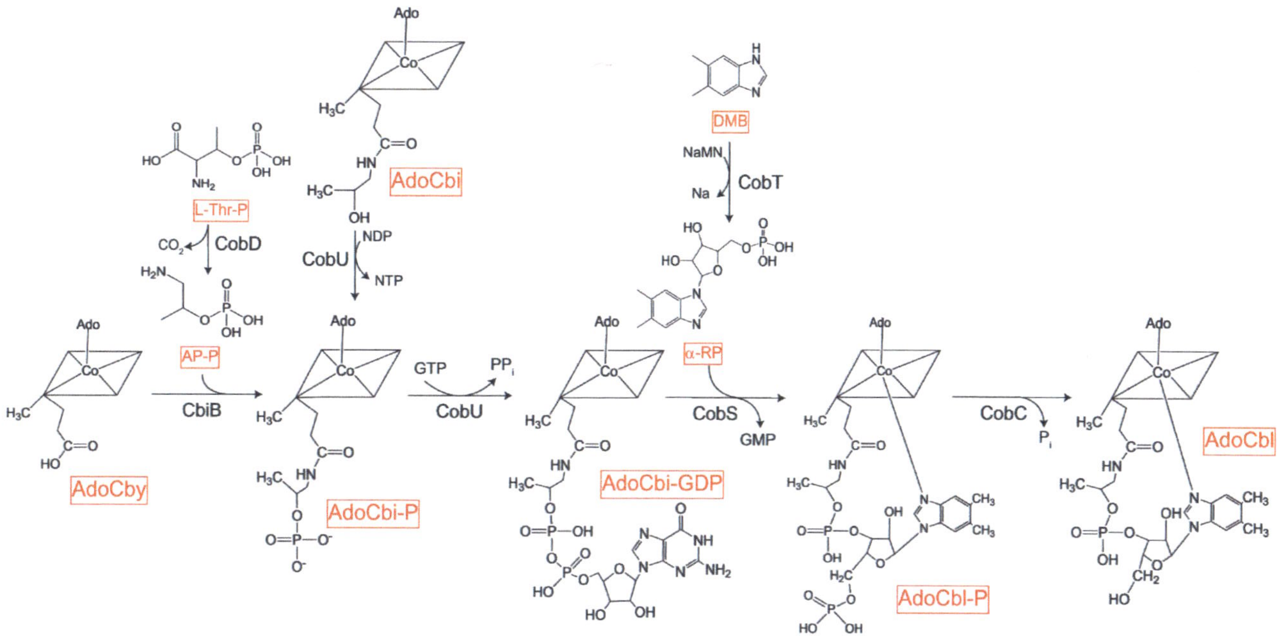

Fig. 1 The NLA pathway in *S. enterica*. Abbreviations: AdoCby, adenosylcobyric acid; L-Thr-P, L-threonine-phosphate; AP-P, aminopropanol-phosphate; AdoCbi, adenosylcobinamide; AdoCbi-P, adenosylcobinamide-phosphate; AdoCbi-GDP, adenosylcobinamide-GDP; AdoCbl-P, adenosylcobalamin-5′-phosphate; AdoCbl, adenosylcobalamin.

(3) β-Nicotinate D-ribonucleotide + 5.6-dimethylbenzimidazole = nicotinate + α-ribazole 5′-phosphate (α-RP).

The reaction is catalyzed by the nicotinate-nucleotide—dimethylbenzimidazole phosphoribosyltransferase (EC 2.4.2.21), a.k.a. CobT.

(4) AdoCbi-GDP + α-RP = GMP + adenosylcobalamin 5′-phosphate (AdoCbl-P)

The reaction is catalyzed by the AdoCbl-P synthase (EC 2.7.8.26), a.k.a. CobS.

(5) AdoCbl-P + H_2O = AdoCbl + phosphate

The reaction is catalyzed by the AdoCbl-P phosphatase (EC 3.1.3.73), a.k.a. CobC

The CobU and CobT enzymes have been well-studied (O'Toole & Escalante-Semerena, 1995; O'Toole, Trzebiatowski, & Escalante-Semerena, 1994; Trzebiatowski & Escalante-Semerena, 1997), however, extensive studies of CobS are limited. In the genomes of all B_{12} producers sequenced to date, CobS is an integral membrane protein whose over-production compromises cell survival (Maggio-Hall, Claas, & Escalante-Semerena, 2004; Maggio-Hall & Escalante-Semerena, 1999) and thus purification efforts have proven difficult. Here we describe a useful scheme for purification of CobS using detergent to maintain solubility throughout the purification process. Subsequent reconstitution of CobS into proteoliposomes for further study has revealed differences in enzyme efficiency when compared to protein solubilized in detergent. This method is of use to improve our understanding of the CobS enzyme and could prove useful in the purification and study of other membrane proteins.

2. Cloning, overexpression and purification of cobamide synthase (CobS)

The CobS enzyme from *S. enterica* used in the following method was purified as described elsewhere (Jeter & Escalante-Semerena, 2021). This method could be implemented to study CobS from other organisms as well. These vectors are recommended and are available upon request. Buffers, strains, and plasmids used in this method are detailed in Table 1.

Table 1 Buffers, strains and plasmids needed for the described procedures.

Buffer Name	Composition	Recommended for
Buffer A	Tris·HCl (0.1 M, pH 7.9)	CobS purification—reconstitution
Buffer B	Tris HCl (0.1 M) pH 7.9, NaCl (0.5 M), imidazole (20 mM)	CobS purification—bind buffer
Buffer C	Tris·HCl (0.1 M, pH 7.9), NaCl (0.5 M), imidazole (20 mM), DHPC (15 mM)	CobS purification—bind buffer
Buffer D	Tris·HCl (0.1 M, pH 7.9), NaCl (0.5 M), imidazole (60 mM), DHPC (15 mM)	CobS purification—wash buffer
Buffer E	[Tris·HCl (0.1 M, pH 7.9), NaCl (0.5 M), imidazole (0.5 M), DHPC (15 mM)	CobS purification—elution buffer
Buffer F	4-(2-hydroxyethyl)-1-piperazineethanesulfonic acid buffer (HEPES-NaOH, pH 7.4), NaCl (500 mM), glycerol [10% (v/v)], 3-[(3-cholamidopropyl)dimethylammonio]-1-propanesulfonate (CHAPS, 40 mM)	Lipid reconstitution, dialysis
Buffer G	Histodenz (80% w/v), HEPES-NaOH, (20 mM, pH 7.4), NaCl (150 mM), glycerol (10% (v/v)	Proteoliposome isolation—separation
Buffer H	Histodenz (30% w/v), HEPES-NaOH, (20 mM, pH 7.4), NaCl (150 mM), glycerol (10% v/v)	Proteoliposome isolation—separation
Buffer I	HEPES-NaOH, (20 mM, pH 7.4), NaCl (150 mM), glycerol (10% v/v)	Proteoliposome isolation—separation
Buffer J	HEPES buffer (20 mM, pH 7.4)	Cobamide synthase activity assay

Strain	Genotype	Source
JE6663 *E. coli* C41 (λDE3)	F *–ompT hsdSB (rB-mB-) gal dcm* (λDE3)	Avanti
JE8248	*metE205 ara-9* Δ*cobS1313*	Laboratory collection

Plasmid	Genotype	Source
pCobS5	*S. enterica cobS*$^+$ cloned into pET15b. The plasmid encodes a H_6-CobS protein (i.e., N-terminally tagged)	Maggio-Hall and Escalante-Semerena (1999)

2.1 Cloning and overexpression of *cobS*

2.1.1 *cobS cloning.* Plasmid pCOBS5 encoding *S. enterica cobS* was constructed by cloning *S. enterica cobS* into pET-15b using *Nco*I and *Hind*III restriction sites. This resulted in the encoded CobS fused to an N-terminal hexahistidine (H_6) tag (Maggio-Hall & Escalante-Semerena, 1999)

2.1.2 *cobS overexpression. Day 1.* Electroporate plasmid pCOBS5 encoding *cobS* into your preferred strain. We have used *E. coli* C41 (λDE3) with success. Plate to selective medium and incubate at 37 °C overnight

Day 2. Inoculate a 10 mL overnight culture (~16 h), grow overnight at 37 °C with shaking.

Day 3. Inoculate a 2-L culture of Terrific Broth (tryptone, (24 g/L), yeast extract (24 g/L), glycerol (4 mL/L, or 54 mM), KH_2PO_4 (17 mM), $MgSO_4$ (2 mM) supplemented with appropriate antibiotic. Use of a baffled flask is recommended to improve aeration. Grow to mid-log (approximate optical density (OD) at 600 nm = 0.6), then induce *cobS* expression by adding the inducer appropriate for your vector of choice. Allow induced culture to incubate for 3 h at 37 °C with shaking. Remove a sample (500 μL) of the culture to verify over-expression. Analyze the 500 μL sample with SDS-PAGE (Laemmli, 1970) to verify that CobS has been overproduced, indicated by a band at ~26 kDa. If CobS overproduction is successful, harvest the remainder of the culture by centrifugation at 4 °C. Discard supernatant and store cell pellet at −80 °C.

2.2 CobS purification

2.2.1 CobS purification was described elsewhere (Jeter & Escalante-Semerena, 2021). Thaw cell pellets and resuspend in 30 mL of Buffer A (Tris·HCl (0.1 M), pH 7.9) and lyse the cells using a cell disruptor or French press. We have used a cell disruptor set to 1.72×10^5 kPa. Following lysis, add 0.5 mL of Protease Inhibitor Cocktail (Sigma), then centrifuge lysate at $5000 \times g$ for 15 min at 4 °C.

2.2.2 Remove supernatant into a new tube and centrifuge at 75,000 × *g* for 90 min to separate cell membranes. Using a glass homogenizer, resuspend membranes in 10 mL of Buffer A. Once resuspended, add 3-((3-cholamidopropyl) dimethylammonio)-1-propanesulfonate (CHAPS) to a final concentration of 20 mM. Incubate on ice for 1 h, then centrifuge at 75,000 × *g* for 30 min at 4 °C.

2.2.3. Resuspend insoluble fraction containing CobS in Buffer B [Tris·HCl (0.1 M) pH 7.9, NaCl (0.5 M), imidazole (20 mM)] using a glass homogenizer, then slowly add 1,2-dihexanoyl-*sn*-glycero-3-phosphocholine (DHPC) to a final concentration of 15 mM. Incubate on ice for 1 h, then centrifuge at 75,000 × *g* for 30 min at 4 °C.

Tip: During this time, equilibrate a 2-mL bed volume of affinity purification resin by first washing with 10 volumes of water and then 10 volumes of Buffer B. We have used HisPur Ni-NTA resin (Fisher Scientific) with success.

2.2.4 Mix supernatant containing solubilized H_6-CobS with resin and incubate at 4 °C using a tube rotator for at least 2 h, then pour slurry into a column. Allow the cell-free extract (CFE) to flow through, then wash with five column volumes of Buffer C [Tris·HCl (0.1 M), pH 7.9, NaCl (0.5 M), imidazole (20 mM), DHPC (15 mM)] followed by five column volumes of Buffer D [Tris·HCl (0.1 M), pH 7.9, NaCl (0.5 M), imidazole (60 mM), DHPC (15 mM)]. Elute protein with five column volumes of Buffer E [Tris·HCl (0.1 M), pH 7.9, NaCl (0.5 M), imidazole (0.5 M), DHPC (15 mM)], and collect flowthrough of each column volume in a separate glass tube.

2.2.5 Identify fractions containing CobS by loading each fraction and resolving proteins on a 12% SDS-PAGE gel.

2.2.6 Remove imidazole by adding fractions containing CobS to a desalting column such as Zeba™ Spin Desalting Column (Fisher Scientific). Add glycerol to a final concentration of 10% (v/v), and flash-freeze in liquid N_2. Store frozen protein at −80 °C (Fig. 2).

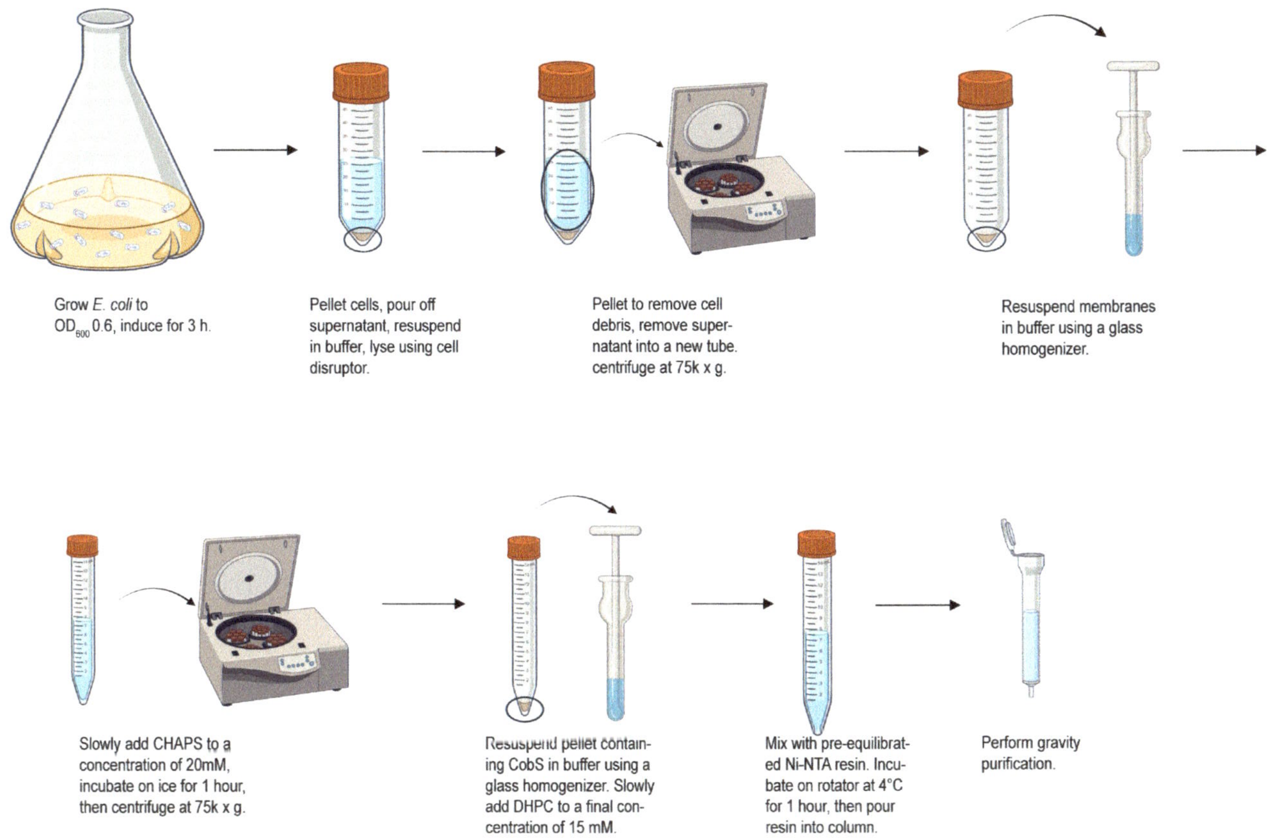

Fig. 2 Overexpression and purification scheme for CobS. Created using BioRender.com.

3. Reconstitution of CobS into liposomes

3.1 Liposome formation

3.1.1 Combine solubilized lipids (Avanti Polar Lipids) in a small glass tube (recommended 10 × 75 mm (Fisher)) and dry under a stream of nitrogen or argon gas. We have used the following composition with success: 1-palmitoyl-2-oleoyl-glycero-3-phosphocholine (POPC, 80% mol/mol), 1-palmitoyl-2-oleoyl-*sn*-glycero3-phospho-L-serine (POPS, 10% mol/mol), 1-palmitoyl-2-oleoyl-*sn*-glycero-3 phospho-ethanolamine (POPE, 9.5% mol/mol), Lissamine™ Rhodamine B 1,2-dihexadecanoyl-*sn*-glycero-phosphoethanolamine (Rh-DHPE, 0.5% mol/mol). A total of 1 μmol lipid at this point will yield a final concentration of 2 mM after drying and reconstituting in buffer.

Tip: Rotate tube under nitrogen stream to ensure that lipids dry in an even layer and avoid "clumping." Once lipids are dry, cover tube with Parafilm immediately to prevent introduction of moisture.

3.1.2 Dry lipids further in a vacuum centrifuge at 30 °C uncovered for a minimum of 1 h or up to overnight. Dried lipids will appear as a film on the bottom of the glass tube. Add 500 μL of Buffer F [4-(2-hydroxyethyl)-1-piperazineethanesulfonic acid buffer (HEPES-NaOH, pH 7.4), NaCl (500 mM), glycerol [10% (v/v)], CHAPS (40 mM)] and mix by vortex until dried lipids are reconstituted. This may take several minutes per tube. Transfer suspended lipids to 1.7 mL Eppendorf tubes and keep on ice.

3.1.3 Add protein to reconstituted lipids at the desired concentration. A ratio of 2 μM protein: 2 mM lipids has been used with success. Gently mix protein and lipids by pipetting, and incubate tube on a tube rotator at 4 °C for 1 h, then transfer lipids to desired dialysis membrane. We have used 20 MWCO Slide-A-Lyzer dialysis cassette with success. Dialyze thrice in 1 L of Buffer F at 4 °C for a minimum of 3 h each or up to overnight to allow proteoliposomes to form. If detergent removal is necessary, dialysis may be extended over several days.

3.2 Proteoliposome isolation

3.2.1 Remove protein-lipid mixture from dialysis cassette into a 15 mL conical tube. Add an equal volume of Buffer G [Histodenz (80%

w/v), HEPES-NaOH, pH 7.4 (20 mM), NaCl (150 mM), glycerol (10% (v/v)] and gently invert at least 20 times to mix. Transfer mix to ultracentrifuge tube. Overlay lipid-protein mix with Buffer H [Histodenz (30% w/v), HEPES-NaOH, pH 7.4 (20 mM), NaCl (150 mM), glycerol (10% v/v)]. We recommend use of polyallomer tubes ultracentrifuge tubes (13 × 51 mm, Beckman Coulter) and a total volume of 4 mL. To the top of each tube, add 200 μL of Buffer I [HEPES-NaOH, pH 7.4 (20 mM), NaCl (150 mM), glycerol (10% v/v)].

3.2.2 Separate proteoliposomes by ultracentrifugation at 200,000 × *g* at 4 °C for at least 3 h or up to overnight. Following ultracentrifugation, remove the top pink layer containing proteoliposomes into a separate tube. Quantify lipid concentration by measuring fluorescence of rhodamine at 540 nm (excitation)/586 nm (emission). Protein concentration can be calculated using 12% SDS-PAGE and Western blot using known concentrations of pure protein as standards. Store liposomes at 4 °C and use within 1 week.

Tip: Commercially available anti-6 × His antibodies (Sigma) can be used to detect tagged CobS protein in Western blot.

4. Proteoliposome quality controls

4.1 Population dispersal

Homogeneity of the proteoliposome population should be assessed by measuring particle size. A monodispersed population is ideal. Varying the ratio between protein and lipid used to form proteoliposomes can lead to changes in particle size, and should be optimized to determine the optimal ratio to yield the most homogeneous population. Dynamic light scattering (DLS) is a useful technique to assess particle size and homogeneity in proteoliposome populations, and employs changes in light intensity as a measurement of aggregation.

4.2 Protein orientation

Proteins can insert into the lipid bilayer in two orientations, starting with the *N*-terminus on either the outside or inside of the proteoliposome structure [Fig. 3; modified from a figure published elsewhere (Jeter & Escalante-Semerena, 2021)]. To assess protein function in the proteoliposome, it is necessary to know the percentage of the population in which

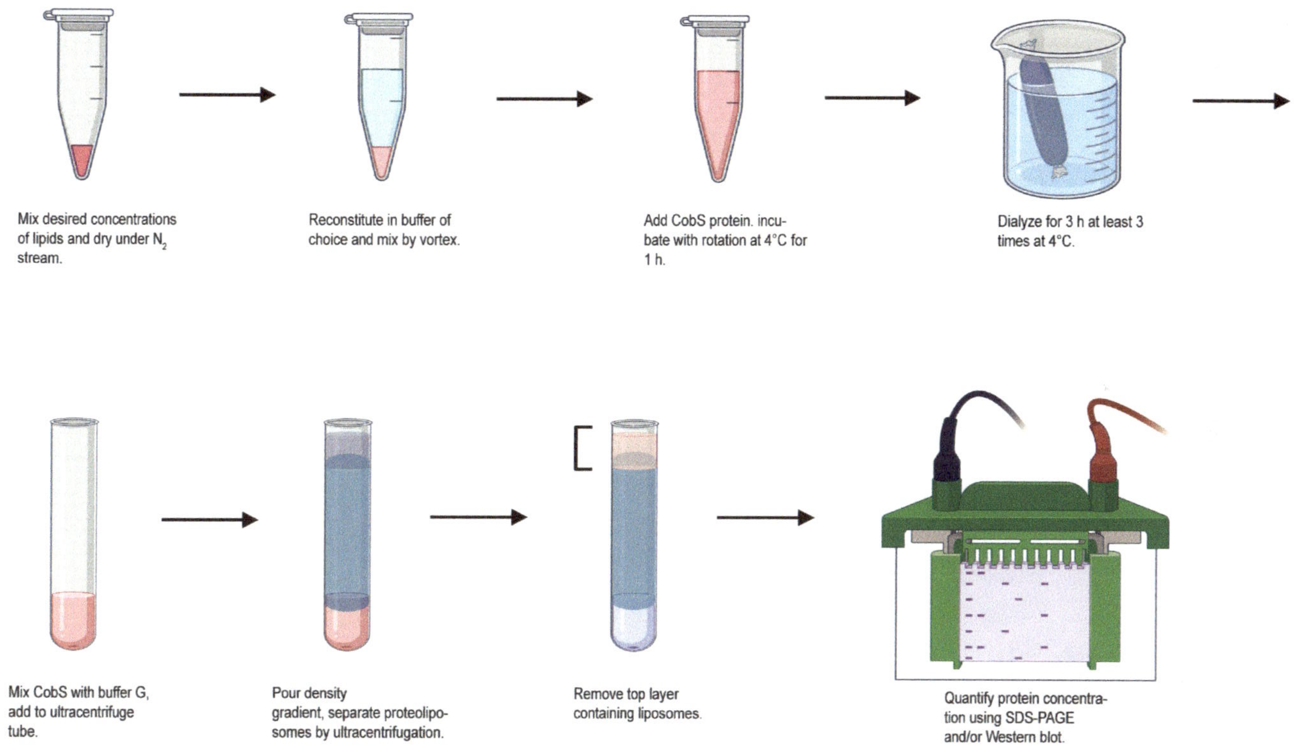

Fig. 3 CobS-containing proteoliposome reconstitution and separation. Created using BioRender.com.

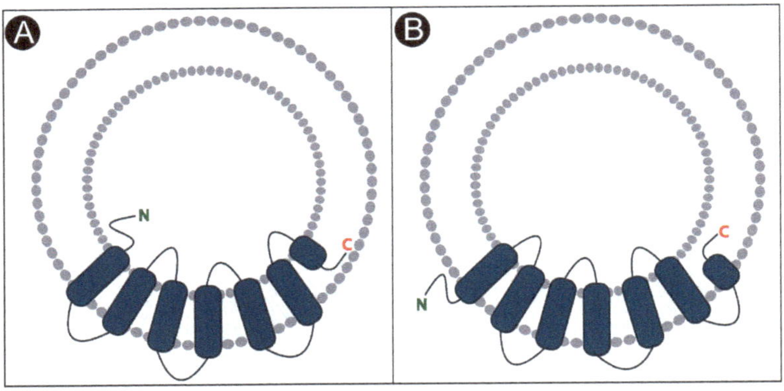

Fig. 4 CobS can insert into the liposome in two orientations. The N-terminus is shown in green, the C-terminus is shown in red. (A) Insertion with the N-terminus in the lumen of the liposome. (B) Insertion with the N-terminus located on the outside of the lipid bilayer. Created using BioRender.com. *Modified from Jeter, V. L., & Escalante-Semerena, J. C. (2021). Insights into the relationship between obamide synthase and the cell membrane.* MBio, 12*(2). doi:10.1128/mBio.00215-21.*

the active site is exposed and accessible to substrate. We have employed a proteolytic digest to assess protein orientation and allow accurate quantification of accessible protein. Protease selection should be tailored to the protein to be studied. For *S. enterica*, trypsin (Sigma-Aldrich), chymotrypsin (Sigma-Aldrich), and protease K (Fisher Scientific) were used. Reaction products can be analyzed by SDS-PAGE and subsequent Western blotting with antibodies against the protein of study (Fig. 4).

5. Activity assays

The substrates of CobS are not commercially available and thus must be synthesized (Maggio-Hall & Escalante-Semerena, 1999; O'Toole & Escalante-Semerena, 1995).

5.1 Preparation of corrinoid substrates

5.1.1 Synthesis of AdoCbi-GDP

AdoCbi-GDP was prepared as described elsewhere. Briefly, commercially-available CN_2Cbi was first incubated with CobA from *S. enterica*, the housekeeping adenosyltransferase, and ATP to yield AdoCbi. This was then incubated with *S. enterica* CobU and GTP to yield AdoCbi-GDP. Reaction conditions are further detailed elsewhere (Blanche, Thibaut, Couder, & Muller, 1990; O'Toole & Escalante-Semerena, 1995). Cbi-GDP may be separated from Cbi by the following HPLC method, using mobile phase (MP) A

[KH_2PO_4 (0.1 M), pH 6.5, KCN (10 mM)] and mobile phase B [KH_2PO_4 (0.1 M), pH 8, KCN (10 mM): acetonitrile in a 1:1 ratio. Equilibrate column to 98% MP A/2% MP B. Follow sample injection with a 5-min linear gradient to 95% A/5% B, then apply a 9-min linear gradient to 80% MP A/20% MP B. Continue to develop the column for 12.5 min at 80% A/20% B. Next, apply an 11.5-min linear gradient to 65% MP A/35% MP B, and finally apply a 7-min linear gradient to 100% MP B.

5.1.2 Adenosylation of Cbi-GDP

Because the Co-C bond of adenosylcobamides is sensitive to light, it is unlikely that the population of Cbi-GDP isolated will be adenosylated following HPLC purification. The adenosylation reaction should be repeated by incubating with CobA and ATP prior to use. To calculate the adenosylation efficiency we refer the reader to chapter "A method for the efficient adenosylation of corrinoids" by Costa and Escalante-Semerena.

5.2 Synthesis of α-ribazole-5′-phosphate (α-RP)

α-RP may be synthesized by incubating *S. enterica* CobT with DMB and NaMN and α-RP can be purified by HPLC (see chapter "A method for the isolation of α-ribazole from vitamin B_{12}, and its enzymatic conversion to α-ribazole 5'-phosphate" by Malalasekara and Escalante-Semerena) (Jeter, Mattes, Beattie, & Escalante-Semerena, 2019). α-RP may be separated from substrate using a Phenomenex Kinetix C18 column. We recommend developing the column with the following mobile phases: MP A, ammonium acetate (20 mM); MP B acetonitrile (100%); MP C H_2O. Equilibrate column to 97% MP A/3% MP B. After injecting sample, allow to develop for 5 min at 97% MP A/3% MP B. Follow with a 12-min linear gradient to 60% MP A/40% MP B. Follow with a 5-min linear gradient to 80% MP B/20% MP C and allow to develop at 80% MP B/20% MP C for 5 min.

5.3 Cobamide synthase activity assays

CobS activity can be assayed in multiple ways. We will describe a continuous manner using spectrophotometric measurements, and an endpoint assay using a bioassay strain for product detection. Recommended reaction conditions are as follows: Buffer J [2-[4-(2-hydroxyethyl)piperazin-1-yl] ethanesulfonic acid (HEPES) buffer (20 mM, pH 7.4)], NaCl (150 mM), glycerol (10% [v/v]), α-RP (200 μM), AdoCbi-GDP (500 μM), MgCl (2 mM) and proteoliposome complex to a final concentration of 0.33 mM lipid.

Tip: Improved activity has been observed in reconstituted CobS (Jeter & Escalante-Semerena, 2021). Pure solubilized protein can be assayed by the described method to compare changes in activity.

5.3.1 Cobamide synthase continuous assay

The activity of CobS can be tested by measuring the decrease in absorbance of the substrate AdoCbi-GDP at 459 nm in a 96-well quartz microtiter plate. Activity of CobS can be calculated using Beer-Lambert's Eq. $A = \varepsilon cL$ with the AdoCbi-GDP extinction coefficient of $\varepsilon_{459} = 3352\,M^{-1}\,cm^{-1}$ (Jeter & Escalante-Semerena, 2021).

5.3.2 Endpoint assay (AdoCbl-P synthesis)

For an endpoint assay, the above reaction components should be incubated at 37 °C for 1 h or desired time point. Stop reactions by adding KCN to a final concentration of 2.4 μM and incubating at 90 °C for 10 min.

Tip: If Cbl is the preferred product, incubate AdoCbl-P with commercially available alkaline phosphatase.

5.3.3 Bioassay

Formation of Cbl-P or Cbl can be assessed by providing reaction components to a *S. enterica metE* and *cobS*-deficient strain. Briefly, cells from an overnight culture should be washed with sterile PBS. An aliquot of 200 μL can be added to 4 mL of melted 0.7% (w/v) agar, mixed, and overlaid on no-carbon E (NCE) medium (Berkowitz, Hushon, Whitfield Jr., Roth, & Ames, 1968) supplemented with Wolfe's trace minerals (Balch & Wolfe, 1976) $MgSO_4$ (1 mM) and glycerol (22 mM). Once cooled, 1 μL of reaction product can be spotted on overlay. Incubate at 37 °C overnight. The following day, assess growth to determine presence of Cbl or Cbl-P (Zayas & Escalante-Semerena, 2007).

Tip: Reaction products can be fractionated using RP-HPLC and products can be verified using mass spectrometry (Jeter et al., 2019).

References

Anderson, P. J., Lango, J., Carkeet, C., Britten, A., Kräutler, B., Hammock, B. D., et al. (2008). One pathway can incorporate either adenine or dimethylbenzimidazole as an alpha-axial ligand of B_{12} cofactors in *Salmonella enterica*. *Journal of Bacteriology*, *190*, 1160–1171.

Balch, W. E., & Wolfe, R. S. (1976). New approach to the cultivation of methanogenic bacteria: 2-Mercaptoethanesulfonic acid (HS-CoM)-dependent growth of *Methanobacterium ruminantium* in a pressurized atmosphere. *Applied and Environmental Microbiology*, *32*, 781–791.

Berkowitz, D., Hushon, J. M., Whitfield, H. J., Jr., Roth, J., & Ames, B. N. (1968). Procedure for identifying nonsense mutations. *Journal of Bacteriology*, *96*, 215–220.
Blanche, F., Thibaut, D., Couder, M., & Muller, J. C. (1990). Identification and quantitation of corrinoid precursors of cobalamin from Pseudomonas denitrificans by high-performance liquid chromatography. *Analytical Biochemistry*, *189*(1), 24–29. https://doi.org/10.1016/0003-2697(90)90038-b.
Chan, C. H., & Escalante-Semerena, J. C. (2011). ArsAB, a novel enzyme from *Sporomusa ovata* activates phenolic bases for adenosylcobamide biosynthesis. *Molecular Microbiology*, *81*, 952–967. https://doi.org/10.1111/j.1365-2958.2011.07741.x.
Costa, F. G., Greenhalgh, E. D., Brunold, T. C., & Escalante-Semerena, J. C. (2020). Mutational and functional analyses of substrate binding and catalysis of the *Listeria monocytogenes* EutT ATP:Co(I)rrinoid adenosyltransferase. *Biochemistry*, *59*(10), 1124–1136. https://doi.org/10.1021/acs.biochem.0c00078.
Jeter, V. L., & Escalante-Semerena, J. C. (2021). Insights into the relationship between obamide synthase and the cell membrane. *MBio*, *12*(2), e00215–21. https://doi.org/10.1128/mBio.00215-21.
Jeter, V. L., Mattes, T. A., Beattie, N. R., & Escalante-Semerena, J. C. (2019). A new class of phosphoribosyltransferases involved in cobamide biosynthesis is found in methanogenic archaea and cyanobacteria. *Biochemistry*, *58*, 951–964. https://doi.org/10.1021/acs.biochem.8b01253.
Johnson, M. G., & Escalante-Semerena, J. C. (1992). Identification of 5,6-dimethylbenzimidazole as the *Coα* ligand of the cobamide synthesized by *Salmonella typhimurium*. Nutritional characterization of mutants defective in biosynthesis of the imidazole ring. *The Journal of Biological Chemistry*, *267*(19), 13302–13305.
Laemmli, U. K. (1970). Cleavage of structural proteins during assembly of head of bacteriophage-T4. *Nature*, *227*(5259), 680–685. https://doi.org/10.1038/227680a0.
Maggio-Hall, L. A., Claas, K. R., & Escalante-Semerena, J. C. (2004). The last step in coenzyme B(12) synthesis is localized to the cell membrane in bacteria and archaea. *Microbiology*, *150*, 1385–1395.
Maggio-Hall, L. A., & Escalante-Semerena, J. C. (1999). In vitro synthesis of the nucleotide loop of cobalamin by *Salmonella typhimurium* enzymes. *Proceedings of the National Academy of Sciences of the United States of America*, *96*, 11798–11803.
O'Toole, G. A., & Escalante-Semerena, J. C. (1995). Purification and characterization of the bifunctional CobU enzyme of *Salmonella typhimurium* LT2. Evidence for a CobU-GMP intermediate. *The Journal of Biological Chemistry*, *270*, 23560–23569.
O'Toole, G. A., Trzebiatowski, J. R., & Escalante-Semerena, J. C. (1994). The *cobC* gene of *Salmonella typhimurium* codes for a novel phosphatase involved in the assembly of the nucleotide loop of cobalamin. *The Journal of Biological Chemistry*, *269*, 26503–26511.
Trzebiatowski, J. R., & Escalante-Semerena, J. C. (1997). Purification and characterization of CobT, the nicotinate-mononucleotide:5,6-dimethylbenzimidazole phosphoribosyltransferase enzyme from *Salmonella typhimurium* LT2. *The Journal of Biological Chemistry*, *272*, 17662–17667.
Zayas, C. L., & Escalante-Semerena, J. C. (2007). Reassessment of the late steps of coenzyme B_{12} synthesis in *Salmonella enterica*: Evidence that dephosphorylation of adenosylcobalamin-5′-phosphate by the CobC phosphatase is the last step of the pathway. *Journal of Bacteriology*, *189*, 2210–2218. https://doi.org/10.1128/JB.01665-06.

CHAPTER SIX

A method for the isolation of α-ribazole from vitamin B_{12}, and its enzymatic conversion to α-ribazole 5′-phosphate

Lahiru Malalasekara and Jorge C. Escalante-Semerena*

Department of Microbiology, University of Georgia, Athens, GA, United States

*Corresponding author: e-mail address: jcescala@uga.edu

Contents

Abstract

Cobamides (Cbas) are the largest coenzymes known and are used by cells in all domains of life. These molecules are characterized by a central cobalt-containing tetrapyrrole ring with two opposing axial ligands on the α and β faces of the ring. All biologically active forms of Cbas have a 5′-deoxyadenosyl group as the upper (*Coβ*) ligand that is covalently attached to the cobalt ion of the ring. In contrast, the lower ligand is a nucleobase of diverse chemical structure; however, nucleobases are usually derivatives of benzimidazole or purine. Phenol and *p*-cresol can also serve as the nucleobase, but they cannot form a coordination bond with the cobalt ion of the ring because they lack a free pair of electrons. The Cba incorporating 5,6-dimethylbenzimidazole (DMB) is known as cobalamin (Cbl), and the coenzymic form of cobalamin is known as adenosylcobalamin (AdoCbl). A common vitamer of cobalamin has a cyano group as the upper ligand. This vitamer is known as cyanocobalamin (CNCbl), which is commercially marketed

Methods in Enzymology, Volume 668
ISSN 0076-6879
https://doi.org/10.1016/bs.mie.2021.11.024

as vitamin B_{12}. Here, we describe a combination of chemical hydrolysis of cobalamin with the enzymatic dephosphorylation of the resulting α-R-3′-phosphate to yield α-R, which we enzymically convert to the pathway intermediate α-R-5′-phosphate (α-RP). The methods describe herein can be readily scaled up to generate large amounts of α-RP.

1. Introduction to coenzyme B_{12} and to the activation of its nucleobase

Cobamides (Cbas) are cyclic tetrapyrroles that belong to the family known as the "pigments of life" (e.g., heme, chlorophylls coenzyme F430) (Battersby, 2000). Many eukaryotic and prokaryotic enzymes require adenosyl-Cbas (AdoCba) to perform either radical chemistry or alkyl transfers. Reactions that require Ado radicals include carbon skeleton rearrangements and functional group eliminations. Despite being an essential coenzyme for many forms of lives, only few prokaryotes have the ability to biosynthesize Cbas *de novo* (Escalante-Semerena & Warren, 2008).

Unlike any other member of the family of biologically active cyclic tetrapyrroles, Cbas contain a central cobalt ion equatorially coordinated to the nitrogen atoms of the pyrrole ring, plus two axial ligands, one on the upper (β) face of the ring, the other one on the lower (α) face. Adenosylcobalamin (AdoCbl) is the only biologically active Cbas used by human cells Notably, in AdoCbl the lower ligand nucleobase is 5,6-dimethylbenzimidazole (DMB).

The so-called late steps of AdoCbl biosynthesis attach the nucleobase (DMB) to the activated ring. In *Salmonella*, the nucleobase-activating enzyme CobT transfers a phosphoribosyl group from nicotinate mononucleotide (NaMN) to one of the nitrogens of the imidazole moiety of DMB *via* a mechanism that inverts the configuration of the N glycosidic bond, yielding an α-ribotide known as α-ribazole-5′- phosphate (α-R-5′-P, α-RP). In the penultimate step of this pathway, α-RP is condensed with the activated ring by the cobamide synthase (CobS, EC 2.7.8.26) enzyme yielding AdoCbl-5′-P, which is in turn dephosphorylated by the CobC (3.1.3.73) phosphatase yielding biologically active AdoCbl (Fig. 1) (Zayas & Escalante-Semerena, 2007).

Some *Firmicutes* lack CobT homologs and thus cannot directly synthesize α-RP directly from NaMN and DMB. Instead, they have evolved an α-ribazole (α-R) transporter and α-R kinase known as CblT and CblS, respectively (Fig. 1). Once taken up, α-R is phosphorylated by CblS.

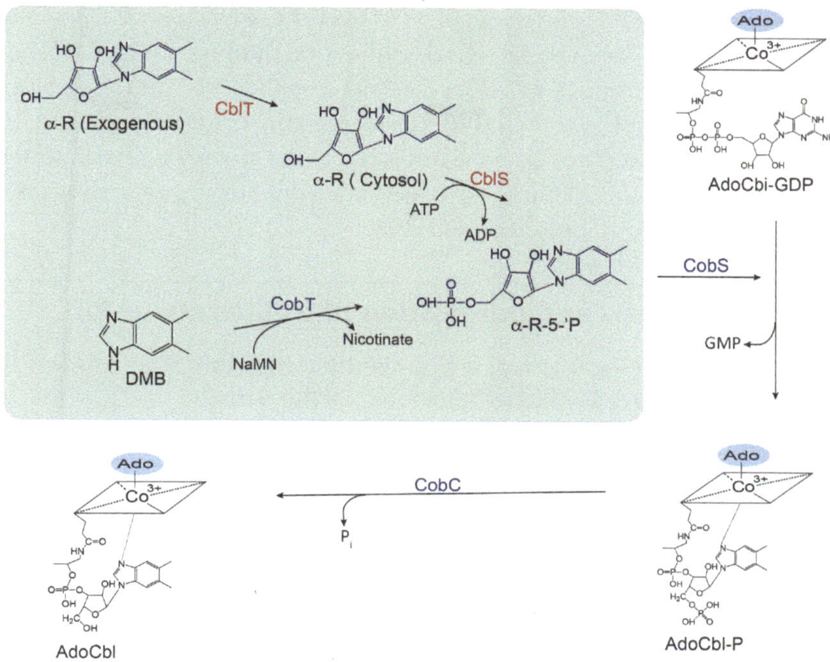

Fig. 1 α-Ribazole-5′-phosphate biosynthesis and its integration into the coenzyme B_{12}. Strategies for α-ribazole-5′-phosphate (α-RP) biosynthesis are depicted in the green box. Abbreviations: α-R, α-ribazole; α-RP, α-ribazole-5′-phosphate; DMB, 5,6-dimethylbenzimidazole; NaMN, nicotinate mononucleotide; AdoCbi-GDP, adenosylcobinamide-GDP; AdoCbl-P, adenosylcobalamin-phosphate; AdoCbl, adenosylcobalamin.

Heterologous expression of CblST has been used to study Cba biosynthesis in *Salmonella* (Gray & Escalante-Semerena, 2010; Mattes & Escalante-Semerena, 2017).

As mentioned above, α-R is a substrate for CblS *in vivo* and *in vitro* (Mattes & Escalante-Semerena, 2017; Mattes, Malalasekara, & Escalante-Semerena, 2021). In addition, once phosphorylated α-RP can be used as substrate for CobS in *in vitro* experiments (Jeter & Escalante-Semerena, 2021; Maggio-Hall & Escalante-Semerena, 1999). However, neither α-R nor α-R-5′-P are commercially available to perform these experiments. Here, we describe a method to synthesize α-R in relatively large quantities by hydrolysis of vitamin B_{12}. In this method, we used boronate-affinity chromatography to purify α-R from the B_{12} hydrolysate. This method is more efficient than conventional chromatography or reverse-phase high performance liquid chromatography (RP-HPLC) (Mattes & Escalante-Semerena, 2018). In addition, we describe a method to purify *Geobacillus kaustophilus* CblS from *E. coli*. The method described later in this protocol can be used to synthesize α-RP by the enzymatic action of purified CblS using α-R as its substrate. Using vitamin B_{12} as the starting material is more cost effective for the synthesis of α-RP than the enzymatic activation of DMB by an NaMN-consuming enzyme.

2. Synthesis and purification of α-Ribazole (α-R)

In the following method, α-R is synthesized in four steps as described elsewhere (Mattes & Escalante-Semerena, 2018), namely:

1. Hydrolysis of vitamin B_{12} to release α-R under high pH and high temperature conditions
2. Dephosphorylation of α-R-3′-P by commercially available alkaline phosphatase
3. Purification of α-R using boronate-affinity chromatography
4. Concentration of α-R by lyophilization

This method has been optimized for the hydrolysis and successive purification of α-R form 50–100 mg of vitamin B_{12}. Typically, hydrolysis of 50 mg of vitamin B_{12} using this method would yield up to ~7 mg (67%) of α-R.

Tip: We recommend quantifying α-R concentration after each of these steps using the method described in step 3.1.2 to estimate ultimate yield.

2.1 Vitamin B_{12} hydrolysis

Alkaline lysis of vitamin B_{12} at higher temperatures is used in this method to release a mixture of α-R and α-R-3′-P (Mattes & Escalante-Semerena, 2018; Pakin, Bergaentzle, Aoude-Werner, & Hasselmann, 2005).

The procedure starts by resuspending 50 mg of vitamin B_{12} in 5 mL of ultra-pure water and transferring this solution into a screw-capped 250-mL Erlenmeyer flask. Add 5 mL of 5 M NaOH and heat the mixture for 75 min in a water bath pre-heated to 85–90 °C. After cooling for 30 min at 4 °C, neutralize the solution by adding 5 mL of 5 M NaOH. To determine the efficiency of hydrolysis, initial vitamin B_{12} concentration and the synthesized α-R concentration can be calculated by the following methods.

2.1.1. *Quantification of vitamin B_{12} concentration.* To determine the initial vitamin B_{12} concentration more accurately, make 500-fold and 1000-fold dilutions into ultrapure water and measure the absorbance at 367 nm. Calculate the concentration using the molar absorptivity of vitamin B_{12} in water ($\varepsilon_{367} = 21{,}800\,M^{-1}\,cm^{-1}$).

2.1.2. *Quantification of α-R concentration by fluorescence*. This quantification is based on the fluorescence emission spectrum of α-R, as described elsewhere (Mattes & Escalante-Semerena, 2018; Pakin et al., 2005). The linear range of this method is from 320 nM to 6.4 μM α-R. To get measurements within this range prepare dilutions in Tris·HCl buffer (1 M, pH 8) as necessary. Load 200 μL of the samples into an opaque black 96-well microtiter plate. Measure the fluorescence at 312 nM after exciting at 250 nM using a fluorescence microplate reader and calculate concentrations using the quantum yield of α-R ($\Phi 250{,}312 = 3.62 \times 10^{13}$ RFU/mol).

Based on the molecular weights of vitamin B_{12} and α-R (1355.37 and 278.30 g/mol, respectively), the expected maximum yield of α-R after hydrolysis of 50 mg vitamin B_{12} is ~10.3 mg. We obtained ~79% (~8.2 mg) maximum release of α-R after hydrolysis.

Tip: Autoclaving vitamin B_{12} as an alternative hydrolysis method generates products that interfere with boronate-affinity chromatography, and could result in a low α-R yield (Mattes & Escalante-Semerena, 2018).

2.2 De-phosphorylation of α-ribazole-3′-phosphate

Hydrolysis of vitamin B_{12} at high temperature under alkaline conditions also releases α-R-3′-P in addition to α-R (Mattes & Escalante-Semerena, 2018; Pakin et al., 2005) (Fig. 2). This α-R-3′-P could be dephosphorylated into α-R with an alkaline phosphatase treatment.

To dephosphorylate α-R-3′-P, add 15 mL of a solution containing ammonium acetate (0.3 M, pH 8.8) and $MgCl_2$ (2 mM) to the hydrolysate, then add alkaline phosphatase in excess (>1000 DEA units; 1 DEA unit hydrolyzes 1 mmol of 4-nitrophenyl phosphate per minute at pH 9.8 at 37 °C). We use lyophilized alkaline phosphatase derived from bovine intestinal mucosa (Millipore Sigma, >10 DEA units per mg). Mix the solution and incubate at least 16 h at 37 °C. Following the incubation transfer the content into a 50 mL conical centrifuge tube and centrifuge at 3000 × *g* for 15 min. Collect the supernatant after filtering through a syringe filter (0.45 μm polyethersulfone membrane).

Tip: Verify the alkaline conditions in the reaction mixture. Alkaline phosphatase, from bovine intestinal mucosa, is most stable in the pH range 7.5–9.5.

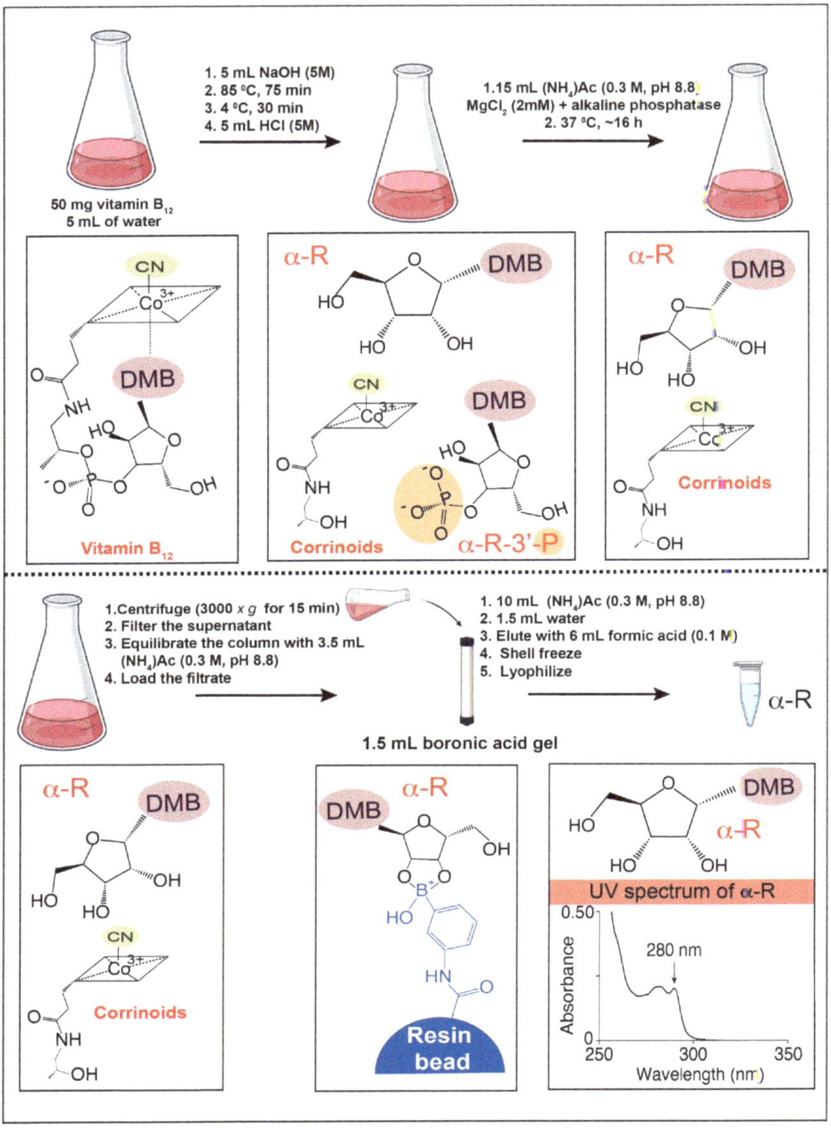

Fig. 2 Scheme for synthesis and purification of α-ribazole. Possible molecules present under each condition are depicted in boxes. Abbreviations: α-R, α-ribazole; α-R-3′-P, α-ribazole-3′-phosphate; DMB, 5,6-dimethylbenzimidazole. UV spectrum of α-R is adapted from Mattes ànd Escalante-Semerena (2018). The figure was created with BioRender.com.

2.3 Purification of α-R using boronate affinity chromatography

In this method, the affinity of boronate for the vicinal *cis*-diol groups of α-R is used to isolate it from the rest of the compounds in the hydrolysate (Mattes & Escalante-Semerena, 2018; Pfadenhauer & Tong, 1979) (Fig. 2). We used immobilized boronic acid gel (Thermo Fisher Scientific) with ≥99% binding and recovery of 110 μmol AMP per mL of gel.

Mix the boronic acid gel to get a homogenized slurry and pack 1.5 mL of the gel into an empty column. Equilibrate the packed column with 3.5 mL of ammonium acetate (0.3 M, pH 8.8). Use gravity to pass the filtered hydrolysate (obtained as described in Section 2.2) through the column and collect the flowthrough. The column may appear slightly pink at this point due to non-specific binding of corrinoids from the hydrolysate. Wash the column with 10 mL of ammonium acetate (0.3 M, pH 8.8) followed with 1.5 mL of ultrapure water and collect the flowthroughs separately. Elute off α-R bound to the column using 6 mL of formic acid (0.1 M) and collect the flowthrough into few separate fractions.

Tip: the vicinal *cis*-diol groups of α-R make a reversible five-member ring complex with immobilized boronic acid in the column under alkaline conditions (Fig. 2). Lowering the pH by adding formic acid destabilizes this complex causing α-R to desorb. Verify the pH of the hydrolysate making sure it is 8.5–9 before loading onto the column. This pH will enhance complex formation of α-R with boronic acid and minimize loss of α-R.

The column can be regenerated and reused. To regenerate the column, wash it with five additional column volumes (CV) of formic acid (0.1 M) flowed by 5 CV of ammonium acetate (0.3 M, pH 8.8). The resin can be used for up to 10 cycles but should be discarded if it remains colored after regeneration. The resin can be stored up to 2 months on a neutral pH buffer at 4 °C, or indefinitely if it is dried down.

The concentration of α-R in collected fractions can be determined by measuring the fluorescence (step 2.1.2). In addition, obtain the 200–400 nM UV–visible spectrum of the fractions containing α-R using a UV–visible spectrophotometer. Pool the fractions with the unique UV spectrum of α-R as shown in Fig. 2 and proceed to the next step.

2.4 Concentration of α-R by lyophilization

Transfer the pooled fractions into a 50 mL conical tube (Avantor™, from VWR). Shell-freeze the content into a thin layer on the wall of the tube by slowly swirling over a mixture of ethanol and dry ice. The purpose of this step is to increase the surface area to speed up sublimation.

Tip: Slowly turning the tube at an angle would help freeze the content over a larger surface.

Replace the cap of the tube with parafilm and puncture it 5–6 times. Keep the tube on dry ice to prevent melting of the shell while in transit to the lyophilizer. Lyophilize the sample until white crystalline powder forms. If necessary, seal the tube with lid and spin down in a centrifuge. Add 200 μL of dimethyl sulfoxide (DMSO) and resuspend the powder by pipetting. Prepare proper dilutions and determine the concentration of α-R by fluorescence (step 2.1.2). Store α-R stocks at −20 °C until further use.

2.4.1 *Bioassay to verify that the isolated material is in fact α-ribazole*

In addition to the UV spectrum and the fluorescence assay, a bioassay can be used to verify that the isolated material is α-R. To do this, we used *S. enterica* strain JE21364 (*metE cobT cobB*/p*cblS*$^+$) as an indicator strain. Under the assay condition described herein, this strain is auxotrophic for α-R.

Bioassay procedure. Spin down cells of an overnight culture of *S. enterica* strain JE21364 grown in nutrient broth (Difco), then resuspend the cell pellet using sterile phosphate-buffered saline (PBS). Add 200 μL cells to 4 mL of melted 0.7% (*w/v*) agar, mix, and overlay it onto a base agar of no-carbon E (NCE) medium (Berkowitz, Hushon, Whitfield, Roth, & Ames, 1968) supplemented with Wolfe's trace minerals (Balch & Wolfe, 1976), $MgSO_4$ (1 mM), ribose (22 mM), L(+)-arabinose (500 μM), ampicillin (50 μM), dicyanocobinamide [(CN)2Cbi] (0.25 nM). Once the soft agar has solidified, 1–5 μL of diluted α-R concentrate can be spotted onto a sterile paper disc placed on the agar. Incubate at 37 °C overnight. The presence of α-R in the sample is detected as a halo of growth.

3. Expression and purification of CblS from *Geobacillus kaustophilus* in *E. coli*

This section describes the expression of the *G. kaustophilus* kinase CblS in *E. coli* and the subsequent purification of the enzyme used to phosphorylate α-R.

3.1 Production of CblS

The method described below uses plasmid pGKCBLS2 to overexpress *cblS*$^+$ in strain JE13607 [*E. coli* BL21 (λDE3) *cobT762::kan*$^+$ / pGKCBLS2].

Plasmid pGKCBLS2 was constructed using the polymerase incomplete primer extension (PIPE) method as described elsewhere (Mattes & Escalante-Semerena, 2017). The alluded plasmid results in the *G. kaustophilus* HTA426 *cblS*$^+$ coding sequence being inserted between the *Nhe*I and *Xho*I sites of overexpression vector pTEV5 which encodes CblS with a N-terminal, rTEV-cleavable hexahistidine tag (H_6-*Gk*CblS) (Rocco, Dennison, Klenchin, Rayment, & Escalante-Semerena, 2008).

Transform 100 ng of plasmid pGKCBLS2 into chemically competent *E. coli* BL21 (λDE3) *cobT762::kan*$^+$. Plate cells on lysogeny broth (LB) agar (10 g/L tryptone, 5 g/L NaCl, 5 g/L yeast extract, 15 g/L agar) supplemented with ampicillin (100 μg/mL), and incubate at 37 °C overnight. *Use an isolated colony start a culture* in LB supplemented with ampicillin (100 μg/mL); incubate the culture overnight (~16 h) at 37 °C with shaking.

Use 1 mL of the overnight culture to inoculate 50 mL of Terrific Broth [tryptone, (24 g/L), yeast extract (12 g/L), glycerol (4 mL/L), KH_2PO_4 (17 mM), $MgSO_4$ (2 mM)] plus ampicillin (100 μg/mL); the use of a baffled flask is recommended to improve aeration. Grow the cultures overnight (~16 h) at 37 °C with shaking. After incubation, remove a sample of the culture (500 μL) and analyze the protein content of the cells using SDS-PAGE to verify CblS production. If CblS overproduction is successful (indicated by a prominent band at ~28.5 kDa), harvest the remainder of the culture by centrifugation at 6000 × *g* for 15 min at 4 °C. Discard supernatant and store the cell pellet at −80 °C. We observed a higher expression of CblS in relatively smaller cultures and we recommend pooling the pellets from several 50 mL cultures if larger amounts of CblS are needed.

3.2 Purification of CblS

Resuspend the cell pellet in binding buffer containing potassium phosphate dibasic (50 mM, pH 7.2 at 25 °C), potassium chloride (350 mM), imidazole (10 mM), glycerol (5%, *v*/v), *tris*(2-carboxyethyl)phosphine (TCEP, 0.5 mM). To effectively resuspend the pellet of cells we recommend using ~1:10 cell pellet (g) to binding buffer (mL) ratio. Add Triton X-100 (0.5%), lysozyme (1.0 mg/mL), and DNase (0.1 mg/mL) to the cell resuspension. Lyse the cells using a cell disruptor or French press. We routinely pass cells twice through a cell disruptor at 1.72×10^5 kPa. We recommend to add a protease inhibitor after lysis. One can use protease inhibitor phenylmethanesulfonyl fluoride (PMSF) to a 1 mM final concentration or commercially available protease inhibitor cocktails. Remove the cell debris by centrifugation at 15,000 × *g*. Pass the supernatant over a 0.45 μm polyethersulfone syringe filter to remove remaining debris. Resuspend the cell pellet in 5 mL binding buffer and centrifuge again at 15,000 × *g* at 4 °C. Pass the supernatant again over a 0.45 μm filter and pool with the first supernatant.

Load the cell-free lysate onto a 5 mL of Ni-nitrilotriacetic acid (NTA) affinity resin. Wash the column with three column volumes of binding buffer and washing buffer [potassium phosphate (50 mM), potassium chloride (350 mM), imidazole (150 mM), glycerol (5%, *v*/v), TCEP (0.5 mM), pH 7.2 at 25 °C]. Elute H_6-*Gk*CblS with four column volumes of elution buffer [potassium phosphate (50 mM), potassium chloride (350 mM), imidazole (750 mM), glycerol (5%), TCEP (0.5 mM), pH 7.2 at 25 °C] and collect multiple fractions. Analyze samples of the collected fractions using SDS-PAGE to determine purity of the protein. Pool the fractions with highest CblS purity and dialyze against buffer containing potassium phosphate (50 mM), glycerol (5%, v/v), TCEP (0.5 mM), pH 7.2 at 25 °C. It is not necessary to use homogeneous CblS; an enriched preparation of the protein is sufficient for the purpose of converting α-R to α-RP. The KCl concentration of the dialysis buffer can be gradually decreased from 350 to 100 mM in a few successive dialyses and the first dialysis may also be supplemented with EDTA, (10 mM) to remove residual nickel ions. The final protein concentration may be quantified by Bradford protein assay. Freeze the purified protein in liquid nitrogen and store at −80 °C until further use.

Tip: Cleavage of the H_6 tag is not necessary for CblS activity (Mattes et al., 2021).

4. Synthesis and purification of α-ribazole-5′-phosphate

Here, we describe phosphorylation of α-R into α-RP using CblS, and the purification of α-RP by RP-HPLC.

4.1 Synthesis of α-ribazole-5′-phosphate

This section describes a method to synthesize α-RP by enzymatic phosphorylation of α-R using CblS. The conditions for this *in vitro* reaction were optimized as described elsewhere (Mattes et al., 2021; Mattes & Escalante-Semerena, 2017).

Set up an *in vitro* reaction mixture containing HEPES buffer (50 mM, pH 7.5), $MgCl_2$ (5 mM), KCl (800 mM), TCEP (1.25 mM), ATP (2 mM), α-ribazole (1 mM) and CblS (>4 μg/100 μL). Incubate at 37 °C for at least for 3 h and stop the reaction by heating for 5 min at 95 °C. Dilute the reaction mixture 1:1 with ammonium acetate (20 mM, pH 4.5) and centrifuge at 15,000 × *g* for 10 min to remove any precipitate. Pass the supernatant through a 0.45 μM cellulose acetate filter and place samples into HPLC autosampler vials. We used Costar® Spin-X® Centrifuge Tube Filters (Corning; 0.45 μM cellulose acetate) and 12 mm × 32 mm MT-IT autosampler vials (Shimadzu).

4.2 HPLC purification of α-ribazole-5′-phosphate (α-RP)

α-RP can be resolved by RP-HPLC using a method described elsewhere (Gray & Escalante-Semerena, 2010; Mattes & Escalante-Semerena, 2017). We used Shimadzu Prominence UFLC equipped with a Kinetex 5 μ C18 column (150 mm × 4.6 mm; Phenomenex) and a SecurityGuard ULTRA guard column (C18 cartridge; Phenomenex). Set up the column according to the manufacturer's instruction and equilibrate at a flow rate of 0.8 mL min^{-1} with 3% (*v*/v) acetonitrile/97% (v/v) of ammonium acetate (20 mM, pH 4.5). Inject the filtered reaction and develop the column for 12 min with a linear gradient to 40% (v/v) acetonitrile/60% (v/v) ammonium acetate (20 mM, pH 4.5). Follow this step with a 5-min linear gradient to 80% (v/v) acetonitrile/20% (v/v) of ammonium acetate (20 mM, pH 4.5). Detect the compounds by their absorbance at 287 nm using a photodiode array detector.

α-RP elutes at ∼8.9 min while ATP and α-R elute at ∼2–3 min and 10.2 min, respectively (Mattes & Escalante-Semerena, 2017). Collect the all the fractions containing α-RP (∼8.9 min) and pool them together. α-RP in these fractions can be recovered and concentrated using a Waters Sep-Pak C18 column. We use a Sep-Pak C18 Plus Short Cartridge (360 mg sorbent per Cartridge, 55–105 μm). Pass the fractions through a C18 column previously conditioned with 10 mL of 100% methanol followed by 10 mL of deionized H_2O. Elute with 100% methanol and vacuum centrifuge for ∼16 h to evaporate the solvent. Resuspend α-RP in DMSO to obtain the desired concentration and store at −20 °C. The concentration of purified α-RP can be determined using the molar absorptivity of the nucleobase DMB in α-RP at 280 nm (ε_{280} of DMB in DMSO = 5091 M^{-1} cm^{-1}) (Mattes et al., 2021).

Synthesis of α-RP can be verified using electrospray ionization mass spectrometry. Theoretical mass of α-RP is 358.28 amu.

References

Balch, W. E., & Wolfe, R. S. (1976). New approach to the cultivation of methanogenic bacteria: 2-mercaptoethanesulfonic acid (HS-CoM)-dependent growth of *Methanobacterium ruminantium* in a pressurized atmosphere. *Applied and Environmental Microbiology*, *32*, 781–791.

Battersby, A. R. (2000). Tetrapyrroles: The pigments of life. *Natural Product Reports*, *17*, 507–526.

Berkowitz, D., Hushon, J. M., Whitfield, H. J., Jr., Roth, J., & Ames, B. N. (1968). Procedure for identifying nonsense mutations. *Journal of Bacteriology*, *96*, 215–220.

Escalante-Semerena, J. C., & Warren, M. J. (2008). Biosynthesis and use of cobalamin (B_{12}). In A. Böck, R. Curtiss III,, J. B. Kaper, P. D. Karp, F. C. Neidhardt, T. Nyström, & C. L. Squires (Eds.), *EcoSal - Escherichia coli and Salmonella: Cellular and molecular biology* ASM Press. http://www.ecosal.org.

Gray, M. J., & Escalante-Semerena, J. C. (2010). A new pathway for the synthesis of alpha-ribazole-phosphate in *Listeria innocua*. *Molecular Microbiology*, 77, 1429–1433. https://doi.org/MMI7294 [pii]10.1111/j.1365-2958.2010.07294.x.

Jeter, V. L., & Escalante-Semerena, J. C. (2021). Insights into the relationship between cobamide synthase and the cell membrane. *MBio*, *12*(2). https://doi.org/10.1128/mBio.00215-21.

Maggio-Hall, L. A., & Escalante-Semerena, J. C. (1999). In vitro synthesis of the nucleotide loop of cobalamin by *Salmonella typhimurium* enzymes. *Proceedings of the National Academy of Sciences of the United States of America*, *96*, 11798–11803.

Mattes, T. A., & Escalante-Semerena, J. C. (2017). *Salmonella enterica* synthesizes 5,6-dimethylbenzimidazolyl-(DMB)-alpha-riboside. Why some *Firmicutes* do not require the canonical DMB activation system to synthesize adenosylcobalamin. *Molecular Microbiology*, *103*, 269–281. https://doi.org/10.1111/mmi.13555.

Mattes, T. A., & Escalante-Semerena, J. C. (2018). Facile isolation of alpha-ribazole from vitamin B_{12} hydrolysates using boronate affinity chromatography. *Journal of Chromatography. B, Analytical Technologies in the Biomedical and Life Sciences*, *1090*, 52–55. https://doi.org/10.1016/j.jchromb.2018.05.022.

Mattes, T. A., Malalasekara, L., & Escalante-Semerena, J. C. (2021). Functional studies of alpha-riboside activation by the alpha-ribazole kinase (CblS) from *Geobacillus kaustophilus*. *Biochemistry*. https://doi.org/10.1021/acs.biochem.1c00119.

Pakin, C., Bergaentzle, M., Aoude-Werner, D., & Hasselmann, C. (2005). α-Ribazole, a fluorescent marker for the liquid chromatographic determination of vitamin B_{12} in foodstuffs. *Journal of Chromatography. A*, *1081*, 182–189. (http://www.ncbi.nlm.nih.gov/pubmed/16038208?dopt=Citation).

Pfadenhauer, E. H., & Tong, S.-D. (1979). Determination of inosine and adenosine in human plasma using high-performance liquid chromatography and a boronate affinity gel. *Journal of Chromatography. B, Analytical Technologies in the Biomedical and Life Sciences*, *162*, 585–590. https://doi.org/https://doi.org/https://doi.org/10.1016/S0378-4347(00)81838-1.

Rocco, C. J., Dennison, K. L., Klenchin, V. A., Rayment, I., & Escalante-Semerena, J. C. (2008). Construction and use of new cloning vectors for the rapid isolation of recombinant proteins from *Escherichia coli*. *Plasmid*, *59*, 231–237. (http://www.ncbi.nlm.nih.gov/pubmed/18295882?dopt=Citation).

Zayas, C. L., & Escalante-Semerena, J. C. (2007). Reassessment of the late steps of coenzyme B_{12} synthesis in *Salmonella enterica*: Evidence that dephosphorylation of adenosylcobalamin-5′-phosphate by the CobC phosphatase is the last step of the pathway. *Journal of Bacteriology*, *189*, 2210–2218. https://doi.org/10.1128/JB.01665-06.

CHAPTER SEVEN

The human B_{12} trafficking chaperones: CblA, ATR, CblC and CblD

Zhu Li, Harsha Gouda, Shubhadra Pillay, Madeline Yaw, Markus Ruetz, and Ruma Banerjee
Department of Biological Chemistry, University of Michigan Medical School, Ann Arbor, MI, United States

Contents

Abstract

Mammals rely on an elaborate intracellular trafficking pathway for processing and delivering vitamin B_{12} to two client enzymes. CblC (also known as MMACHC) is postulated to receive the cofactor as it enters the cytoplasm and converts varied B_{12} derivatives to a common cob(II)alamin intermediate. CblD (or MMADHC) reacts with CblC-bound cob(II) alamin forming an interprotein thiolato-cobalt coordination complex and, by a mechanism that remains to be elucidated, transfers the cofactor to methionine synthase. In the mitochondrion, CblB (also known as MMAB or adenosyltransferase) synthesizes AdoCbl from cob(II)alamin and ATP in the presence of an electron donor. CblA (or MMAA), a GTPase, gates cofactor loading from CblB to methylmalonyl-CoA mutase and off-loading of cob(II)alamin in the reverse direction. This chapter focuses on assays for measuring the activities of the four B_{12} chaperones CblA-D.

Methods in Enzymology, Volume 668
ISSN 0076-6879
https://doi.org/10.1016/bs.mie.2021.12.009

1. Introduction

Cobalamin (known generically as vitamin B_{12}) is a complex organometallic cofactor that is required by two mammalian enzymes: cytoplasmic methionine synthase (MS), which uses methylcobalamin (MeCbl) and mitochondrial methylmalonyl-CoA mutase (MCM), which uses 5′-deoxyadenosylcobalamin (AdoCbl) (Banerjee & Ragsdale, 2003). Unable to synthesize this essential cofactor, mammals rely on a dietary supply of cobalamin and an intracellular B_{12} trafficking pathway comprising a multitude of chaperones for its processing and delivery to MS and MCM (Banerjee, Gherasim, & Padovani, 2009; Banerjee, Gouda, & Pillay, 2021; Gherasim, Lofgren, & Banerjee, 2013).

CblC, also known as MMACHC (for methylmalonic aciduria type C and homocystinuria), is an early component in the B_{12} trafficking pathway (Fig. 1).

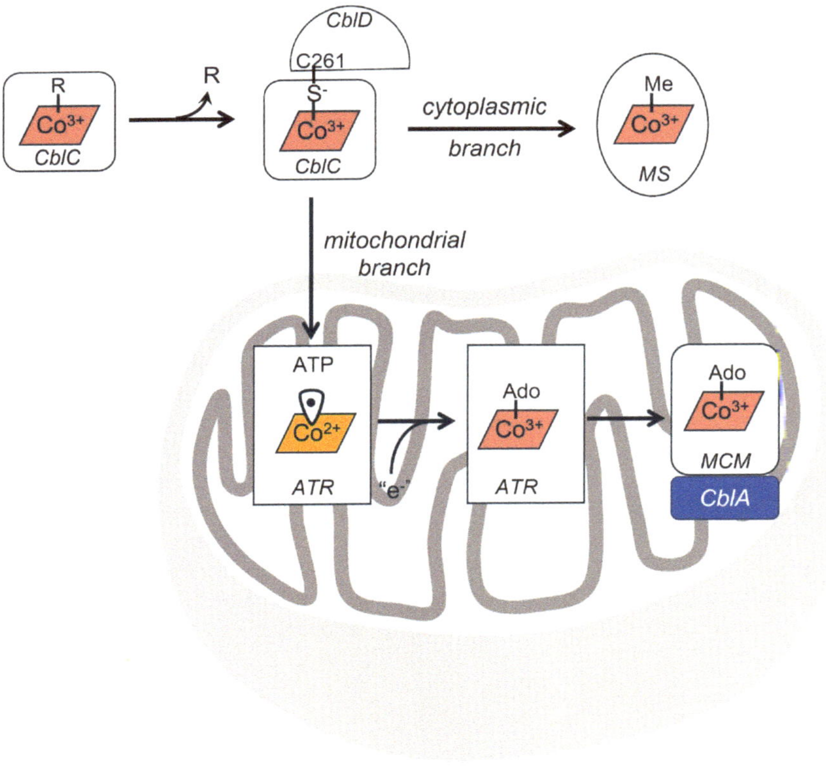

Fig. 1 Functions of CblC, CblD, ATR and CblA in the B_{12} trafficking pathway. R represents an alkyl group.

It binds various B_{12} derivatives and, via its β-ligand transferase activity (Kim et al., 2009; Kim, Gherasim, & Banerjee, 2008), removes the upper or β-axial ligand, forming cob(II)alamin under aerobic conditions. Once the β-axial position has been vacated, CblD (also known as MMADHC) provides a thiolato ligand to the cobalt ion, forming an interprotein thiolato-cobalamin complex with CblC (Li et al., 2020). The function and fate of the CblC•CblD complex await further studies. However, clinical data suggest that CblD is important for delivery of cobalamin to MS (in the cytoplasmic branch) and MCM (in the mitochondrial branch) (Coelho et al., 2008).

In mitochondria, CblB, which is an ATP:cob(I)alamin transferase (hereafter referred to as ATR) converts cob(II)alamin to AdoCbl (Yamanishi, Labunska, & Banerjee, 2005). This reaction requires the injection of an electron from a donor that remains to be identified, to convert cob(II)alamin to cob(I)alamin (Leal et al., 2004). The AdoCbl product is then transferred directly from ATR to MCM without release into solution, in a process that is gated by CblA, a G-protein chaperone (Padovani et al., 2008; Padovani & Banerjee, 2009a, 2009b). The GTPase activity of CblA leads to the nucleotide-sensitive conformational changes that are important for signaling to MCM (Lofgren et al., 2013; Ruetz et al., 2019). MCM in turn, exhibits GAP (GTPase activating protein) activity and regulates CblA (Padovani, Labunska, & Banerjee, 2006). MCM is a radical enzyme that is vulnerable to occasional inactivation leading to loss of the deoxyadenosine moiety and accumulation of inactive cob(II)alamin (Padovani & Banerjee, 2006, 2009b). CblA and ATR engage in repairing MCM in a process that involves off-loading cob(II)alamin, reforming AdoCbl on ATR and re-loading MCM (Campanello et al., 2018; Padovani & Banerjee, 2009b; Ruetz et al., 2019).

2. Assays for chaperones in the cytoplasmic branch

CblC houses a β-ligand transferase/eliminase activity that converts incoming cobalamins to cob(II)alamin under aerobic conditions. In the glutathione (GSH)-dependent dealkylation reaction, the thiolate group of the co-substrate GSH, displaces the alkyl-group in alkylcobalamins (e.g., MeCbl or AdoCbl), forming the corresponding thioether of GSH and cob(I)alamin (Eq. 1) (Kim et al., 2009). The latter is rapidly oxidized in air to aquocobalamin (H_2OCbl) (Eq. 2). CblC also catalyzes the reductive elimination of cyanocobalamin (CNCbl) (Eq. 3) (Kim et al., 2008), H_2OCbl (Li et al., 2017) and nitritocobalamin (NO_2Cbl) (Mascarenhas et al., 2020).

Flavin or GSH can serve as an election donor in these reactions and the cob(II)alamin product is observed under anaerobic conditions but is oxidized to H_2OCbl under aerobic conditions.

$$\mathrm{R-Cbl + GSH \rightarrow Cob(I)alamin + GSR + H^+} \quad (1)$$

$$\mathrm{Cob(I)alamin} + O_2 + \mathrm{H_2}O + 2\mathrm{H}^+ \rightarrow \mathrm{H_2OCbl} + \mathrm{H_2}O_2 \quad (2)$$

$$\mathrm{CNCbl + ``e"^- \rightarrow Cob(II)alamin + CN^-} \quad (3)$$

Following removal of the β-ligand by CblC, CblD forms an interprotein Co—S coordination complex, donating a thiolato ligand from Cys-261 (Fig. 1). The resulting thiolato-cob(II)alamin product (Eq. 4) is subsequently oxidized to thiolato-cob(III)alamin in air (Eq. 5). Additionally, the thiolate group of Cys-261 in CblD can react directly with H_2OCbl in solution (Eq. 6). The low affinity of CblD for H_2OCbl however, makes the physiologically relevance of this interaction unlikely (Li et al., 2020). The vulnerability of CblC-bound cob(I)alamin (formed via dealkylation, (Eq. 1)) and cob(II)alamin (formed via reductive elimination (Eq. 3)) to air oxidation, suggests that they must occur in situ in the CblC•CblD complex for the thiolato complex to form (Eq. 4).

$$\mathrm{CblD - S^- + CblC{\bullet}cob(II)alamin \leftrightarrow CblD - thiolato - cob(II)alamin{\bullet}CblC} \quad (4)$$

$$\mathrm{CblD - thiolato - Cob(II){\bullet}CblC + O_2 \rightarrow CblD - thiolato - Cob(III){\bullet}CblC + O_2^{\bullet -}} \quad (5)$$

$$\mathrm{CblD} - S^- + \mathrm{H_2OCbl} \rightarrow \mathrm{CblD} - S^- - \mathrm{Cob(III)} + H_2O \quad (6)$$

In this chapter, the purification and assays for CblC, CblD, ATR and CblA are described.

3. Assays for CblC

3.1 Expression and purification of CblC

3.1.1 Recombinant expression of human CblC

Materials

- *E. coli* expression strain BL21 (DE3)
- Luria-Bertani (LB) medium
- LB agar plates containing 100 μg/mL ampicillin
- 100 mg/mL ampicillin
- 1 M Isopropyl-β-D-thiogalactopyranoside (IPTG)

Procedure

1. The human CblC cDNA was cloned into a ligation-independent vector with a His_6-tag followed by a TEV-protease site at the N-terminus of CblC (Li et al., 2020).

2. *E. coli* (BL21/DE3) cells are transformed with the CblC expression vector and plated on an LB agar plate containing 100 µg/mL of ampicillin. The plate is incubated overnight at 37 °C.
3. The next day, a single colony from the plate is used to inoculate 100 mL of LB containing 100 µg/mL ampicillin in a 250 mL culture flask. The flask is incubated overnight at 37 °C and shaken at 220 rpm.
4. The overnight inoculate (10 mL) and 1 mL ampicillin (100 µg/mL final concentration) are added to each 2.8 L culture flask containing 1 L LB. The six flasks are incubated at 37 °C with shaking at 220 rpm until the OD_{600} = 0.5–0.6 (~4 h).
5. The incubator is cooled to 15 °C with shaking at 220 rpm for 1 h.
6. IPTG (100 µL) is added and incubation continued overnight at 15 °C with shaking at 220 rpm.
7. After 16–18 h, the culture is collected in six 1 L centrifuge bottles and centrifuged at 5500 × *g* for 30 min. The pelleted cells is stored at −80 °C until further use or used immediately to purify CblC.

3.1.2 Purification of human CblC

Equipment.

Misonix Sonicator (Model XL, 1.2 cm diameter probe).
Temperature-controlled high speed centrifuge.
FPLC system.
Nickel-NTA-agarose (Ni-NTA) column (Qiagen).
Superdex 200 (S200, 100 mL, GE Healthcare).
Amicon Ultra-15 30 kDa centrifugal filters.

Materials

- Buffer A: 50 mM Tris-HCl, pH 8.0, 300 mM KCl, 15 mM imidazole
- Buffer B: 50 mM Tris-HCl, pH 8.0, 300 mM KCl, 400 mM imidazole
- Lysozyme from chicken egg white
- EDTA-free cOmplete Protease Inhibitor Cocktail (Roche)
- Wash Buffer: 50 mM Tris-HCl, pH 8.0, 300 mM KCl, 30 mM imidazole
- Dialysis Buffer: 0.1 M Tris-HCl, pH 8.0, 150 mM KCl, 10% glycerol
- Storage Buffer: 0.1 M HEPES, pH 7.4, 150 mM KCl, 10% glycerol
- Purified His_6-tagged tobacco etch virus (TEV) protease (1 mg/mL)
- Dithiothreitol (DTT), 1 M
- Bio-Rad Protein Assay Kit

Procedure

1. The cell pellet is thawed on ice and resuspended in 200 mL Buffer A supplemented with 20 mg lysozyme and 1 tablet of EDTA-free

cOmplete Protease Inhibitor Cocktail. The mixture is transferred to a 300 mL beaker and stirred for 15–20 min at 4 °C until the suspension is homogeneous.

2. Cells are sonicated on ice at 30% power with cycles of 15 s on and 60 s off for a total of 8 min.
3. Cell lysates are transferred to 50 mL centrifuge tubes and centrifugated at 38,000 × *g* at 4 °C for 30 min.
4. The supernatant is collected in a 250 mL beaker and loaded onto a Ni-NTA column (2.5 × 5 cm) equilibrated with Buffer A, followed by washing with 100 mL of Wash Buffer.
5. His_6-tagged CblC is eluted by an imidazole gradient using a gradient mixer containing 150 mL Buffer A and 150 ml Buffer B and 10 mL fractions are collected. The fractions containing CblC are determined by measuring the 595 nm absorbance using 5 μL of each fraction and 200 μL of Bio-Rad Protein Assay Reagent in a 96-well plate.
6. The fractions containing CblC (with $A_{595} > 0.15$) are combined and concentrated to 15 mL using Amicon Ultra-15 (MWCO = 30 kDa) filters. The concentrated protein is transferred to an appropriate length of SnakeSkin Dialysis Tubing (MWCO = 10 kDa) and 1 mL of TEV-protease is added. The protein mixture is dialyzed overnight with stirring in a 1 L beaker containing 1 L Dialysis Buffer supplemented with 1 mL DTT at 4 °C.
7. The dialyzed mixture is transferred to a 50 mL conical tube and clarified by centrifugation at 2998 × *g* and 4 °C for 10 min. The supernatant is loaded onto a Ni-NTA column (2.5 × 2 cm) and washed with Dialysis Buffer +25 mM imidazole. The flow-through containing tagless CblC is collected.
8. CblC is concentrated in Amicon Ultra-15 (MWCO = 30 kDa) filters to 5 mL and dialyzed in SnakeSkin Dialysis Tubing against 1 L of the Storage Buffer at 4 °C overnight.
9. Dialyzed CblC is clarified by centrifugation at 20,000 × *g* and 4 °C for 10 min and loaded onto a Superdex 200 column (S200, 100 mL, GE Healthcare) equilibrated with the Storage Buffer. Fractions containing CblC are monitored by absorbance at 280 nm.
10. CblC (monomer) is concentrated to 3 mL using Amicon Ultra-15 (MWCO = 30 kDa) filters. The molecular weight of purified CblC is 34 kDa (Fig. 2A). The concentration is determined by the absorbance at 280 nm ($\varepsilon_{280\ nm} = 49.6\ cm^{-1}\ M^{-1}$). The purified protein is then aliquoted (50 μL) in 750 μL microcentrifuge tubes, flash-frozen in liquid nitrogen and stored at −80 °C till further use. A typical purification following this protocol yields 5 mg of CblC per liter of culture.

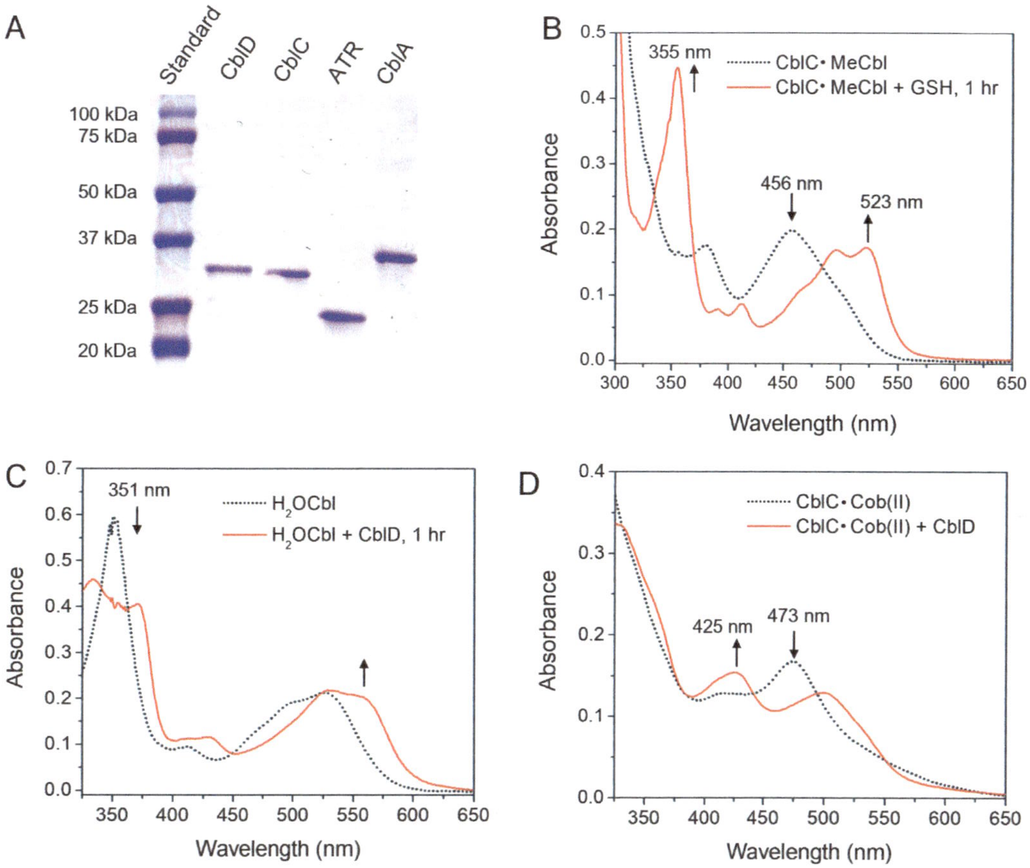

Fig. 2 Protein purity and spectral characterization. (A) SDS-PAGE analysis of purified human CblC, CblD, ATR and CblA. (B) UV–visible spectra of CblC (40 μM) with MeCbl (20 μM), before (*dotted black line*) and after incubation with 1 mM GSH for 1 h (*red solid line*). (C) Spectral changes induced by H_2OCbl (20 μM) (*dotted black line*) binding to CblD (40 μM), after 1 h incubation (*red solid line*). (D) UV–visible spectra of CblC (40 μM) and cob(II) alamin (20 μM) in in before (*dotted black line*) and after incubation with CblD (40 μM) for 200 s aerobically (*red solid line*). The buffer used in B–D was 100 mM HEPES-KOH, pH 7.4, 150 mM KCl, 10% glycerol.

3.1.3 CblC activity assay: GSH-dependent dealkylation

Reagents:

- 100 mM HEPES-KOH buffer, pH 7.4, 150 mM KCl, 10% glycerol
- 1 mM MeCbl or 1 mM AdoCbl
- 400 μM purified human CblC in 100 mM HEPES-KOH buffer, pH 7.4, 150 mM KCl, 10% glycerol
- 100 mM GSH

Method: In a 150 μL quartz cuvette, mix 130 μL HEPES buffer and 3 μL MeCbl (or AdoCbl) and place the cuvette in a spectrophotometer at 20 °C. Add 15 μL CblC (40 μM final concentration) to the cuvette and mix well. Initiate the dealkylation reaction by adding 1.5 μL GSH (1 mM final concentration) and record the absorption spectrum every minute between 300 and 700 nm for a total of 1 h (Fig. 2B). Reaction progress is monitored by H_2OCbl formation and quantified by the change in absorbance at 355 nm using $\Delta\varepsilon_{355\ nm} = 11\ cm^{-1}\ mM^{-1}$ (Kim et al., 2009). Activity is expressed in units of min^{-1} at 20 °C and is typically 0.04 – 0.05 min^{-1} (Li et al., 2020).

3.1.4 CblC activity assay: Reductive elimination of CNCbl by methionine synthase reductase (MSR) and NADPH

Reagents:

- 100 mM HEPES buffer-KOH, pH 7.0, 150 mM KCl, 5% glycerol, anaerobic
- 5 mM NADPH, anaerobic
- 200 μM purified human MSR, anaerobic
- 400 μM purified human CblC, anaerobic
- 1 mM CNCbl, anaerobic

Method: HEPES buffer (40 mL) is purged with nitrogen gas in a sealed falcon tube for 0.5 h to remove dissolved O_2. All reagents are dissolved in the anaerobic buffer. In a quartz cuvette, add 171.5 μL HEPES buffer, 12.5 μL CblC, 4 μL CNCbl, and 8 μL NADPH. Place the cuvette in a spectrophotometer for 5 min at 23 °C and initiate the reaction with 4 μL MSR purified as described in Olteanu et al. (2004). Monitor the decrease in absorbance at 530 nm (disappearance of CNCbl) over 30 min. From the slope of the absorbance change, calculate the reaction rate using $\varepsilon_{530\ nm} = 3700\ M^{-1}\ cm^{-1}$ (Li et al., 2014). Activity is expressed in units of min^{-1} at 23 °C and is estimated to be 0.1 min^{-1} (Koutmos et al., 2011). Alternatively, MSR and NADPH can be replaced with 1.5 μL 100 mM GSH (1 mM final concentration) (Li et al., 2014).

3.2 Assays for CblD

3.2.1 Expression and purification of CblD

3.2.1.1 Recombinant expression of human CblD

Materials

- *E. coli* strains BL21 (DE3)
- LB medium
- LB agar plates containing 100 μg/mL ampicillin
- 100 mg/mL ampicillin
- 1 M IPTG

Procedure

1. Human CblD is expressed as a GST fusion protein, using the pGEX-TEV-CblD vector in which the thrombin cleavage site has been replaced by a TEV-protease site (Li et al., 2020).
2. *E. coli* BL21 (DE3) is transformed with pGEX-TEV-CblD and plated onto an LB agar plate containing 100 μg/mL ampicillin, and incubated overnight at 37 °C.
3. A single colony is selected to inoculate 100 mL of LB medium containing ampicillin (100 μg/mL) in a 200 mL culture flask and grown overnight at 37 °C with shaking at 250 rpm.
4. Culture flasks (6 × 2.8 L) each containing 1 L of LB medium and ampicillin (100 μg/mL) are inoculated with 5 mL of the overnight inoculate and incubated at 37 °C with shaking at 220 rpm until the culture reaches an OD_{600} of 0.5–0.6 (~4–5 h).
5. The incubator is then cooled to 15 °C for 1 h for the temperature to stabilize. Then, 100 μM IPTG (final concentration) is added, and incubation is continued for 16 h with shaking at 220 rpm.
6. The culture is distributed between six 1 L centrifuge bottles and centrifuged at 5500 × *g* for 30 min to pellet cells, which can be used immediately for purification or stored at −80 °C until further use.

3.2.1.2 Purification of human CblD

Equipment

Misonix Sonicator (Model XL, 1.2 cm diameter probe)
Temperature-controlled high-speed centrifuge
FPLC system
Glutathione Sepharose 4B resins (GE Healthcare)
Superdex 200 (S200, 100 mL, GE Healthcare)
Amicon Ultra-15 30 kDa centrifugal filters

Materials

- Buffer A: 50 mM Tris-HCl, pH 7.4, 50 mM KCl
- Buffer B: 50 mM Tris-HCl, pH 8.0, 50 mM KCl, 10 mM GSH
- 14 M β-mercaptoethanol (β-ME)
- Lysozyme from chicken egg white
- EDTA-free cOmplete™ Protease Inhibitor Cocktail (Roche)
- Purified His_6-tagged TEV protease (1 mg/mL)
- DTT, 1 M
- Reaction Buffer: 100 mM HEPES-KOH, pH 7.4, 150 mM KCl and 10% glycerol
- Bio-Rad Protein Assay Kit

Procedure

1. The cell pellet from a 6 L culture is thawed on ice and resuspended in 200 mL Buffer A supplemented with 285 μL β-ME, 20 mg lysozyme and 1 tablet of EDTA-free cOmplete Protease Inhibitor Cocktail. The mixture is transferred to a 300 mL beaker and stirred for 15–20 min at 4 °C till the suspension is homogeneous.
2. The cell suspension is sonicated on ice at 30% power with cycles of 15 s on and 60 s off for a total time of 8 min. The cell lysate is separated from insoluble debris by centrifugation (38,000 × *g*, 30 min, 4 °C).
3. The clarified lysate is filtered using 0.8 μM filters (Sartorius) and loaded on a GSH-agarose column (2.5 × 5 cm) pre-equilibrated with Buffer A, followed by washing with 100 mL of Buffer A.
4. Glutathione-S-transferase (GST)-tagged CblD is eluted with 100 mL Buffer B and 10 mL fractions are collected. CblD containing fractions are detected by their absorbance at 595 nm by mixing 5 μL of the sample with 200 μL of Bio-Rad Protein Assay Reagent in a 96-well plate.
5. The fractions containing GST-CblD are pooled and 1.5 mL TEV-protease is added prior to overnight dialysis at 4 °C against 1 L Buffer A containing 0.7 mL β-ME.
6. The dialyzed proteins containing tag-less CblD and GST are centrifuged at 2998 × *g* for 10 min at 4 °C, and loaded onto a second GSH agarose column (2.5 × 5 cm) pre-equilibrated with Buffer A. The column is washed with 30 mL of Buffer A till CblD is no longer present in the eluate as determined by the Bio-Rad Protein Assay. The flow-through contains CblD and is pooled.
7. CblD is concentrated to 5 mL using an Amicon Ultra-15 (MWCO = 30 kDa) filter and dialyzed overnight at 4 °C in SnakeSkin dialysis tubing against 0.5 L of Reaction Buffer containing 2 mL DTT.
8. CblD is clarified by centrifugation at 20,000 × *g* and 4 °C for 10 min and loaded onto a Superdex 200 column (S200, 100 mL, GE Healthcare)

equilibrated with Reaction Buffer. Fractions containing CblD (monomer) are monitored by their absorbance at 280 nm, pooled and concentrated to 3 mL using an Amicon Ultra-15 10 kDa centrifugal filter.

9. CblD concentration is estimated by the absorbance at 280 nm (ε_{280} = 32,930 M^{-1} cm^{-1}). The purified protein (Fig. 2A) is separated into 50 μL aliquots in 750 μL microcentrifuge tubes, flash-frozen in liquid nitrogen and stored at −80 °C till further use. A typical purification following this protocol yields 5 mg of CblD per liter of culture.

3.2.2 Monitoring H_2OCbl binding to CblD

Reagents

- Reaction Buffer: 0.1 M HEPES-KOH, pH 7.4, 150 mM KCl, 10% glycerol
- 400 μM CblD
- 1 mM H_2OCbl
- 1 M DTT

Method: To ensure that the cysteines are reduced, 100 μL CblD is mixed overnight with 2 μL DTT on ice. The protein is diluted with 400 μL of the Reaction Buffer and centrifuged in an Amicon Ultra-0.5 10 kDa centrifugal filter at 4 °C till the protein is concentrated to 100 μL. This procedure is repeated 6 times to completely remove DTT. The concentration of reduced CblD is estimated by the absorbance at 280 nm (ε_{280} = 32,930 M^{-1} cm^{-1}) and adjusted to 400 μM using the Reaction Buffer.

To monitor H_2OCbl binding to CblD, 132 μL Reaction Buffer and 3 μL H_2OCbl are added to a 150 μL quartz cuvette, mixed and placed in a spectrophotometer at 23 °C. Formation of the thiolato-cobalamin complex is initiated by addition of 15 μL CblD (40 μM final concentration) to the cuvette and the UV–visible spectrum is recorded between 300 and 700 nm every minute for 1 h to monitor reaction progress (Fig. 2C). Formation of the CblD-thiolato-cobalamin complex is quantified by the decrease in absorbance at 351 nm ($\Delta\varepsilon_{351nm}$ = 11.8 mM^{-1} cm^{-1}). The activity is typically 0.09 min^{-1} (Li et al., 2020).

3.2.3 Monitoring formation of the CblC-cob(III)alamin-thiolato-CblD complex

Reagents:

- Reaction buffer: 100 mM HEPES buffer, pH 7.4, 150 mM KCl, 10% glycerol, 500 U bovine superoxide dismutase, aerobic
- 1 mM AdoCbl
- 400 μM purified human CblC, aerobic
- 400 μM purified human CblD, aerobic

Method: Prepare 2 mL of 1 mM AdoCbl stock solution in a glass cuvette sealed with a rubber stopper and parafilm and purge with N_2 gas for 15 min. Then, place the AdoCbl sample on ice 10 cm away from a halogen lamp (38 W, G.E.) After 2–3 h, move the sample into an anaerobic chamber. Complete photolysis of AdoCbl is confirmed by the absorption spectrum of cob(II)alamin ($\lambda_{max} = 473$ nm, $\varepsilon_{473nm} = 9.2\,mM^{-1}\,cm^{-1}$).

CblC•CblD complex formation is monitored using a stopped flow spectrophotometer. For this, CblC (300 μL) and cob(II)alamin (80 μL) are mixed and then diluted with 1.62 mL of aerobic Reaction Buffer. In a separate syringe, 300 μL CblD is mixed with 1.7 mL aerobic Reaction Buffer. The two solutions are then rapidly mixed in a stopped-flow spectrophotometer at 20 °C and spectra are recorded between 300 and 700 nm for 200 s at 0.2 s intervals (Fig. 2D). The reaction is estimated by the decrease in absorbance at 473 nm ($\Delta\varepsilon_{473\ nm} = 4.1\,mM^{-1}\,cm^{-1}$) and fitting the trace to a double exponential equation $A_{473nm} = y_0 + a_1 \times e^{-t \times k_{app1}} + a_2 \times e^{-t \times k_{app2}}$. The estimated k_{app1} of the fast phase with the major absorption change is 4 min^{-1} (Li et al., 2020).

4. Assays for B_{12} chaperones in the mitochondrial branch

Human ATR catalyzes the synthesis of AdoCbl and also functions as a chaperone for loading AdoCbl directly to MCM and off-loading cob(II) alamin from MCM. Following binding of ATP and cob(II)alamin, reduction to cob(I)alamin (Eq. 7), sets up a nucleophilic attack on the C5' carbon of ATP, forming AdoCbl and triphosphate (PPPi) (Eq. 8). The electron donor to ATR is not known.

$$\text{Cob(II)alamin} + \text{“}e\text{”}^{-} \rightarrow \text{Cob(I)alamin} \tag{7}$$

$$\text{Cob(I)alamin} + \text{ATP} \rightarrow \text{AdoCbl} + \text{PPPi} \tag{8}$$

Human CblA is a P-loop GTPases (Leipe et al., 2002), which gates the translocation of cobalamin between MCM and ATR (Ruetz et al., 2019). CblA communicates with MCM using the classical G-protein “switch” elements and exhibits very low intrinsic GTPase activity that is enhanced by the GAP function of MCM (Campanello et al., 2017; Lofgren et al., 2013; Lofgren, Koutmos, & Banerjee, 2013). Assays for ATR and for measuring the intrinsic and GAP-activated GTPase activity of CblA are described in this section.

4.1 Assays for ATR

4.1.1 Expression and purification of ATR

4.1.1.1 Recombinant expression of His_6-tagged ATR

Materials

- *E. coli* expression strain BL21 (DE3)
- LB medium
- LB agar plates containing 50 μg/mL of kanamycin
- Kanamycin, 50 mg/mL in sterilized water
- IPTG, 50 mM solution in sterilized water
- Erlenmeyer flasks (250 ml and 2.8 L), temperature-controlled shaker

Procedure:

1. The expression construct for human ATR uses the pET28b vector and provides an N-terminal hexahistidine (His_6) tag that can be cleaved by thrombin (Campanello et al., 2018).
2. Transform BL21(DE3) cells with the ATR expression vector and select on LB agar plates following overnight incubation at 37 °C.
3. Inoculate a single colony into a 50 mL of LB medium containing 50 μg/mL of kanamycin in a 250 ml flask. Incubate overnight at 37 °C with shaking at 220 rpm.
4. Inoculate 6 × 2.8 L flasks each containing 1 L LB medium with 5 mL of the overnight culture and incubate at 37 °C with shaking at 220 rpm until OD_{600} = 0.6, (~2.5–3 h). Then, cool to 20 °C over 40 min.
5. Add 1 mL IPTG to each flask (50 μM final concentration) and continue incubation overnight (14–16 h) at 20 °C with shaking at 220 rpm.
6. Harvest cells by centrifugation at 5500 × *g* for 20 min using 1 L centrifuge bottles. Transfer cell pellets in a 50 mL pre-weighed conical tube, and freeze at −80 °C until further use.

4.1.1.2 Purification of ATR

Equipment.

Misonix Sonicator (Model XL, 1.2 cm diameter probe)
Temperature-controlled centrifuge
Ni-NTA resins (Qiagen)
Amicon Ultra-15 30 kDa centrifugal filters
HiTrap Benzamidine FF column (5 mL, GE Healthcare)

Materials

- Lysis Buffer: 50 mM Tris-HCl pH 8.0, 300 mM NaCl, 10 mM imidazole and 1 mM TCEP (tris (2-carboxyethyl)phosphine)

- Elution Buffer: 50 mM Tris-HCl pH 8.0, 300 mM NaCl, 300 mM imidazole and 1 mM TCEP
- Thrombin Buffer: 50 mM Tris-HCl pH 8.0, 300 mM NaCl and 1 mM TCEP
- 1000 units/mL thrombin
- EDTA-free cOmplete™ Protease Inhibitor Cocktail (Roche)
- Benzamidine Buffer: 25 mM HEPES pH 7.4, 500 mM NaCl and 1 mM TCEP
- Buffer A: 50 mM HEPES pH 7.4, 150 mM KCl, 2 mM TCEP, 2 mM $MgCl_2$, 5% glycerol
- Bio-Rad Protein Assay Kit

Procedure:

1. Thaw the cell pellet and resuspend in lysis buffer (6–8 mL per gm of cell pellet) in a 500 mL beaker. Add one tablet of EDTA-free cOmplete™ Protease Inhibitor Cocktail and stir at 4 °C for 30 min.
2. Lyse cells by sonication at 30% power with cycles of 15 s on and 45 s off for a total of 8 min.
3. Clarify the cell lysate by centrifugation at 38,500 × *g* for 20 min at 4 °C.
4. Load the supernatant onto a Ni-NTA column (2.5 × 5 cm) washed with 150 mL of Lysis Buffer. Elute ATR using a linear gradient comprising 150 mL Lysis Buffer and 150 mL Elution Buffer. Collect 8–10 mL fractions.
5. Detect protein-containing fractions by monitoring the 595 nm absorbance in a 10 μL aliquot of the fraction mixed with 990 μL of Bio-Rad Protein Assay in a plastic cuvette. Pool and concentrate protein-containing fractions to ~20 mL using Amicon Ultra-15 30 kDa centrifugal filters at 4 °C.
6. Add thrombin (5 unit/mg of ATR) to the protein in a dialysis bag and dialyze overnight against 1 L of Thrombin Buffer at 4 °C to cleave the His_6-tag.
7. Load the protein on to a second Ni-NTA column (2.5 × 5 cm) equilibrated with Thrombin buffer. Wash column with 150 mL of Thrombin Buffer until no further protein is detected by Bradford assay. Collect the flow-through and concentrate it to ~20 mL.
8. Remove thrombin by loading the protein solution on a benzamidine column (1.6 × 2.5 cm). Wash the column with 50 mL Benzamidine Buffer and collect the flow-through containing ATR.
9. Concentrate ATR to ~2 mL at 4 °C (Fig. 2A), using Amicon Ultra-15 (MWCO = 30 kDa) filters, and dialyze overnight against 1 L of

Buffer A at 4 °C. Aliquot 50 µL of the protein in 750 µL microcentrifuge tubes, flash-freeze in liquid nitrogen and store at −80 °C till further use.

10. Calculate the concentration of ATR (trimer) using $\varepsilon_{280nm} = 43{,}200\,cm^{-1}\,M^{-1}$. A typical purification yields 150–180 mg of ATR from 6 L of culture.

4.1.2 ATR-dependent AdoCbl synthesis assay

Reagents:

- Buffer A: 50 mM HEPES pH 7.4, 150 mM KCl, 2 mM TCEP, 2 mM $MgCl_2$, 5% glycerol, anaerobic
- 2 mM cob(II)alamin, anaerobic
- 20 mM Ti(III)citrate, anaerobic (prepared as described in Zehnder & Wuhrman, 1976)
- 100 mM ATP, anaerobic
- 100 µM ATR (trimer)

Method: Preparation of cob(II)alamin is described in Section 3.2.3. Prepare all reagents anaerobically prior to the start of the assay. Place the ATR solution in an anaerobic chamber (with <0.2 ppm O_2) and incubate at 4 °C for 2 h with gentle mixing every 30 min. In a quartz cuvette, add 188 µL Buffer A, 2 µL ATP and 4 µL cob(II)alamin and place the cuvette in the spectrophotometer for 10 min at 20 °C. Add 4 µL of Ti(III)citrate and observe the increase in 386 nm absorbance corresponding to cob(I)alamin formation. Estimate the concentration of cob(I)alamin after stabilization of the 386 nm absorbance using $\varepsilon = 28.0\,cm^{-1}\,mM^{-1}$ (Campanello et al., 2018). Initiate the reaction by adding 2 µL of ATR (1 µM final concentration) and monitor the increase in absorbance at 522 nm corresponding to AdoCbl formation. From the rate of change at 522 nm, calculate the specific activity using $\Delta\varepsilon_{522nm} = 5.04\,cm^{-1}\,mM^{-1}$ (Campanello et al., 2018). The specific activity of ATR is typically $0.16 \pm 0.02\,\mu mol\,min^{-1}\,mg^{-1}$ protein (k_{cat} is $11.5 \pm 1.6\,min^{-1}$).

4.2 Assays for CblA

4.2.1 Expression and purification of CblA

4.2.1.1 Recombinant expression of His_6 fusion CblA

Materials

- *E. coli* expression strain BL21 (DE3)
- LB medium and TB medium
- LB agar plates containing 50 µg/mL of kanamycin
- Kanamycin, 50 mg/mL in sterilized water
- IPTG, 500 mM solution in sterilized water
- 6 Erlenmeyer flasks (2.8 L) and temperature-controlled shaker

Procedure:

1. The *E. coli* codon optimized construct for human CblA cloned into pET28b(+) vector with a TEV cleavable His-tag at the N-terminus (Ruetz et al., 2019) is used.
2. Transform BL21(DE3) cells with the expression vector for CblA and select on LB agar plates incubated overnight at 37 °C overnight.
3. Inoculate 7 mL of LB medium containing 7 µL kanamycin in a 15 mL conical tube with a single colony and incubate overnight at 37 °C with shaking at 220 rpm.
4. Inoculate 6 × 2.8 L culture flasks each containing 1 L TB medium and 1 mL kanamycin with 1 mL of the overnight culture and incubate at 37 °C with shaking at 220 rpm until OD_{600} = 1.0–1.2 (~5–6 h).
5. Add 1 mL IPTG (final concentration of 500 µM) to each flask and incubate at 37 °C overnight (18–20 h) with shaking at 220 rpm.
6. Harvest cells by centrifugation at 5500 × *g* for 20 min and transfer the cell pellet to a pre-weighed 50 mL conical tube, and store at −80 °C until further use.

4.2.1.2 Purification of CblA

Equipment.

Misonix Sonicator (Model XL, 1.2 cm diameter probe).
Temperature-controlled centrifuge
FPLC system
Ni-NTA resins (Qiagen)
Amicon Ultra-15 30 kDa centrifugal filters
Superdex 200 (S200, 100 mL, GE Healthcare)

Materials

- EDTA-free cOmplete™ Protease Inhibitor Cocktail (Roche)
- Purified His_6-tagged TEV protease (1 mg/mL)
- Lysis Buffer: 50 mM Tris-HCl pH 8.0, 500 mM NaCl, 10 mM Imidazole and 5% glycerol
- Elution Buffer: 50 mM Tris-HCl pH 8.0, 500 mM NaCl, 200 mM Imidazole and 5% glycerol
- TEV buffer: 50 mM Tris-HCl pH 8.0, 300 mM NaCl, 0.5 mM EDTA, 1 mM DTT and 5% glycerol
- Buffer A: 50 mM HEPES pH 7.5, 150 mM KCl, 2 mM TCEP, 2 mM $MgCl_2$, 5% glycerol
- 100 mM Phenylmethylsulfonyl fluoride (PMSF) in DMSO
- Bio-Rad Protein Assay Kit

Procedure:

1. Thaw the cell pellet and resuspend in Lysis Buffer (6–8 mL per g of cell pellet) in a 500 mL beaker. Add 0.1 mL PMSF per 100 mL of cell suspension (100 μM final concentration) and one tablet of EDTA-free cOmplete™ Protease Inhibitor Cocktail to the suspension and stir at 4 °C for 30 min.
2. Disrupt cells by sonication at 30% power with cycles of 15 s on and 45 s off for a total of 8 min.
3. Transfer the cell lysate to 50 mL tubes and centrifuge at 38,500 × *g* for 20 min at 4 °C.
4. Load the supernatant onto a Ni-NTA column (2.5 × 5 cm) pre-equilibrated with Lysis Buffer. Wash column with 100 mL Lysis Buffer to remove unbound proteins. Elute the protein with a linear gradient containing 150 mL Lysis Buffer and 150 mL Elution Buffer. Collect 8–10 mL fractions.
5. Determine the protein concentration in fractions by mixing 10 μL aliquots and 990 μL Bio-Rad Protein Assay reagent in a plastic cuvette and read the absorbance at 595 nm. Pool protein containing fractions and concentrate to ~20 mL using Amicon Ultra-15 30 kDa centrifugal filters.
6. Add TEV protease (0.02 mg/mg of CblA) to the concentrated protein in a dialysis bag and dialyze overnight at 4 °C against 1 L TEV Buffer to cleave the His_6 tag.
7. Load the protein onto a second Ni-NTA column (2.5 × 5 cm) and wash with 100 mL TEV Buffer until no protein is detected by Bradford assay. Collect the flow-through containing CblA, concentrate to ~5 mL, and dialyze overnight at 4 °C against 1 L of Buffer A.
8. Load CblA onto a Superdex200 column connected to an FPLC system. Elute the column with 100% Buffer A at a flow rate of 1 ml/min. Monitor the absorbance at 280 nm and collect ~5 mL fractions.
9. Pool fractions containing CblA (Fig. 2A) and concentrate using Amicon Ultra-15 30 kDa centrifugal filters and flash freeze 100 μL aliquots in liquid N_2 and store at −80 °C until further use. Estimate the concentration of the CblA (dimer) using $\varepsilon_{280nm} = 62{,}000\,M^{-1}\,cm^{-1}$. A typical purification yields 70–80 mg of CblA from a 6 L of culture.

4.2.2 GTPase activity of human CblA with and without MCM

Materials

- 75 mM GTP
- 1 M Trichloroacetic acid (TCA)

- 100 μM CblA
- 150 μM MCM
- Buffer R: 50 mM HEPES pH 7.5, 150 mM KCl, 2 mM $MgCl_2$, 2 mM DTT and 5% glycerol
- Buffer 1: 50 mM KH_2PO_4 pH 6.0, 4 mM tetrabutyl ammonium bromide (Sigma)
- Buffer 2: 50% (*v*/v) methanol, 50%(v/v) Buffer 1. Adjust pH to 7.2 using KOH
- C-18 HPLC Column (Beckman Ultrasphere C-18, 4.6 × 25 cm)

Method: In a 1.5 mL Eppendorf tube, add 3 μL CblA, and 0 or 10 μL MCM and Buffer R to bring the final volume to 115 μL. Place the tube in a thermostat for 10 min at 30 °C. To initiate the GTP hydrolysis reaction, add 5 μL GTP and incubate at 30 °C. Remove 50 μL aliquots after 10 and 20 min and quench by immediately adding 2.5 μL TCA. Add 50 ul of Buffer 1 and incubate on ice for 10 min. Remove pellet by centrifuging at 15,870 × *g* for 10 min at 4 °C. Collect the supernatant carefully and transfer 80 μL into an autosampler vial for HPLC analysis. The K_{act} for MCM is determined in the same assay by varying the volume of MCM added to the mixture from 1 to 20 μL with the corresponding change in the volume of Buffer R to maintain a final volume of 120 μL.

Load samples onto an Ultrasphere C-18 HPLC column and elute at 1 ml/min using the following gradient: 0–4.2 min: 100% Buffer 1; 4.2–8.3 min: 0–20% Buffer 2; 8.3–16.7 min: 20–40% Buffer 2; 16.7–21.7 min: 100% Buffer 2; 21.7–28.3: isocratic 100% Buffer 2; 28.3–30 min: 100–0% Buffer 2; 30–35 min: isocratic 100% Buffer 1. Monitor eluant at 260 nm. GDP and GTP elute at 15.9 and 19.8 min, respectively under these conditions. Measure the area under the peaks and determine the concentration of GDP formed using a calibration curve generated with standards of known concentrations (20–500 μM GDP). The GTPase activity of CblA (k_{cat}) is reported in the presence or absence of MCM with typical values of 2.8 ± 0.4 and $0.06 \pm 0.02\ min^{-1}$, respectively. The concentration of MCM needed to achieve half maximal GAP activation (K_{act}) is determined by plotting the dependence of the GTPase activity (min^{-1}) on the concentration of MCM and using the Michaelis-Menten equation.

Acknowledgment

This work was supported in part by a grant from the National Institutes of Health RO1-DK045776.

References

Banerjee, R., Gherasim, C., & Padovani, D. (2009). The tinker, tailor, soldier in intracellular B_{12} trafficking. *Current Opinion in Chemical Biology, 13*, 484–491.

Banerjee, R., Gouda, H., & Pillay, S. (2021). Redox-linked coordination chemistry directs vitamin B12 trafficking. *Accounts of Chemical Research, 54*(8), 2003–2013.

Banerjee, R., & Ragsdale, S. W. (2003). The many faces of vitamin B_{12}: Catalysis by cobalamin-dependent enzymes. *Annual Review of Biochemistry, 72*, 209–247.

Campanello, G. C., et al. (2017). Switch I-dependent allosteric signaling in a G-protein chaperone-B12 enzyme complex. *The Journal of Biological Chemistry, 292*(43), 17617–17625.

Campanello, G. C., et al. (2018). Sacrificial cobalt-carbon bond homolysis in coenzyme B_{12} as a cofactor conservation strategy. *Journal of the American Chemical Society, 140*(41), 13205–13208.

Coelho, D., et al. (2008). Gene identification for the cblD defect of vitamin B_{12} metabolism. *The New England Journal of Medicine, 358*(14), 1454–1464.

Gherasim, C., Lofgren, M., & Banerjee, R. (2013). Navigating the B_{12} road: Assimilation, delivery and disorders of cobalamin. *The Journal of Biological Chemistry, 288*, 13186–13193.

Kim, J., Gherasim, C., & Banerjee, R. (2008). Decyanation of vitamin B_{12} by a trafficking chaperone. *Proceedings. National Academy of Sciences. United States of America, 105*(38), 14551–14554.

Kim, J., et al. (2009). A human vitamin B_{12} trafficking protein uses glutathione transferase activity for processing alkylcobalamins. *The Journal of Biological Chemistry, 284*(48), 33418–33424.

Koutmos, M., et al. (2011). Structural basis of multifunctionality in a vitamin B_{12}-processing enzyme. *The Journal of Biological Chemistry, 286*(34), 29780–29787.

Leal, N. A., et al. (2004). Human ATP: Cob(I)alamin adenosyltransferase and its interaction with methionine synthase reductase. *The Journal of Biological Chemistry, 279*(46), 47536–47542.

Leipe, D. D., et al. (2002). Classification and evolution of P-loop GTPases and related ATPases. *Journal of Molecular Biology, 317*(1), 41–72.

Li, Z., et al. (2014). Glutathione-dependent one-electron transfer reactions catalyzed by a B12 trafficking protein. *The Journal of Biological Chemistry, 289*(23), 16487–16497.

Li, Z., et al. (2017). Coordination chemistry controls the thiol oxidase activity of the B_{12}-trafficking protein CblC. *The Journal of Biological Chemistry, 292*(23), 9733–9744.

Li, Z., et al. (2020). An Interprotein Co-S coordination complex in the B12-trafficking pathway. *Journal of the American Chemical Society, 142*(38), 16334–16345.

Lofgren, M., Koutmos, M., & Banerjee, R. (2013). Autoinhibition and signaling by the switch II motif in the G-protein chaperone of a radical B_{12} enzyme. *The Journal of Biological Chemistry, 288*(43), 30980–30989.

Lofgren, M., et al. (2013). A switch III motif relays signaling between a B12 enzyme and its G-protein chaperone. *Nature Chemical Biology, 9*(9), 535–539.

Mascarenhas, R., et al. (2020). The human B_{12} trafficking protein CblC processes nitrocobalamin. *The Journal of Biological Chemistry, 295*(28), 9630–9640.

Olteanu, H., et al. (2004). Kinetic and thermodynamic characterization of the common polymorphic variants of human methionine synthase reductase. *Biochemistry, 43*(7), 1988–1997.

Padovani, D., & Banerjee, R. (2006). Assembly and protection of the radical enzyme, methylmalonyl-CoA mutase, by its chaperone. *Biochemistry, 45*(30), 9300–9306.

Padovani, D., & Banerjee, R. (2009a). A rotary mechanism for coenzyme B_{12} synthesis by adenosyltransferase. *Biochemistry, 48*, 5350–5357.

Padovani, D., & Banerjee, R. (2009b). A G-protein editor gates coenzyme B_{12} loading and is corrupted in methylmalonic aciduria. *Proceedings of the National Academy of Sciences of the United States of America*, *106*(51), 21567–21572.

Padovani, D., Labunska, T., & Banerjee, R. (2006). Energetics of interaction between the G-protein chaperone, MeaB and B_{12}-dependent methylmalonyl-CoA mutase. *The Journal of Biological Chemistry*, *281*, 17838–17844.

Padovani, D., et al. (2008). Adenosyltransferase tailors and delivers coenzyme B_{12}. *Nature Chemical Biology*, *4*(3), 194–196.

Ruetz, M., et al. (2019). Allosteric regulation of oligomerization by a B_{12} trafficking G-protein is corrupted in methylmalonic aciduria. *Cell Chemistry & Biology*, *26*(7), 960–969 e4.

Yamanishi, M., Labunska, T., & Banerjee, R. (2005). Mirror "base-off" conformation of coenzyme B_{12} in human adenosyltransferase and its downstream target, methylmalonyl-CoA mutase. *Journal of the American Chemical Society*, *127*, 526–527.

Zehnder, A. J. B., & Wuhrman, K. (1976). Titanium(III) citrate as a nontoxic oxidation-reduction buffering system for the culture of obligate anaerobes. *Science*, *194*, 1165–1166.

CHAPTER EIGHT

Antivitamins B_{12}: Synthesis and application as inhibitory ligand of the B_{12}-tailoring enzyme CblC

Markus Ruetz[a], Markos Koutmos[b,*], and Bernhard Kräutler[c,*]
[a]Department of Biological Chemistry, University of Michigan Medical School, Ann Arbor, MI, United States
[b]Department of Chemistry, Program in Biophysics, Program in Chemical Biology, University of Michigan, Ann Arbor, MI, United States
[c]Institute of Organic Chemistry and Center of Molecular Biosciences, University of Innsbruck, Innsbruck, Austria
*Corresponding authors: e-mail address: mkoutmos@umich.edu; bernhard.kraeutler@uibk.ac.at

Contents

Abstract

Antivitamins B_{12} are non-natural corrinoids that have been designed to counteract the metabolic effects of vitamin B_{12} and related cobalamins (Cbls) in humans and other mammals. A basic structure- and reactivity-based concept typifies antivitamins B_{12} as close structural mimics of vitamin B_{12} that are not transformed by the cellular

Methods in Enzymology, Volume 668
ISSN 0076-6879
https://doi.org/10.1016/bs.mie.2021.12.016

metabolism into organometallic B_{12}-cofactors. Antivitamins B_{12} have the correct structure for efficient take-up and transport via the natural mammalian pathway for cobalamin assimilation. Thus they can be delivered to every cell in the body, where they are proposed to target and inhibit the Cbl tailoring enzyme CblC. Antivitamins B_{12} may be specifically inert Cbls or isostructural Cbl-analogues that carry a metal centre other than a cobalt-ion. The syntheses of two antivitamins B_{12} are detailed here, as are biochemical and crystallographic studies that provide insights into the crucial binding interactions of Cbl-based antivitamins B_{12} with the human B_{12}-tailoring enzyme CblC. This key enzyme binds genuine antivitamins B_{12} as inert substrate mimics and enzyme inhibitors, effectively repressing the metabolic generation of the B_{12}-cofactors. Hence, antivitamins B_{12} induce the diagnostic symptoms of (functional) B_{12}-deficiency, as observed in healthy laboratory mice.

Abbreviations

_a_VitB$_{12}$ antivitamin B_{12}
Cbl cobalamin
CNCbl vitamin B_{12}
ITC isothermal titration calorimetry
Metbl metbalamin

1. Introduction

Cobalamins (Cbls) and the related cobamides (Cbas) are unique and very complex cobalt-containing molecules (Eschenmoser, 1988; Hodgkin et al., 1956). Such 'complete' corrinoids are biosynthesized only by some prokaryotes and archaea (Leeper, Warren, Kelly, & Lawrence, 2012), but are strikingly indispensable for a functioning metabolism of organisms in most spheres of life, including humans and other mammals (Banerjee, 1999; Green et al., 2017; Kräutler, Arigoni, & Golding, 1998). The uptake and the further metabolism of the very scarce Cbls is highly effective and strictly structure-selective in mammalian cells. In humans, the B_{12}-binding proteins intrinsic factor, transcobalamin and haptocorrin manage uptake and intercellular transport of the Cbls, in concert with corresponding membrane bound receptors (Fedosov, 2012). When Cbls are finally internalized in cells, they are 'tailored' by the enzyme CblC (also known as MMACHC) (Banerjee, Gouda, & Pillay, 2021), a'Cbl-deligase', that binds natural Cbls in their dimethylbenzimidazole de-coordinated (or 'base-off') form and that is also remarkably effective in removing the cobalt-bound beta-ligand of natural Cbls (Gherasim, Lofgren, & Banerjee, 2013). Among the known substrates of CblC are vitamin B_{12} (CNCbl), which is converted to cob(*II*)alamin (Cbl^{II}) via reductive de-coordination of the cyanide ligand (Kim, Gherasim, & Banerjee, 2008) and methylcobalamin (MeCbl), from which the methyl group is abstracted in a nucleophilic displacement reaction by protein-bound glutathione (Hannibal et al., 2009). The de-ligated enzyme product Cbl^{II} is the corrinoid substrate of other cellular enzymes that (re)generate the proper B_{12}-cofactors, MeCbl and coenzyme B_{12} (AdoCbl) (Banerjee et al., 2021). Thus, the crucial and unique tailoring capacity of CblC paves the further intracellular pathway to the two active organometallic B_{12}-cofactors. When CblC is impaired or inhibited (by mutations or by antivitamins B_{12}, respectively), the required cellular supply of MeCbl and AdoCbl is shut off and cells become B_{12}-deficient (Lerner-Ellis et al., 2006).

2. Structure-based concept of antivitamins B_{12}

Constructed, primarily, for their effect in humans and other mammals, genuine antivitamins B_{12} (*a*VitB_{12}s) are designed as close structural mimics of the natural Cbls that (i) are inert to the tailoring by CblC and (ii) do not themselves harbor the organometallic reactivity of the active B_{12}-cofactors (Kräutler, 2015). Thanks to their chemical and structural features, *a*VitB_{12}s may also interfere with the basic metabolism in B_{12}-dependent microorganisms (Guzzo et al., 2016), inviting their consideration as alternative antibiotics (Kräutler, 2020).

As close structural imitations of the natural Cbls, *a*VitB_{12}s should be taken up and transported by the mammalian B_{12}-transporters in a similar way to CNCbl itself. *A*VitB_{12}s need a design that renders them inert against the two known modes of tailoring by CblC (removal of the beta-ligand, either by one-electron reduction or by nucleophilic displacement). Such properties were proposed to be found in some non-natural organometallic aryl-Cbls, presumably because they featured very negative reduction potentials and carryed specifically substitution-inert aromatic carbon centres of their aryl-moiety (Kräutler, 2015). The aryl-Cbl 4-ethylphenyl-cobalamin (EtPhCbl) (Fig. 1) was prepared as a potential *a*VitB_{12} (Ruetz, Gherasim, et al., 2013), as further detailed below. Indeed, EtPhCbl bound well to the three human B_{12}-transporter proteins and bound to CblC in its 'base-off' form but resisted the removal of its organometallic aryl-moiety (Ruetz, Gherasim, et al., 2013). EtPhCbl was tested as a metabolically bona fide *a*VitB_{12} in studies with healthy laboratory mice. These latter experiments indicated impairment of the uptake of CNCbl by EtPhCbl, and the development within 4 weeks of strong metabolic signs of functional B_{12}-deficiency in these mice (Mutti, Ruetz, Birn, Kräutler, & Nexo, 2013).

EtPhCbl is a light sensitive, 'photo-conditional' *a*VitB_{12}, from which the active B_{12}-vitamer aquocobalamin (AquoCbl) is produced (with low

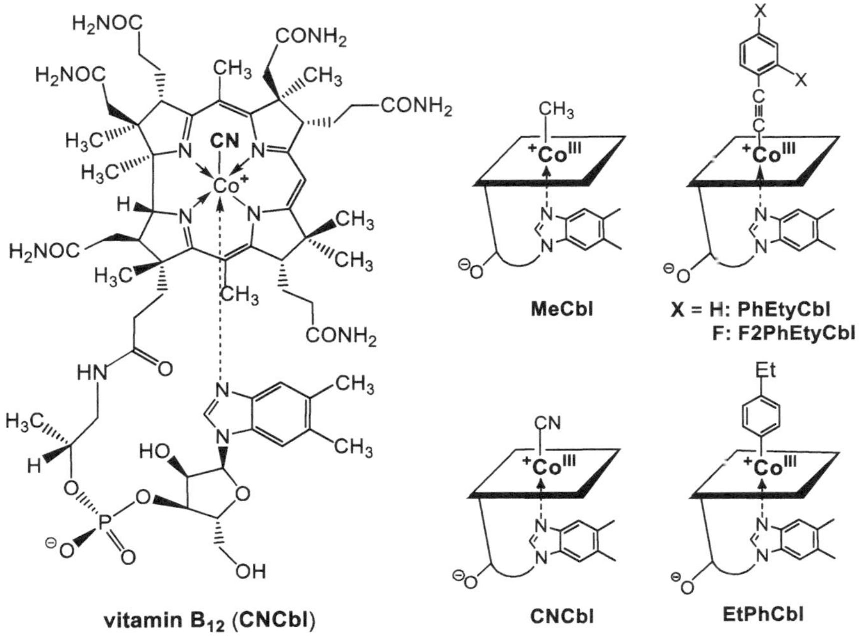

Fig. 1 Structural formula of vitamin B_{12} (CNCbl) and graphical symbols for CNCbl and methylcobalamin (MeCbl), and for the antivitamins B_{12} 4-ethylphenyl-cobalamin (EtPhCbl), phenylethynyl-cobalamin (PhEtyCbl) and 2(2,4-difluoro)phenylethynyl-cobalamin (F2PhEtyCbl).

quantum yield) in aerobic solution upon absorption of visible light at wavelengths of about 600nm or shorter (Miller et al., 2016; Ruetz, Gherasim, et al., 2013). This rather undesirable photo-reactivity in a fully functional *a*VitB_{12}, was circumvented in photo-stable alkynyl-Cbls, a 'next generation' type of *a*VitB_{12}s, such as phenylethynyl-cobalamin (PhEtyCbl) (Ruetz, Salchner, Wurst, Fedosov, & Kräutler, 2013). The alkynyl-Cbl PhEtyCbl exhibited slow acid-induced conversion into aquocobalamin (AquoCbl). This stability problem was successfully addressed by the introduction of fluorine-substituents in the 2(2,4-difluorophenyl)ethynyl-cobalamin (F2PhEtyCbl), which also served as an effective inhibitor of human CblC, useful for a first X-ray crystal structure analysis of glutathione loaded CblC (Ruetz et al., 2017).

Cbl-based *a*VitB_{12}s are, potentially, chemical sources of active B_{12}-vitamers. This latent drawback may limit their general use and broad application in studies requiring extended periods of time (Kräutler, 2020). To address this problem, transition metal analogues of the cobalt-containing

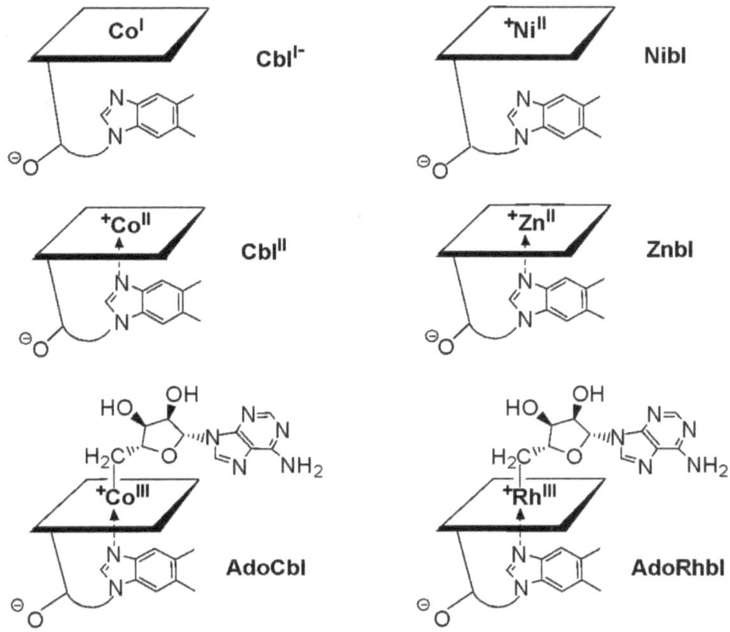

Fig. 2 Graphical symbols (*left*) for the representative cobalamins coenzyme B_{12} (adenosylcobalamin, AdoCbl), cob(*II*)alamin (Cbl^{II}) and cob(*I*)alamin (Cbl^{I-}), and (*right*) for the corresponding isostructural metbalamins (Metbls) adenosylrhodibalamin (AdoRhbl), zincobalamin (Znbl) and nibalamin (Nibl).

Cbls were considered. They provide the basis for an alternative basic design for corrinoid B_{12}-mimics, among them *a*VitB_{12}s. Metal-analogues of Cbls, or metbalamins (Metbls), would not convert into Cbls when exposed to the conditions of any biological or biomedical experiment. Therefore, suitably structured Metbls could represent potentially powerful *a*VitB_{12}s (Fig. 2). These expectations were in line with earlier qualitative studies (Koppenhagen, 1982) and were precisely supported experimentally with some well characterized rhodibalamins (Rhbls), rhodium analogues of the Cbls (Widner et al., 2016). Adenosylrhodibalamin (AdoRhbl), the nearly isostructural Rh-analogue of coenzyme B_{12} (AdoCbl) was prepared in a combined chemical and biological synthesis sequence starting with the biosynthetic metal-free AdoCbl-precursor hydrogenobyric acid a,c-diamide (Widner et al., 2016). Related strategies, starting with biosynthetic hydrogenobyric acid (Kieninger, Deery, et al., 2019), opened up a rational synthesis path to the stable and inert Ni- and Zn-analogues of the highly reactive reduced Cbls Cbl^{I-} and Cbl^{II}, respectively (Kieninger et al., 2020;

Kieninger, Baker, et al., 2019), as well as to further Rhbls (Widner et al., 2021). Hence, Metbls with 4-, 5- and 6-coordinate metal centres are now available (Kräutler, 2020), mimicking the three basic coordination forms of the corresponding isostructural Cbls (Cbl^{I-}, Cbl^{II} and Cbl (Co(*III*)Cbl)), some of them deduced to have the properties of *a*VitB_{12}s.

3. Synthesis of antivitamins B_{12} from vitamin B_{12}

The preparation of the organometallic antivitamins B_{12} (*a*VitB_{12}s) described here led to the introduction of methods differing from those in the traditional synthetic tool box (Kräutler, 2019) in order to specifically couple an unsaturated organic ligand to the Cbl-core (Ruetz, Gherasim, et al., 2013). The *a*VitB_{12} 4-ethylphenyl-cobalamin (EtPhCbl), the first aryl-Cbl made, was prepared via a proposed radical coupling of the 4-ethylphenyl radical with Cbl^{II} (see details below) (Ruetz, Gherasim, et al., 2013). From this reaction crystalline EtPhCbl was obtained as a single isomer in nearly 60% yield. The crystal structure of EtPhCbl confirmed the structural similarity of this 'base-on' Cbl with CNCbl. The related phenylcobalamin (PhCbl) was synthesized from Cbl^{II} and diphenyliodonoium chloride. This reaction produced PhCbl in over 70% yield, as well as about 20% of alpha-phenylcobalamin (αPhCbl), the 'base-off' coordination isomer of PhCbl (Brenig, Ruetz, Kieninger, Wurst, & Kräutler, 2017). For the syntheses of the novel alkynylcobalamins 2-phenylethynylcobalamin (PhEtyCbl) and 2(2,4-difluoro-phenyl)ethynyl-cobalamin (F2PhEtyCbl) presumed radical processes were likewise employed to join the organometallic and the corrin moieties (see details below) (Ruetz et al., 2017; Ruetz, Salchner, et al., 2013). However, a non-radical ('reduction-free') synthesis method for PhEtyCbl and other alkynylcobalamins using CNCbl, an alkyne and Cu^{I}-acetate as a catalyst, has also been reported (Chrominski, Lewalska, & Gryko, 2013; Chrominski, Lewalska, Karczewski, & Gryko, 2014).

4. Study of antivitamins B_{12} as inhibitors of the B_{12} tailoring enzyme CblC

4.1 Biochemical experiments with antivitamins B_{12} and CblC

In the early steps of intracellular vitamin B_{12} trafficking, all cobalamins entering the cell are processed by CblC into an intermediate that can be used by the downstream enzymes, methionine synthase and mitochondrial methylmalonyl CoA mutase. (Banerjee and Ragsdale (2003), Banerjee, Gherasim, and Padovani (2009)) CblC is able to remove the axial ligands of alkylcobalamins such as Me- or AdoCbl via a SN_2 reaction using GSH as nucleophile. (Kim, Hannibal, Gherasim, Jacobsen, & Banerjee, 2009) The resulting cob(*I*)alamin is rapidly oxidized under aerobic conditions. Cbls with inorganic ligands (e.g., H_2O, CN, NO_2) (Kim et al., 2008; Li et al., 2017; Mascarenhas, Li, Gherasim, Ruetz, & Banerjee, 2020) are processed via reductive de-ligation using GSH or flavins as electron donor. (See also chapter "The human B_{12} trafficking chaperones: CblA, ATR, CblC and CblD"). The promiscuity of CblC towards the axial ligands in Cbls allowed the development of synthetic Cbls with axial ligands unable to be removed by CblC. EtPhCbl as well as F2PhEtyCbl bind in the base-off form to CblC (Fig. 3) like other natural occurring cobalamins (Ruetz et al., 2017). However, the addition of GSH does not result in removal of the upper axial ligand even after prolonged incubation time. The resistance towards GSH-induced ligand removal allowed determination of the binding constant (K_D) in the presence of GSH by isothermal titration calorimetry (ITC). The K_D's of EtPhCbl, of F2PhEtyCbl and of MeCbl are remarkably similar. The inactivity of the ternary CblC•F2PhEtyCbl•GSH complex also enabled its crystallization and provided information on the location of the GSH co-substrate in holo-CblC.

4.2 Crystallographic studies with antivitamins B_{12} and CblC

Multiple structures of CblC from human and *C.elegans* CblC (Froese et al., 2012; Koutmos, Gherasim, Smith, & Banerjee, 2011; Li et al., 2017) have provided a structural framework to understand aspects of cobalamin recognition and processing by CblC. CblC has been crystallized as a binary complex with either MeCbl (Koutmos et al., 2011), AdoCbl (Froese et al., 2012), CNCbl and H_2OCbl (data not shown). These studies underscore CblC's promiscuity in accommodating ready to be processed cobalamins in a base-off form and with variable axial ligands. Initial unsuccessful attempts to obtain structural information on ternary cobalamin and substrate/product CblC complexes could not allow us to closely examine the differential chemical mechanisms that CblC employs. Co-crystallization or soaking of cobalamin loaded CblC with substrates like GSH had been largely unsuccessful despite multiple attempts with combinations of various CblC and Cbl forms. CblC's reactivity, even *in crystallo*, leading for example to de-alkylation of alkyl-Cbls in the presence of the co-substrate GSH, prevented capture of any ternary complexes. We therefore turned to using antivitamins B_{12} (aVitB_{12}s) such as F2PhEtyCbl, not only to structurally characterize how these aVitB_{12}s are accommodated by Cbl trafficking and processing proteins like CblC, but also to trap ternary CblC complexes with substrates like GSH bound (Figs. 4 and 5). CblC's inability to process F2PhEtyCbl provides us with a tool to stabilize abortive enzyme ternary complexes. Our chemical trap method allowed us to gain an unprecedented structure-based glimpse into a ternary CblC complex (CblC•F2PhEtyCbl/GSH) in the presence of both the antivitamin B_{12} and GSH, as well as into the mode of action of an antivitamin B_{12} inhibiting CblC (Ruetz et al., 2017).

5. Methods

5.1 Synthesis of 4-ethylphenyl-cobalamin (EtPhCbl)

5.1.1 Synthesis of 4-ethylphenyldiazonium tetrafluoroborate

Materials and equipment

- 100 mL 3-necked round bottom flask
- Dropping funnel
- Magnetic stirrer
- Buchner funnel
- 4-Ethylaniline, freshly distilled
- Tetrafluoroboric acid 50% (HBF_4)
- Sodium nitrite $NaNO_2$
- Water, distilled

Procedure

1. In a 100 mL 3-necked round bottom flask place 20 mL of distilled water and 2.0 mL of 4-ethylaniline (1.87 g, 15.8 mmol) and mix at room temperature.
2. Slowly add 8.9 mL of 50% HBF_4 (71.1 mmol) and cool the solution to −5 °C.
3. Add dropwise a saturated solution of 2.19 g (32.2 mmol) $NaNO_2$ in water within 15 min.
4. Stir the reaction for additional 15 min at −5 °C.
5. Filter off the white crystals using a Buchner funnel and wash the crystals with a small amount of ice-cold water.
6. Dry the crystals under high vacuum overnight. A typical yield is 2.14 g (9.7 mmol, 62%) 4-ethylphenyldiazonium tetrafluoroborate.

Note: Diazonium salts are best stored cold and are potentially explosive!

5.1.2 Synthesis of 4-ethylphenyl-cobalamin (Ruetz, Gherasim, et al., 2013)

Materials and equipment

- Aquocobalamin hydrochloride
- Sodium formate
- 4-Ethylphenyldiazonium tetrafluoroborate
- Acetone
- LiChroprep RP-18, particle size 25–40 μm (Merck) (15 g in a glass column with 2 cm diameter)
- Water, distilled
- Acetonitrile
- Methanol
- 50 mM potassium phosphate buffer pH 7.0
- 1 g RP-18 cartridge (Waters)
- 10 mL two necked round bottom flask
- Magnetic stirrer
- Rotary evaporator
- Argon

Procedure

1. Dissolve 50.0 mg of aquocobalamin chloride (36.2 μmol) in 2 mL of distilled water in a 10 mL two-necked round bottom flask. Purge the solution with argon for 15 min to remove oxygen.
2. Add 43.5 mg sodium formate (0.65 mmol) under argon and stir the solution for 30 min at room temperature. The color of the solution turns from red to brown.
3. Add 12.2 mg 4-ethylphenyldiazonium tetrafluoroborate (55.5 μmol) to the reaction solution under argon and stir the mixture for another hour. The flask is covered with aluminum foil and all further operations are done under protection from light.
4. Pour the reaction mixture into 15 mL of acetone to precipitate the product.
5. Remove the supernatant and dry the orange-red precipitate under high vacuum for 2 h.
6. Dissolve the raw 4-ethylphenyl-cobalamin (EtPhCbl) in 1 mL of water and apply to a RP-18 column (15 g RP-18 in a 2 cm diameter glass column) for purification by chromatography using a stepwise 50 mM

potassium phosphate buffer pH 7.0/acetonitrile gradient from 5 to 30% acetonitrile (v/v) in 2% steps. The product elutes at 16% acetonitrile.

7. Remove the solvents using a rotary evaporator at room temperature.
8. Dissolve the red residue in 1 mL water.
9. Load the solution onto a 1 g RP-18 cartridge. Before loading, wash the cartridge with 10 mL methanol followed by 10 mL water. After loading of the sample wash the column with 20 mL water.
10. Elute the red product with 5–7 mL methanol.
11. Remove solvent on a rotary evaporator.
12. Dissolve the residue in 0.3 mL water and precipitate in 3 mL acetone. Remove the supernatant.
13. Dissolve the precipitate in 0.4 mL of water and crystallize the product by addition of 5 mL acetone.
14. Wash the crystals first with 0.5 mL (1:9) mixture of water: acetone (v/v), then with 0.5 acetone. Dry the red crystals under high vacuum overnight. A typical yield is 29.1 mg (20.3 μmol) 4-ethylphenyl-cobalamin (EtPhCbl) (56%).

    Note: For safe handling and storage, the *a*VitB_{12} EtPhCbl must be protected from light (photodecomposition of EtPhCbl generates B_{12} vitamin forms)!

5.2 Synthesis of 2(2,4-difluorophenyl)ethynyl-cobalamin (F2PhEtyCbl)

5.2.1 Synthesis of 1-iodo-2-(2,4-difluorophenyl)ethyne

Materials and equipment

- 5 mL round bottom flask
- Dropping funnel
- Magnetic stirrer
- Buchner funnel
- Rotary evaporator
- 4-iodomorpholin-4-ium iodide (Sigma-Aldrich)
- 2,4-difluorophenylethyne (Sigma-Aldrich)
- Tetrahydrofuran (THF) distilled
- Methylene chloride (CH_2Cl_2)
- Neutral aluminum oxide (activated)
- *n*-Hexane

*Note:*4-iodomorpholin-4-ium iodide can be prepared by the method described by Hein, Tripp, Krasnova, Sharpless, and Fokin (2009).

Procedure

1. In a 5 mL round bottom flask mix 2 mL of distilled THF, 301.3 mg 4-iodomorpholin-4-ium iodide (0.88 mmol), 110.5 mg 2,4-difluorophenylethyne (0.80 mmol) and 7.6 mg 2,4-difluorophenylethyne (0.040 mmol) and stir at room temperature for 1 h. A white precipitate will form during the reaction.
2. Pour the reaction mixture on a pad of activated neutral aluminum oxide (3.6 g) inside a Buchner funnel and filter off the solution. Wash four times with 2 mL CH_2Cl_2 and combine filtrates.
3. Evaporate the solvents using a rotary evaporator until a yellow oil remains.
4. Drop the yellow oil into 10 mL of n-hexane to precipitate residual morpholine.
5. Filter off the brown precipitate and dry the filtrate on a rotary evaporator. A typical yield is 177.9 mg 1-iodo-2-(2,4-difluorophenyl) ethyne (0.674 mmol, 84.3% yield) as yellowish crystals.

5.2.2 Synthesis of 2(2,4-diflourphenyl)ethynyl-cobalamin (Ruetz et al., 2017)

Materials and equipment

- Aquocobalamin hydrochloride (AquoCbl)
- Formic acid
- Triethylamine
- 1-iodo-2-(2,4-difluorophenyl)ethyne
- Acetone
- RP-18 column chromatography: Merck LiChroprep RP-18, particle size 25–40 μm, (15 g in a glass column with 2 cm diameter)
- Water, distilled
- Methanol
- Methylene chloride (CH_2Cl_2)
- Ethyl acetate
- 10 mL two necked round bottom flask
- Rotary evaporator
- Magnetic stirrer
- Argon

Procedure

1. Place 100.0 mg of AquoCbl (72.3 μmol) in a 10 mL two necked round bottom flask and dissolve in 2.5 mL of methanol. Purge the solution with argon for 15 min to remove oxygen.

2. Add 50 μL formic acid (1.3 mmol) and 187 μL triethylamine (1.3 mmol) under argon and stir the solution for 30 min at room temperature. The color of the solution turns from red to brown.
3. Add 158.4 mg 1-iodo-2-(2,4-difluorophenyl)ethyne (222.0 μmol), dissolved in 25 μL CH_2Cl_2, to the solution under argon and stir the mixture for another hour. The flask is covered with aluminum foil and all further operations are done under protection from light.
4. Pour the reaction mixture into 15 mL of ethyl acetate to precipitate the product.
5. Remove the supernatant and dry the orange-red precipitate under high vacuum for 2 h.
6. Dissolve the raw product in 2 mL of water and purify it by RP-18 column chromatography using a stepwise water/methanol gradient from 10% to 70% methanol (v/v) in 10% steps. AquoCbl elutes first at 30% methanol, the product elutes at 50% methanol.
7. Remove the solvents using a rotary evaporator at room temperature.
8. Dissolve the red residue in 0.5 mL water and crystallize by addition of 5 mL acetone. Crystals are formed by storage (light protected) at room temperature overnight.
9. Wash the crystallized F2PhEtyCbl first with 0.5 mL (1:9) mixture of water: acetone (v/v), then with 0.5 mL acetone. 10. Dry the red crystals under high vacuum overnight. A typical yield of F2PhEtyCbl is 61.7 mg (42.1 μmol, 58.2% yield) of red crystalline solid.

5.3 Spectroscopic characterization of antivitamin B_{12}

EtPhCbl exhibits a UV/visible spectrum that is similar to organometallic cobalamin derivatives, such as methyl- or adenosyl-cobalamin, and displays absorption maxima at 518, 373 and 343 nm (see Fig. 3). This novel Cbl has been characterized by other spectroscopic methods (Ruetz, Gherasim, et al., 2013), in particular by 2-dimensional homo- and hetero-nuclear nuclear magnetic resonance spectroscopy (2D-NMR) (Konrat, Tollinger, & Kräutler, 1998; Summers, Marzilli, & Bax, 1986) and by electrospray ionization mass spectrometry (ESI-MS) (Fenn, Mann, Meng, Wong, & Whitehouse, 1989).

The UV/visible spectrum of F2PhEtyCbl (see Fig. 3) displays absorption maxima at 550, 527 and 367 nm and resembles the spectrum of vitamin B_{12} (CNCbl). This new Cbl has been characterized further (Ruetz et al. 2017) by 2-dimensional homo- and hetero-nuclear nuclear magnetic resonance spectroscopy (2D-NMR) and by electrospray ionization mass spectrometry (ESI-MS), in particular.

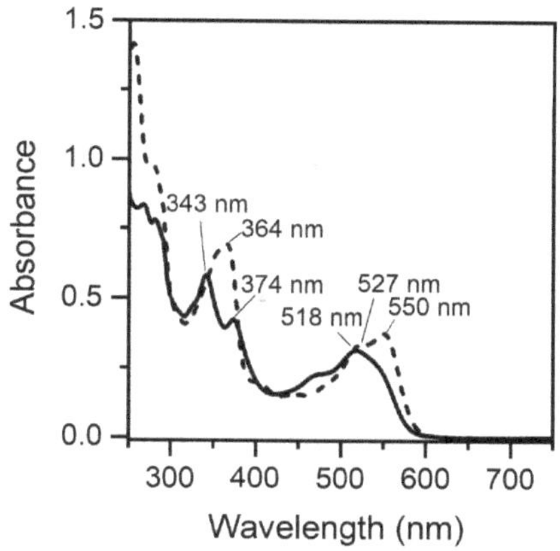

Fig. 3 UV/visible spectra of 50 μM EtPhCbl (solid line) and 50 μM F2PhEtyCbl (dashed line) in H_2O.

5.4 Preparation of CblC•F2PhEtyCbl/GSH ternary complex sample

Crystals of the CblC•F2PhEtyCbl/GSH ternary complex were obtained through co-crystallization of a hCblC protein sample loaded with both the F2PhEtyCbl and GSH ligands. The hCblC protein samples used for crystallization were based on a C-terminal truncation lacking the last 44 residues that were predicted to be disordered. The expression and purification of hCblC(1–238)ΔC44 has been previously described (Koutmos et al., 2011). This truncated construct increased protein stability, crystal formation frequency and crystallization reproducibility.

Materials and equipment

- Size exclusion buffer (SEC buffer): 50 mM Tris pH 7.8, 0.5 mM TCEP
- Protein sample: 12 mg/mL human CblC(1–238)ΔC44 in the presence of 0.25 mM of F2PhEtyCbl and 1 mM of GSH in SEC buffer
- Crystallization solution: 1.26 M NH_4SO_4, 200 mM NaCl, 100 mM CHES:NaOH pH 9.5
- Cryoprotective solution: 1 M NH_4SO_4, 15% glycerol, 150 mM NaCl and 75 mM CHES:NaOH at pH 9.5
- AKTA Purifier (GE Healthcare)
- Phoenix crystallization robot (Art Robbins)
- Stereomicroscope (Olympus)

Procedure

1. Filter all solutions through a 0.2 μm filter.
2. Run protein sample for crystallization through a size exclusion column (S200; GE Healthcare). Note that TCEP is required since it prevents CblC oxidation and aggregation over the crystallization timeframe.

3. Reconstitute CblC with 0.25 mM of F2PhEtyCbl and 1 mM of GSH.
4. Concentrate reconstituted sample to 12 mg/mL in SEC buffer.
5. Pass concentrated sample through a Spin-X centrifuge filter (Corning Costar) with a 0.22 μm cutoff to remove any insoluble aggregates formed during sample concentration.
6. Dispense protein and commercially available (Hampton Research, Molecular Dimensions) precipitation cocktails into 96 well sitting drop crystallization plates (Art Robbins, INTELLI-PLATE).
7. Store duplicate plates in both 4 °C and 20 °C.
8. Identify conditions affording best diffracting crystals. Crystals of CblC•F2PhEtyCbl/GSH were obtained at 4 °C by mixing a 2 μL of CblC•F2PhEtyCbl/GSH solution with a 1 μL of crystallization solution.
9. Harvest crystals in the dark using a red filter to cover the microscope light needed to illuminate the crystal trays and visualize crystals.
10. Cryoprotect, if necessary, crystals. In this case cryoprotection was accomplished by transferring CblC•F2PhEtyCbl/GSH crystals to the cryoprotective solution. Don't leave crystals in cryoprotective solution for a prolonged time.
11. Plunge freeze crystals in liquid N_2.

6. Biochemical experiments with antivitamins B_{12} and human CblC

6.1 De-ligation assay with antivitamins B_{12} and CblC

The assay is identical to the de-alkylation assay of CblC described in the chapter in this volume "The human B_{12} trafficking chaperones: CblA, ATR, CblC and CblD" (see also (Kim et al., 2009)). Unlike MeCbl, which is rapidly de-alkylated by GSH, antivitamins B_{12} are resistant toward the removal of the upper axial ligand. However, GSH enhances the affinity of CblC towards Cbls as can be observed by spectroscopic changes following addition of GSH to antivitamin B_{12} bound CblC that indicate a complete switch into the base-off form (Fig. 4).

Procedure

Substitute MeCbl with 4-EtPhCbl or with F2PhEtyCbl and perform the assay as described.

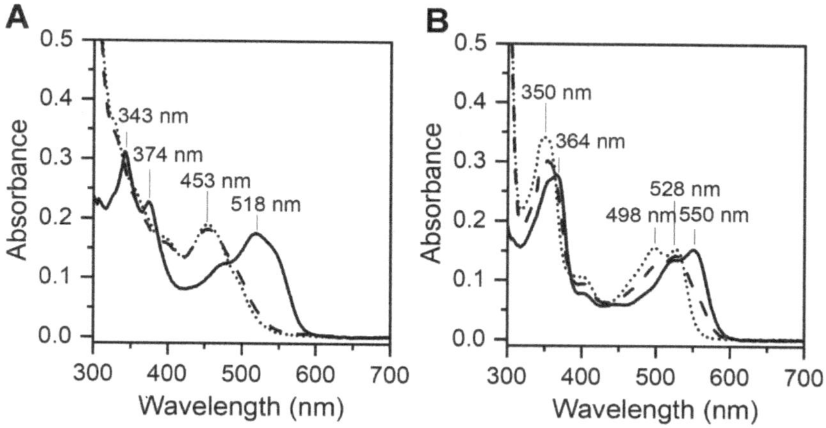

Fig. 4 Binding of antivitamins B_{12} to human CblC. (A) UV/visible spectrum of 20 μM EtPhCbl in 100 mM HEPES, 150 mM KCl, 10% glycerol pH 7.4 at 20 °C before (*solid line*) and 60 min after addition of 50 μM human CblC (*dashed line*). CblC induced a blue shift of the UV/visible spectrum indicating binding of EtPhCbl in the base-off form on CblC. The addition of 5 mM GSH caused only minor spectroscopic changes (*dotted line*) after 60 min incubation. (B) Same as in panel (A), but with 20 μM F2PhEtyCbl (*solid line*). The addition of CblC resulted in a blue shift of the UV/visible spectrum (dashed line). A further hypsochromic shift of the absorbance maxima (*dotted line*) is observed following incubation with 5 mM GSH for 60 min.

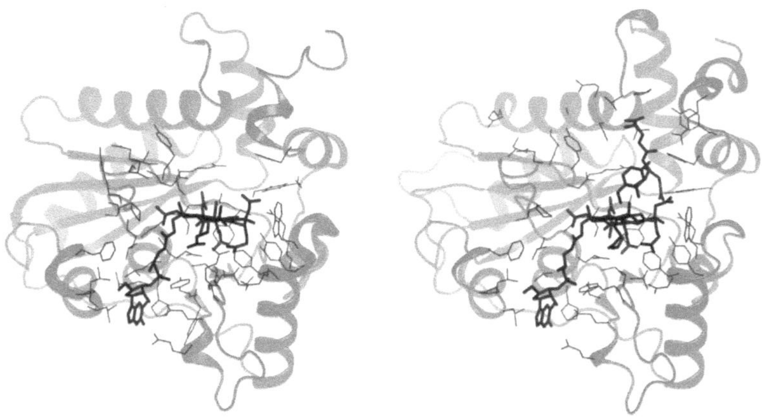

Fig. 5 Structures of CblC•MeCbl (*left*, see Koutmos et al., 2011) and of the ternary complex CblC•F2PhEtyCbl/GSH (right, see Ruetz et al., 2017) are shown as cartoon models with protein parts depicted in *gray*. Protein-bound ligands (Cbls, GSH, 2(2,4-difluorophenyl)ethynyl-cobalamin) are displayed as dark stick models, and selected amino acid side chains of the protein as lines. Graphics were generated with Pymol.

6.2 Binding of antivitamins B_{12} to CblC

The affinity of antivitamins B_{12} (*a*VitB_{12}s) to CblC was determined by isothermal titration calorimetry (ITC). The resistance of *a*VitB_{12}s toward axial ligand displacement by CblC allows the determination of the dissociation constant in the presence of GSH. In using ITC, high affinity binding of EtPhCbl ($K_D = 29.7 \pm 7.8$ nM) and F2PhEtCbl ($K_D = 130 \pm 13.0$ nM) to human CblC was revealed. The structurally related PhEtyCbl also displays high affinity binding to CblC with a $K_D = 30.0 \pm 5.6$ nM (Ruetz et al., 2017; Ruetz, Gherasim, et al., 2013).

Materials and equipment

- Buffer A: 100 mM HEPES pH 7.4, 150 mM KCl, 10% glycerol
- 10 μM human CblC in the presence of 1 mM GSH in Buffer A
- 200 μM antivitamin B_{12} in the presence of 1 mM GSH in Buffer A
- Microcal VP-ITC (GE Healthcare)
- ThermoVac sample degasser

Procedure

1. Filter all solutions through a 0.2 μm filter and degas under vacuum at 15 °C for 5 min.
2. Place CblC solution in the sample cell.
3. Fill reference cell with Buffer A.
4. Fill syringe with antivitamin B_{12} solution.
5. Perform the ITC experiment at 20 °C with 37 aliquots of 8 μL antivitamin B_{12} solution.
6. Analyze with Microcal ORIGIN software using a single-site-binding model to determine the equilibrium association constant, $K_A = (1/K_D)$ and the binding enthalpy ΔH°. Representative ITC titration curves and the corresponding binding isotherm of F2PhEtyCbl and PhEtyCbl binding to CblC are presented in graphical from in Ruetz et al. (2017).

7. Crystallographic structure determination of a CblC•F2PhEtyCbl/GSH ternary complex

Crystallographic data were collected at GM/CA-CAT 23-IDB (Advanced Photon Source, Argonne National Laboratory) using synchrotron derived X-rays. Data to 2.5 Å resolution for CblC•F2PhEtyCbl/GSH crystals were collected at 100 K using a Mar 300 detector. Data are typically processed with either HKL2000 or xds. Crystals were determined to belong to $P6_122$ space group ($a = b = 114.5$ Å, $c = 151.0$ Å, $\alpha = \beta = 90°$, $\gamma = 120°$) with one monomer in the asymmetric unit. The space group and unit cell differed from previously determined hCblC structures therefore molecular replacement was used to obtain an initial model. The initial molecular replacement model was subsequently refined. Detailed data processing and refinement metrics are provided in (Ruetz et al., 2017).

Procedure

1. Collect crystallographic data. Note that when intense synchrotron radiation is used for data collection metalloenzymes can be susceptible to photo reduction. Moreover, care needs to be taken to prevent excessive radiation damage that would not allow the collection of complete data sets. If needed, multiple data sets should be collected either on multiple and well separated spots in the same big crystal, and/or using multiple crystals.
2. Index data to determine space group (HKL2000 (Otwinowski & Minor, 1997) or xds (Kabsch, 2010)).
3. Integrate (HKL2000 or xds) and use a data reduction program (such as Aimless (Evans & Murshudov, 2013)) to scale and assess the collected crystallographic data.
4. Use a closely related search model (such as model with PDB accession number 3SC0 in this case) for obtaining initial phases through Molecular Replacement (Phaser (McCoy et al., 2007)). Remove all ligands and most loops (especially with high B-factors) from the search model for faster results and for removing model bias.
5. Refine resulting molecular replacement model with either Refmac5 (Murshudov et al., 2011) in the CCP4i suite (Potterton, Briggs, Turkenburg, & Dodson, 2003) or PHENIX (Adams et al., 2002). During first refinement, simulated annealing is recommended to remove model bias. Refinement strategies including coordination minimization, simulated annealing in torsional and Cartesian space, and restrained individual B-factor adjustment with maximum-likelihood targets were used for the CblC•F2PhEtyCbl/GSH data.
6. Load the resulting electron density and refined model in a program such as Coot. Use the electron density (both Fo-Fc and 2Fo-Fc type of maps) as guide to correct, modify and update the existing model. Assign electron density corresponding to protein residues as needed.
7. Perform iterative cycles of steps 5–6 till all, if possible, electron density corresponding to protein residues is assigned. For the CblC•F2PhEtyCbl/GSH structure Refmac5 was used during step 7 and restrained refinement was performed using isotropic individual

B-factors with maximum-likelihood targets, while the babinet model was used for bulk solvent scaling.

8. Assign electron density for water molecules and proceed with refinement as in step 7.

9. Assign electron density corresponding to ligands such as F2PhEtyCbl and GSH. Ligand restraints and parameters need to be prepared for any new ligand such as F2PhEtyCbl before proceeding with refinement. Use a program like PRODRG server (Schüttelkopf & van Aalten, 2004) or eLBOW (Moriarty, Grosse-Kunstleve, & Adams, 2009) in the PHENIX program suite to generate ligand restraint libraries necessary for refinement. Note that due to occasional photo-reduction in the X-Ray beam of Cbl ligands from Co(III) to Co(II), the axial ligand above the Co atom may no longer be covalently bound or the Co—C bond may increase. This was the case with CblC•F2PhEtyCbl. As a result, cobalamin ligands should be refined both as a single ligand with the Co—C bond intact but also as two separate fragments generated from the dissociation of the axial ligand from Co, such as Co(*II*)Cbl and 2,4-difluorophenyl-ethynyl (F2PhEty). No or weak restraints between the Co atom and the C atom of the axial ligand should be used during refinement. For the CblC•F2PhEtyCbl structure the electron density supported the assignment of the CblC•F2PhEtyCbl ligand as two separate fragments. The ligand was photo-reduced and was therefore modeled separately as B_{12} and F2PhEty ('8FS').

10. Check and assess the geometric quality of the final model and its agreement with the experimental structure factors using MolProbity (Davis et al., 2007). Correct any issues (e.g., fix rotamers and Ramachandran outliers, address major clashes) to improve as much as possible the clash score and a molprobity score.

11. Create unbiased maps such as composite omit maps (in PHENIX or in CCP4i using FFT) or polder maps (PHENIX) to check the quality of elector density assigned to ligands.

12. Visualize final structure and create figures using PyMOL (Schrödinger., 2015) or Chimera (Pettersen et al., 2004).

8. Summary

In this chapter the syntheses of two Cbl-based antivitamins B_{12} (*a*VitB_{12}s) are detailed, as are basic and essential biochemical experiments for testing *a*VitB_{12}s as inhibitors of the human B_{12}-tailoring enzyme CblC (hCblC). The resistance of *a*VitB_{12}s against the degradation by hCblC has allowed the crystallization of a ternary complex (hCblC• *a*VitB_{12}/GSH), fostering understanding of the role of protein residues critical for the physiological activity of this enzyme. Clearly, these first detailed insights into the interaction of *a*VitB_{12}s with hCblC also invite further biochemical, structural biological and biomedical experiments with *a*VitB_{12}s, in order to gain deeper knowledge of the intricate structure-based mechanisms of B_{12}-dependent processes in animals. *A*VitB_{12}s may open-up further precise structure-based insights into the roles of cobamides in fundamental life processes in B_{12}-dependent organisms and, as a consequence, help find alternative ways to control and to exploit their metabolism.

Acknowledgments

Financial support by the National Science Foundation (CAREER Award 1945174 to MK) and by the Austrian Science Foundation (FWF, project P-33059 to BK) is gratefully acknowledged.

References

Adams, P. D., Grosse-Kunstleve, R. W., Hung, L.-W., Ioerger, T. R., McCoy, A. J., Moriarty, N. W., et al. (2002). PHENIX: Building new software for automated crystallographic structure determination. *Acta Crystallographica, Section D: Biological Crystallography*, *58*(11), 1948–1954.

Banerjee, R. (Ed.). (1999). *Chemistry and biochemistry of B_{12}*. New York, Chichester: John Wiley & Sons.

Banerjee, R., Gherasim, C., & Padovani, D. (2009). The tinker, tailor, soldier in intracellular B_{12} trafficking. *Current Opinion in Chemical Biology*, *13*, 484–491.

Banerjee, R., Gouda, H., & Pillay, S. (2021). Redox-linked coordination chemistry directs vitamin B-12 trafficking. *Accounts of Chemical Research*, *54*(8), 2003–2013.

Banerjee, R., & Ragsdale, S. W. (2003). The many faces of vitamin B_{12}: Catalysis by cobalamin-dependent enzymes. *Annual Review of Biochemistry*, *72*, 209–247.

Brenig, C., Ruetz, M., Kieninger, C., Wurst, K., & Kräutler, B. (2017). Alpha- and Beta-Diastereoisomers of Phenylcobalamin from cobalt-Arylation with Diphenyliodonium chloride. *Chemistry--A European Journal*, *23*(41), 9726–9731.

Chrominski, M., Lewalska, A., & Gryko, D. (2013). Reduction-free synthesis of stable acetylide cobalamins. *Chemical Communications*, *49*, 11406–11408.

Chrominski, M., Lewalska, A., Karczewski, M., & Gryko, D. (2014). Vitamin B_{12} derivatives for orthogonal functionalization. *The Journal of Organic Chemistry*, *79*(16), 7532–7542.

Davis, I. W., Leaver-Fay, A., Chen, V. B., Block, J. N., Kapral, G. J., Wang, X., et al. (2007). MolProbity: All-atom contacts and structure validation for proteins and nucleic acids. *Nucleic Acids Research*, *35(Web Server)*. W375–W383.

Eschenmoser, A. (1988). Vitamin-B_{12} - experiments concerning the origin of its molecular-structure. *Angewandte Chemie, International Edition*, *27*, 5–39.

Evans, P. R., & Murshudov, G. N. (2013). How good are my data and what is the resolution? *Acta Cryst*, *D69*, 1204–1214.

Fedosov, S. N. (2012) Physiological and molecular aspects of cobalamin transport. In J. R. Harris (Series Ed.), *Subcellular biochemistry*: Vol. 56. *Water soluble vitamins* (pp. 347-368): Springer.

Fenn, J., Mann, M., Meng, C., Wong, S., & Whitehouse, C. (1989). Electrospray ionization for mass spectrometry of large biomolecules. *Science*, *246*(4926), 64–71.

Froese, D. S., Krojer, T., Wu, X., Shrestha, R., Kiyani, W., von Delft, F., et al. (2012). Structure of MMACHC reveals an arginine-rich pocket and a domain-swapped dimer for its B12 processing function. *Biochemistry*, *51*(25), 5083–5090.

Gherasim, C., Lofgren, M., & Banerjee, R. (2013). Navigating the B_{12} road: Assimilation, delivery, and disorders of cobalamin. *The Journal of Biological Chemistry*, *288*(19), 13186–13193.

Green, R., Allen, L. H., Bjorke-Monsen, A. L., Brito, A., Gueant, J. L., Miller, J. W., et al. (2017). Vitamin B_{12} deficiency. *Nature Reviews. Disease Primers*, *3*, 17040.

Guzzo, M. B., Nguyen, H. T., Pham, T. H., Wyszczelska-Rokiel, M., Jakubowski, H., Wolff, K. A., et al. (2016). Methylfolate trap promotes bacterial Thymineless death by sulfa drugs. *PLoS Pathogens*, *12*, e1005949.

Hannibal, L., Kim, J., Brasch, N. E., Wang, S. H., Rosenblatt, D. S., Banerjee, R., et al. (2009). Processing of alkylcobalamins in mammalian cells: A role for the MMACHC (cblC) gene product. *Molecular Genetics and Metabolism*, *97*(4), 260–266.

Hein, J. E., Tripp, J. C., Krasnova, L. B., Sharpless, K. B., & Fokin, V. V. (2009). Copper(I)-catalyzed cycloaddition of organic azides and 1-iodoalkynes. *Angewandte Chemie, International Edition*, *48*, 8018–8021.

Hodgkin, D. C., Kamper, J., Mackay, M., Pickworth, J., Trueblood, K. N., & White, J. G. (1956). Structure of vitamin-B_{12}. *Nature*, *178*(4524), 64–66.

Kabsch, W. (2010). XDS. *Acta Cryst*, *D66*, 125–132.

Kieninger, C., Baker, J. A., Podewitz, M., Wurst, K., Jockusch, S., Lawrence A. D., et al. (2019). Zinc Substitution of Cobalt in Vitamin B_{12} : Zincobyric acid and Zincobalamin as Luminescent Structural B_{12}-Mimics. *Angewandte Chemie, International Edition*, *58*(41), 14568–14572.

Kieninger, C., Deery, E., Lawrence, A. D., Podewitz, M., Wurst, K., Nemoto-Smith, E., et al. (2019). The Hydrogenobyric acid structure reveals the Corrin ligand as an Entatic state module empowering B_{12}-cofactors for catalysis. *Angewandte Chemie, International Edition*, *58*, 10756–10760.

Kieninger, C., Wurst, K., Podewitz, M., Stanley, M., Deery, E., Lawrence, A. D., et al. (2020). Replacement of the cobalt-center of vitamin B_{12} by nickel–Nibalamin and Nibyric acid prepared from metal-free B_{12}-ligands hydrogenobalamin and hydrogenobyric acid. *Angewandte Chemie, International Edition*, *59*, 20129–20136.

Kim, J., Gherasim, C., & Banerjee, R. (2008). Decyanation of vitamin B_{12} by a trafficking chaperone. *Proceedings of the National Academy of Sciences of the United States of America*, *105*(38), 14551–14554.

Kim, J., Hannibal, L., Gherasim, C., Jacobsen, D. W., & Banerjee, R. (2009). A human vitamin B12 trafficking protein uses glutathione transferase activity for processing Alkylcobalamins. *The Journal of Biological Chemistry*, *284*(48), 33418–33424.

Konrat, R., Tollinger, M., & Kräutler, B. (1998). New NMR structural and dynamical probes of organometallic B_{12} derivatives. In B. Kräutler, D. Arigoni, & B. T. Golding (Eds.), *Vitamin B_{12} and B_{12}-Proteins* (pp. 349–368). Weinheim: Wiley-VCH.

Koppenhagen, V. B. (1982). Metal-Free Corrinoids and Metal Insertion. In D Dolphin (Ed.), *2. B12* (pp. 105–150). John Wiley & Sons.

Koutmos, M., Gherasim, C., Smith, J. L., & Banerjee, R. (2011). Structural basis of multifunctionality in a vitamin B_{12}-processing enzyme. *The Journal of Biological Chemistry*, *286*, 9780–29787.

Kräutler, B. (2015). Antivitamins B_{12} - a structure- and reactivity-based concept. *Chemistry - A European Journal*, *21*, 11280–11287.

Kräutler, B. (2019). Biological organometallic chemistry of vitamin B_{12}-derivatives. In T. Hirao, & T. Moriuchi (Eds.), *Adv. Bioorganomet. Chem* (pp. 399–429). Cambridge, USA: Elsevier.

Kräutler, B. (2020). Antivitamins B_{12} - Some Inaugural Milestones. *Chemistry - A European Journal, 26*, 15438–15445.
Kräutler, B., Arigoni, D., & Golding, B. T. (Eds.). (1998). *Vitamin B_{12} and B_{12}-Proteins*. Weinheim: John Wiley VCH.
Leeper, F. J., Warren, M. J., Kelly, J. M., & Lawrence, A. D. (2012). Biosynthesis of vitamin B_{12}. In K. M. Kadish, K. M. Smith, & R. Guilard (Eds.), *Vol. 25. Handbook of porphyrin science* (pp. 2–83). World Scientific.
Lerner-Ellis, J. P., Tirone, J. C., Pawelek, P. D., Dore, C., Atkinson, J. L., Watkins, D., et al. (2006). Identification of the gene responsible for methylmalonic aciduria and homocystinuria, cblC type. *Nature Genetics, 38*(1), 93–100.
Li, Z., Shanmuganathan, A., Ruetz, M., Yamada, K., Lesniak, N. A., Kräutler, B., et al. (2017). Coordination chemistry controls the thiol oxidase activity of the B12-trafficking protein CblC. *The Journal of Biological Chemistry, 292*(23), 9733–9744.
Mascarenhas, R., Li, Z., Gherasim, C., Ruetz, M., & Banerjee, R. (2020). The human B12 trafficking protein CblC processes nitrocobalamin. *The Journal of Biological Chemistry, 295*(28), 9630–9640.
McCoy, A. J., Grosse-Kunstleve, R. W., Adams, P. D., Winn, M. D., Storoni, L. C., & Read, R. J. (2007). Phaser crystallographic software. *Journal of Applied Crystallography, 40*(4), 658–674.
Miller, N. A., Wiley, T. E., Spears, K. G., Ruetz, M., Kieninger, C., Kräutler, B., et al. (2016). Toward the design of Photoresponsive conditional antivitamins B_{12}: A transient absorption study of an Arylcobalamin and an Alkynylcobalamin. *Journal of the American Chemical Society, 138*(43), 14250–14256.
Moriarty, N. W., Grosse-Kunstleve, R. W., & Adams, P. D. (2009). Electronic ligand builder and optimization workbench (eLBOW): A tool for ligand coordinate and restraint generation. *Acta Crystallographica, Section D: Biological Crystallography, 65*(10), 1074–1080.
Murshudov, G. N., Skubák, P., Lebedev, A. A., Pannu, N. S., Steiner, R. A., Nicholls, R. A., et al. (2011). REFMAC5 for the refinement of macromolecular crystal structures. *Acta Crystallographica, Section D: Biological Crystallography, 67*(4), 355–367.
Mutti, E., Ruetz, M., Birn, H., Kräutler, B., & Nexo, E. (2013). 4-Ethylphenyl-cobalamin impairs tissue uptake of vitamin B_{12} and causes vitamin B_{12} deficiency in mice. *PLoS One, 8*, e75312.
Otwinowski, Z., & Minor, W. (1997). Processing of X-ray diffraction data collected in oscillation mode. *Methods in Enzymology, 276*, 307–326.
Pettersen, E. F., Goddard, T. D., Huang, C. C., Couch, G. S., Greenblatt, D. M., Meng, E. C., et al. (2004). UCSF chimera: A visualization system for exploratory research and analysis. *Journal of Computational Chemistry, 25*(13), 1605–1612.
Potterton, E., Briggs, P., Turkenburg, M., & Dodson, E. (2003). A graphical user interface to theCCP4 program suite. *Acta Crystallographica, Section D: Biological Crystallography, 59*(7), 1131–1137.
Ruetz, M., Gherasim, C., Fedosov, S. N., Gruber, K., Banerjee, R., & Kräutler, B. (2013). Radical synthesis opens access to organometallic aryl-Cobaltcorrins - 4-Ethylphenyl-cobalamin, a potential "antivitamin B_{12}". *Angewandte Chemie, International Edition, 52*, 2606–2610.
Ruetz, M., Salchner, R., Wurst, K., Fedosov, S., & Kräutler, B. (2013). Phenylethynylcobalamin: A light-stable and Thermolysis-resistant organometallic vitamin B_{12} derivative prepared by radical synthesis. *Angewandte Chemie, International Edition, 52*, 11406–11409.
Ruetz, M., Shanmuganathan, A., Gherasim, C., Karasik, A., Salchner, R., Kieninger, C., et al. (2017). Antivitamin B_{12} inhibition of the human B_{12}-processing enzyme

CblC: Crystal structure of an inactive ternary complex with glutathione as the Cosubstrate. *Angewandte Chemie, International Edition, 56*(26), 7387–7392.
Schrödinger. (2015). *The PyMOL molecular graphics system*. Version 1.8.
Schüttelkopf, A. W., & van Aalten, D. M. F. (2004). PRODRG: A tool for high-throughput crystallography of protein–ligand complexes. *Acta Crystallographica, Section D: Biological Crystallography, 60*(8), 1355–1363.
Summers, M. F., Marzilli, L. G., & Bax, A. (1986). Complete proton and carbon-13 assignments of coenzyme B_{12} through the use of new two-dimensional NMR experiments. *Journal of the American Chemical Society, 108*, 4285.
Widner, F. J., Kieninger, C., Wurst, K., Deery, E., Warren, M. J., & Kräutler, B. (2021). Synthesis, spectral characterization and crystal structure of Chloro-Rhodibalamin – A synthesis platform for rhodium analogues of vitamin B_{12} and for Rh-based antivitamins B_{12}. *Synthesis, 53*, 332–337.
Widner, F. J., Lawrence, A. D., Deery, E., Heldt, D., Frank, S., Gruber, K., et al. (2016). Total synthesis, structure, and biological activity of Adenosylrhodibalamin, the non-natural rhodium homologue of coenzyme B_{12}. *Angewandte Chemie, International Edition, 55*(37), 11281–11286.

SECTION II

B_{12}-dependent enzyme reactions

CHAPTER NINE

Coenzyme B_{12}-dependent eliminases: Diol and glycerol dehydratases and ethanolamine ammonia-lyase

Tetsuo Toraya[a,*], Takamasa Tobimatsu[a], Koichi Mori[a], Mamoru Yamanishi[a], and Naoki Shibata[b]

[a]Graduate School of Natural Science and Technology, Okayama University, Tsushima-naka, Kita-ku, Okayama, Japan

[b]Graduate School of Life Science, University of Hyogo, 3-2-1 Koto, Kamigori-cho, Ako-gun, Hyogo, Japan

*Corresponding author: e-mail address: toraya@cc.okayama-u.ac.jp

Contents

Methods in Enzymology, Volume 668
ISSN 0076-6879
https://doi.org/10.1016/bs.mie.2021.11.027

Abstract

Adenosylcobalamin (AdoCbl) or coenzyme B_{12}-dependent enzymes catalyze intramolecular group-transfer reactions and ribonucleotide reduction in a wide variety of organisms from bacteria to animals. They use a super-reactive primary-carbon radical formed by the homolysis of the coenzyme's Co-C bond for catalysis and thus belong to the larger class of "radical enzymes." For understanding the general mechanisms of radical enzymes, it is of great importance to establish the general mechanism of AdoCbl-dependent catalysis using enzymes that catalyze the simplest reactions—such as diol dehydratase, glycerol dehydratase and ethanolamine ammonia-lyase. These enzymes are often called "eliminases." We have studied AdoCbl and eliminases for more than a half century. Progress has always been driven by the development of new experimental methodologies. In this chapter, we describe our investigations on these enzymes, including their metabolic roles, gene cloning, preparation, characterization, activity assays, and mechanistic studies, that have been conducted using a wide range of biochemical and structural methodologies we have developed.

1. Introduction

Since the Barker's discovery of a light-sensitive corrinoid coenzyme (Barker, Weissbach, & Smyth, 1958), adenosylcobalamin (AdoCbl) (Fig. 1A) or coenzyme B_{12} has fascinated many investigators in various fields with its complicated structure and peculiar functions (Banerjee, 1999; Dolphin, 1982). It enables many enzymes to catalyze intramolecular group-transfer reactions or rearrangements (Fig. 1C) and ribonucleotide

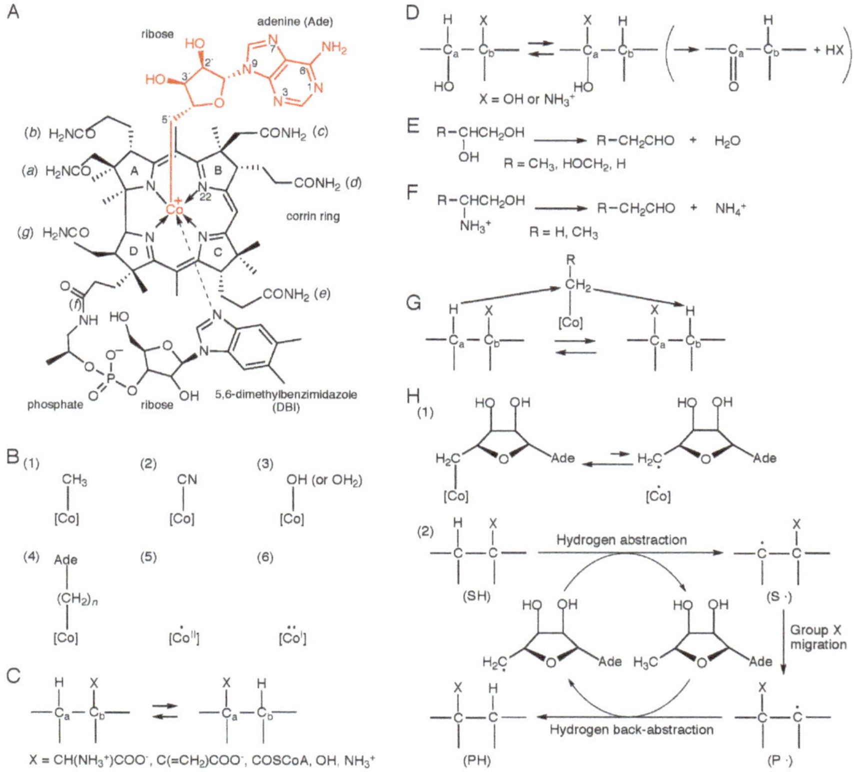

Fig. 1 Objects of this study. (A) Coenzyme B_{12} (AdoCbl). (B) Some analogs of vitamin B_{12}. (1) MeCbl. (2) CN-Cbl. (3) OH-Cbl (or aqCbl). (4) AdePeCbl (n = 5), AdeBuCbl (n = 4), and AdeEtCbl (n = 2). (5) Cob(II)alamin (B_{12r}). (6) Cob(I)alamin (B_{12s}). (C) Common feature of AdoCbl-dependent rearrangements. (D) Common feature of reactions catalyzed by AdoCbl-dependent isomerases (eliminating). (E) Reactions catalyzed by DD and GD. (F) Reactions catalyzed by EAL. (G) AdoCbl as an intermediate hydrogen carrier. (H) Minimal mechanism of AdoCbl-dependent rearrangements: (1) homolysis of coenzyme Co-C bond, and (2) adenosyl radical-catalyzed rearrangements. [Co], cobalamin; SH, substrate; $S^\bullet$, substrate radical; PH, product; $P^\bullet$, product radical; X, generic migrating group; RCH_2, adenosyl group.

reduction using a primary-carbon radical formed by the homolysis of coenzyme's Co-C bond. The former reactions include carbon-skeleton rearrangements, heteroatom eliminations, and intramolecular amino group migrations.

It is well known that many proteins utilize the high reactivity of radicals for catalyzing chemically challenging reactions. For understanding general mechanisms of such radical enzymes, AdoCbl-requiring reactions are prototypical examples. For these reasons, we have concentrated our efforts on establishing the general mechanism of AdoCbl-assisted enzymatic catalysis and radical-catalyzed reactions by using enzymes that catalyze simple isomerization reactions leading to the elimination of water or ammonia (Fig. 1D). These types of enzymes catalyzing heteroatom eliminations are often called "eliminases" and include diol dehydratase (DD) or propanediol hydro-lyase (EC 4.2.1.28) (Fig. 1E) (Abeles, Brownstein, & Randles, 1960; Lee & Abeles, 1963), glycerol dehydratase (GD) or glycerol hydro-lyase (EC 4.2.1.30) (Fig. 1E) (Pawelkiewicz & Zagalak, 1965; Smiley & Sobolov, 1962), and ethanolamine ammonia-lyase (EAL) or ethanolamine deaminase (EC 4.3.1.7) (Fig. 1F) (Bradbeer, 1965b; Kaplan & Stadtman, 1968). The reactions catalyzed by these enzymes were extensively studied in late 1960s and early 1970s (Abeles & Dolphin, 1976; Babior, 1975).

Earlier studies established that all the AdoCbl-dependent rearrangements share a common feature (Banerjee, 1999; Dolphin, 1982)—that is, they are intramolecular group-transfer reactions in which a hydrogen atom migrates from one carbon atom of the substrate to an adjacent carbon atom in exchange for a group, X, that moves in the opposite direction (Fig. 1C). Enzyme-bound AdoCbl serves as an intermediate hydrogen carrier, first accepting a hydrogen from C1 of the substrate to C5′ of the coenzyme's 5′-deoxyadenosyl group and then, in a subsequent step, returning a hydrogen back to C2 of the product (Frey & Abeles, 1966; Frey, Essenberg, & Abeles, 1967; Frey, Essenberg, Abeles, & Kerwar, 1970; Frey, Kerwar, & Abeles, 1967) (Fig. 1G). Hydrogen transfer can be intramolecular or intermolecular as it occurs through an enzyme-bound intermediate in which the hydrogen abstracted from the substrate and the two hydrogens on C5′ of AdoCbl all become equivalent (Abeles & Zagalak, 1966; Essenberg, Frey, & Abeles, 1971; Frey, Essenberg, & Abeles, 1967). Hence, 5′-deoxyadenosine was postulated to be an intermediate. Reversible formation of this nucleoside demonstrated with EAL provided further evidence for its involvement as intermediate in the catalytic process (Babior, Carty, & Abeles, 1974).

Starting from this point, we investigated AdoCbl-dependent eliminases to refine the overall mechanisms and stereochemical course of their reactions. For a deep understanding of enzyme-catalyzed reactions, we believe that combined biochemical, structural, and computational studies are required. In this chapter, we focus on the major biochemical and structural methodologies that we have used. The computational methodology will be discussed elsewhere. In addition, it is also an important question whether and how radical B_{12} enzymes undergo reactivation when inactivated by the extinction of radicals through side reactions. We discovered reactivases or reactivating chaperones that are involved in the activity-maintaining systems for the AdoCbl-dependent eliminases. The methodology for studying these chaperones is discussed in another chapter in this volume.

2. Discovery and metabolic roles

2.1 Diol dehydratase (DD)

Glycerol-grown cells of *Klebsiella oxytoca* (formerly *Aerobacter aerogenes* and *Klebsiella pneumoniae*) ATCC 8724 convert 1,2-propanediol, glycerol, and 1,2-ethanediol to propionaldehyde, β-hydroxypropionaldehyde, and acetaldehyde, respectively (Abeles et al., 1960). It was reported that purified DD from *K. oxytoca* catalyzes the conversion of 1,2-propanediol and 1,2-ethanediol to the corresponding aldehydes in the presence of AdoCbl (Fig. 1E), but glycerol reacts at less than 5% the rate of 1,2-propanediol (Abeles, 1966; Lee & Abeles, 1963). We purified DD in a soluble form from *K. oxytoca* by a conventional method (Poznanskaja, Tanizawa, Soda, Toraya, & Fukui, 1979) and demonstrated that glycerol serves as both a good substrate and a potent inactivator (Toraya, Shirakashi, Kosuga, & Fukui, 1976).

DD and GD are isofunctional enzymes but distinguishable from each other by their substrate-specificity (ratio of glycerol-dehydrating activity to 1,2-propanediol-dehydrating activity) and immunochemical cross-reactivity with rabbit anti-DD antiserum (Toraya & Fukui, 1977). Selected genera of *Enterobacteriaceae*, such as *Klebsiella* and *Citrobacter*, grown anaerobically on 1,2-propanediol and on glycerol produce AdoCbl-dependent DD and GD, respectively (Toraya & Fukui, 1977; Toraya, Kuno, & Fukui, 1980), whereas *K. oxytoca* ATCC 8724, *Propionibacterium freudenreichii*, and *Salmonella typhimurium* produce only DD (Obradors, Badía, Baldomà, & Aguilar, 1988; Toraya, Honda, Kuno, & Fukui, 1978; Toraya et al., 1980). Some other bacteria also express DD (Toraya, 1999), but *Clostridium glycolicum* produces an AdoCbl-independent membrane-associated DD (Hartmanis & Stadtman, 1986). Recently, B_{12}-independent, glycyl radical enzyme-type DD of *Roseburia inulinivorans* (LaMattina et al., 2016; Levin & Balskus, 2018) and GD of *Clostridium butyricum* (O'Brien et al., 2004; Raynaud, Sarçabal, Meynial-Salles, Croux, & Soucaille, 2003) were also discovered and biochemically and structurally characterized. Since the cloning of DD genes (*pddABC* or *pduCDE*) (Fig. 2A) (Bobik et al., 1997; Tobimatsu et al., 1995),

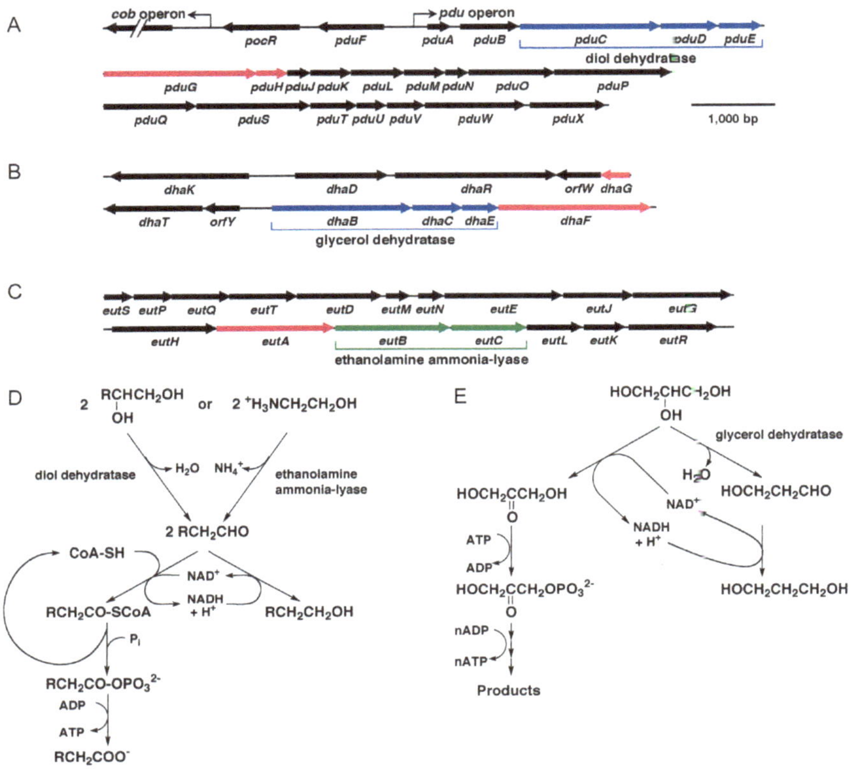

Fig. 2 Metabolic roles of DD, GD, and EAL and gene organizations around their genes. (A) *pdu/cob* regulon of *S. typhimurium* (Bobik, Xu, Jeter, Otto, & Roth, 1997). (B) *dha* regulon of *C. freundii* (Seyfried, Daniel, & Gottschalk, 1996). (C) *eut* operon of *S. typhimurium* (Kofoid, Rappleye, Stojiljkovic, & Roth, 1999). (D) Fermentation of 1,2-diols and ethanolamine. R = CH_3 or H. (E) Fermentation of glycerol.

recombinant enzymes of *K. oxytoca* (Tobimatsu et al., 1997) and *S. typhimurium* (Magnusson & Frey, 2002) have been generally used.

1,2-Propanediol, 1,2-ethanediol, and glycerol are more reduced substrates than the corresponding carbohydrates and thus utilized by only a limited number of bacteria under anaerobic conditions without external electron acceptors (Toraya, Honda, & Fukui, 1979; Toraya et al., 1980). 1,2-Propanediol and 1,2-ethanediol are fermented as illustrated in Fig. 2D in selected genera of *Enterobacteriaceae* and *P. freudenreichii* and finally disproportionated to acids and alcohols via aldehydes (Hosoi, Morimoto, Ozaki, Kitamoto, & Ichikawa, 1978; Toraya, Honda, et al., 1979). DD and other enzymes necessary for the utilization of 1,2-propanediol are encoded by genes in the *pdu* (*p*ropane*d*iol *u*tilization) operon (Daniel, Bobik, & Gottschalk, 1998; Jeter, 1990) (Fig. 2A). DD is induced when bacteria are grown anaerobically in 1,2-propanediol-containing media (Forage & Foster, 1979; Toraya & Fukui, 1977; Toraya et al., 1978). Glycerol is also effective as an inducer for DD, but to a much lesser extent, and the synthesis of DD undergoes catabolite repression reversed by cyclic AMP (Toraya et al., 1978). The metabolic role of DD is to produce aldehydes that serve as both an electron acceptor and a metabolic intermediate for the incomplete oxidation of 1,2-diols (Toraya, Honda, et al., 1979).

2.2 Glycerol dehydratase (GD)

GD, an isofunctional enzyme of DD (Fig. 1E), was found in acrolein-forming *Lactobacillus* sp. (Smiley & Sobolov, 1962) and in *K. pneumoniae* (formerly *A. aerogenes*) ATCC 25955 (Pawelkiewicz & Zagalak, 1965). GD was purified from *K. pneumoniae* (Schneider, Larsen, Jacobson, Johnson, & Pawelkiewicz, 1970; Stroinski, Pawelkiewicz, & Johnson, 1974). It prefers glycerol to 1,2-propanediol as substrate, whereas DD prefers 1,2-propanediol to glycerol (Toraya & Fukui, 1977). They are immunochemically distinct (Toraya & Fukui, 1977) and differ in the monovalent cation-selectivity pattern (Toraya & Fukui, 1977) and the binding affinity for AdoCbl (Forage & Foster, 1979) as well. Some genera of *Enterobacteriaceae*, such as *Klebsiella* and *Citrobacter*, produce GD and DD when grown anaerobically on glycerol and 1,2-propanediol, respectively (Toraya & Fukui, 1977; Toraya et al., 1980), while *Citrobacter freundii* NCIB418 forms only GD (Forage & Foster, 1979). Since the cloning of GD genes (*gldABC* or *dhaBCE*) (Fig. 2B) (Seyfried et al., 1996; Tobimatsu et al., 1996), recombinant enzymes of *K. pneumoniae* (Yamanishi et al., 2002) and *C. freundii* (Poppe & Rétey, 1997) have been usually used.

Glycerol is fermented by bacteria with accumulating 1,3-propanediol (Mickelson & Werkman, 1940; Sobolov & Smiley, 1960). A *dha* (*di*hydroxy*a*cetone) system is responsible for anaerobic dissimilation of glycerol, i.e., two enzymes constitute a branch for oxidation of glycerol to dihydroxyacetone phosphate, while the other two enzymes including GD serve as a branch for the reduction of glycerol to 1,3-propanediol (Fig. 2E) (Forage & Foster, 1982; Forage & Lin, 1982). GD and other enzymes necessary for the utilization of glycerol are encoded by genes in the *dha* regulon (Forage & Foster, 1982; Forage & Lin, 1982; Seyfried et al., 1996) (Fig. 2B) whose expression is induced by glycerol or dihydroxyacetone.

The metabolic role of GD is to produce β-hydroxypropionaldehyde that serves as an electron acceptor in this type of glycerol fermentation (Abeles et al., 1960; Forage & Foster, 1982; Toraya et al., 1980). In some bacteria, GD is substituted by DD that is induced on a low level by glycerol (Forage & Foster, 1982; Toraya et al., 1978, 1980).

2.3 Ethanolamine ammonia-lyase (EAL)

EAL catalyzes the conversion of ethanolamine to acetaldehyde in the presence of AdoCbl (Fig. 1F) (Bradbeer, 1965b). It was discovered in choline-fermenting *Clostridium* sp. (Bradbeer, 1965a) and found in many bacteria, such as *E. coli*, *Klebsiella aerogenes*, *Bacillus megaterium*, and *S. typhimurium*, which require vitamin B_{12} for growth on ethanolamine (Chang & Chang, 1975). EAL was purified from *Clostridium* sp. (Kaplan & Stadtman, 1968) and *E. coli* (Blackwell & Turner, 1978). Since the cloning of EAL genes (*eutBC*) (Fig. 2C) (Faust, Connor, Roof, Hoch, & Babior, 1990), recombinant enzymes of *S. typhimurium* (Faust & Babior, 1992) and *E. coli* (Akita et al., 2010) have been generally used.

In clostridia, the metabolic role of EAL is to produce acetaldehyde that serves as both a metabolic intermediate and an electron acceptor for the incomplete oxidation of ethanolamine (Fig. 2D) (Bradbeer, 1965a). In *E. coli* and other bacteria, its metabolic role is to produce ammonia in the anaerobic dissimilation of ethanolamine that is utilized as a sole nitrogen source (Chang & Chang, 1975; Scarlett & Turner, 1976). In the latter case, both ethanolamine and vitamin B_{12} are required for the induction of EAL (Blackwell, Scarlett, & Turner, 1977; Roof & Roth, 1988). EAL and other enzymes necessary for the utilization of ethanolamine are encoded by genes in the *eut* (*e*thanolamine *ut*ilization) operon (Kofoid et al., 1999) (Fig. 2C).

3. Gene cloning and high-level expression

3.1 Diol dehydratase genes

In the shotgun cloning of DD genes, we digested genomic DNA from *K. oxytoca* ATCC 8724 partially with *Sau*3AI and inserted 6–20 kb fragments into the *Bam*HI site of plasmid pUC119 (Tobimatsu et al., 1995). *E. coli* HB101 was transformed with the plasmids, and ampicillin-resistant transformants were screened using a synthetic 50-mer oligodeoxyribonucleotide probe which is complementary to the DNA sequence deduced from the

N-terminal amino acid sequence of the M_r 60,000 subunit (McGee & Richards, 1981) of DD. Hybridization-positive clones were isolated, cultivated aerobically in a glycerol/1,2-propanediol medium, and examined for DD activity. The clone pUCDD11 showing the highest specific activity was cultivated, and its cell-free extract was analyzed by nondenaturing polyacrylamide gel electrophoresis (PAGE) through protein staining and activity staining (Protocol 9.1). A new protein band was catalytically active and immunochemically identified as DD of *K. oxytoca*. The addition of isopropyl-1-thio-β-D-galactopyranoside (IPTG) did not increase the level of DD activity, suggesting that DD genes in this clone are under control of their own promoter. Nucleotide sequencing revealed that the DD genes (*pddABC*) encode polypeptides of 554, 224, and 173 amino acid residues with predicted molecular weights of 60,348 (α), 24,113 (β), and 19,173 (γ), respectively (Tobimatsu et al., 1995). *pddABC* genes correspond to *pduCDE* of the *pdu/cob* (1,2-*p*ropane*d*iol *u*tilization/*co*balamin *b*iosynthesis) regulon of *S. typhimurium* (Bobik et al., 1997) whose gene organization is depicted in Fig. 2A.

For high-level expression of DD genes, pUSI2E(DD) and other expression plasmids were constructed. In pUSI2E(DD), *pddABC* genes are placed in tandem in this order downstream of the *tac* promoter (Tobimatsu et al., 1995). When *E. coli* JM109 carrying this plasmid was cultivated at 30 °C and induced by IPTG, membrane-associated DD was produced with a specific activity of homogenates 53-fold higher than that of *K. oxytoca* ATCC 8724.

DD is a low-solubility enzyme (Bobik et al., 1997; Tobimatsu et al., 1997). For soluble DDs, expression plasmids encoding DDs with the full-length α subunit, the full-length β subunit or β subunit lacking the residues Gluβ2-Lysβ20, and the full-length γ subunit or γ subunit lacking the residues Asnγ2-Argγ16 were constructed (Tobimatsu, Kawata, & Toraya, 2005). Another expression plasmid encoding DD with the full-length α subunit and the β and γ subunits lacking the residues Gluβ2-Aspβ45 and Asnγ2-Serγ37, respectively, was also constructed (Shibata, Sueyoshi, Higuchi, & Toraya, 2018). In the latter, a hexahistidine (His_6)-tag followed by a human rhinovirus 3C (HRV3C) protease cleavage site was inserted before Glyβ46. These N-terminal truncations do not affect catalytic properties.

3.2 Glycerol dehydratase genes

GD genes were also cloned by shotgun cloning. We digested genomic DNA from *K. pneumonia* ATCC 25955 partially with *Sau*3AI and inserted 6–20 kb fragments into the *Bam*HI site of pUC119 (Tobimatsu et al., 1996). *E. coli* JM109 was transformed with the plasmids, and ampicillin-resistant transformants were screened by cross-hybridization with the 1.6-kb *Pst*I fragment of pUCDD11 (Tobimatsu et al., 1995). This probe carried most of the *pddA* gene that encodes the *K. oxytoca* DD α subunit. Strongly-hybridizing and weakly-hybridizing clones were isolated, analyzed by Southern blot analysis, and examined for DD activity. When grown in a glycerol medium or a glycerol/1,2-propanediol medium, most of the strongly-hybridizing clones showed definite DD activity, whereas none of the weakly-hybridizing clones showed significant activity. Sequence analysis revealed that the former contain the *K. pneumonia* DD genes (*pddABC*) (Tobimatsu, Azuma, Hayashi, Nishimoto, & Toraya, 1998), whereas the latter contain the GD genes. The GD genes (*gldABC*) exist in tandem in this order and encode polypeptides of 555, 194, and 141 amino acid residues with predicted molecular weights of 60,659 (α), 21,355 (β), and 16,104 (γ), respectively (Tobimatsu et al., 1996). *gldABC* genes correspond to *dhaBCE* of the *dha* regulon of *C. freundii* (Seyfried et al., 1996) whose gene organization is illustrated in Fig. 2B.

For high-level expression of GD genes, the expression plasmid pUSI2E(GD) that contain *gldABC* downstream of the *tac* promoter was constructed (Tobimatsu et al., 1996). When *E. coli* JM109 carrying this plasmid was cultivated at 30 °C and induced by IPTG, GD was produced in a soluble form with the specific activity of homogenate 14-fold higher than that of *K. pneumoniae* ATCC 25955. In contrast to DD, GD is a soluble enzyme. For easier purification of GD, expression plasmids encoding GD with the N-terminal His_6-tagged γ subunit followed by a thrombin cleavage site and encoding its N-terminal His_6-tagged β subunit followed by a thrombin cleavage site were also constructed (Yamanishi et al., 2002). The latter plasmid pET19b($H6\beta_G$) was used for supplying the purified β subunit to partially dissociated GD.

3.3 Ethanolamine ammonia-lyase genes

EAL is composed of two kinds of subunits—large (α) and small (β) subunits, and the subunit structures of both clostridial (Wallis, Johnson, & Lappert, 1979) and *S. typhimurium* (Faust & Babior, 1992) EALs are $\alpha_6\beta_6$. The α and β subunits are encoded by *eutBC* genes, respectively, in the *eut* operon of *S. typhimurium* (Sheppard & Roth, 1994) whose gene organization is shown in Fig. 2C. *E. coli* EAL genes (*eutBC*) encode polypeptides of 453 and 295 amino acid residues with predicted molecular weights of 49,402 (α) and 31,781 (β), respectively (Akita et al., 2010).

For high-level expression of *E. coli* EAL genes, we constructed expression plasmids encoding the wild-type and truncated EALs (Akita et al., 2010). The 3.1-kb *Pvu*II fragment containing *eutBC* genes was excised from DNA prepared from clone #421 of the Kohara *E. coli* genomic λ phage library (Kohara, Akiyama, & Isono, 1987), treated with *Taq* DNA polymerase, and cloned into pCR2.1 (Invitrogen). The 630-bp DNA fragment containing the 5′-terminal region of *eutB* with preceding *Nde*I site and following *Eco*RI site was amplified by PCR with the same DNA from Kohara clone #421 as template using *Taq* DNA polymerase and cloned into pCR2.1. The 2.2-kb *Afl*II-*Eco*RI fragment obtained from the former and the 610-bp *Nde*I-*Afl*II fragment obtained from the latter were inserted into the *Nde*I-*Eco*RI region of pUSI2ENd (Tobimatsu et al., 2005). Unexpectedly, resulting plasmid pUSI2ENd(EAL) contained extra 0.6-kb sequence corresponding to the 5′-terminal region of *eutB*, and this extra sequence was removed to generate pUSI2ENd(EAL)2 (Akita et al., 2010).

EAL is a low-solubility enzyme (Akita et al., 2010; Faust & Babior, 1992). For soluble EALs, expression plasmids encoding EALs with the N-terminal His_6-tagged α subunit and the β subunit lacking the residues Aspβ2-Cysβ29 and with the full-length α subunit and the N-terminal His_6-tagged β subunit lacking the residues Aspβ2-Cysβ29 were constructed (Akita et al., 2010). Removing this short N-terminal sequence of the β subunit is sufficient to change the solubility and stability of EAL without affecting the catalytic properties. Expression plasmids encoding EALs with the full-length α subunit and the His_6-tagged β subunits lacking the residues Lysβ4-Alaβ30 and Lysβ4-Cysβ43 were also constructed for X-ray analysis (Shibata et al., 2010).

4. Purification and protein-chemical properties

4.1 Diol dehydratase

Native DD was first purified from *K. oxytoca* grown on glycerol (Lee & Abeles, 1963) or glycerol/1,2-propanediol medium (Abeles, 1966). We have developed a conventional method for purifying the enzyme to homogeneity in the soluble form from *K. oxytoca* (Poznanskaja et al., 1979; Tanizawa, Nakajima, Toraya, Tanaka, & Soda, 1987). The molecular weight obtained by gel filtration and sedimentation equilibrium analytical ultracentrifugation was approximately 230,000. The enzyme consisted of subunits with M_rs of 60,000, 25,500, 23,000, 15,500, and 14,000, and this subunit heterogeneity was due to partial proteolysis during isolation. Upon diethylaminoethyl (DEAE)-cellulose chromatography in the absence of substrate, DD is dissociated into two protein components, F and S, with M_rs of 26,000 and 200,000, respectively (Toraya, Uesaka, & Fukui, 1974). Neither component alone possesses activity, and the activity is restored when the components are combined. Substrates promote the association of the components. A simple purification procedure has been developed based on the finding that DD is a membrane-associated enzyme (McGee, Carroll, Bond, & Richards, 1982; McGee & Richards, 1981). The subunit structure was suggested to be $(60{,}000)_2(51{,}000)_1(29{,}000)_2(15{,}000)_2$, but it turned out later that the M_r 51,000 polypeptide is not a DD subunit but CoA-acylating propionaldehyde dehydrogenase (PduP) (Leal, Havemann, & Bobik, 2003; Sauvageot et al., 2002).

Recombinant wild-type DD of *K. oxytoca* was purified to homogeneity from overexpressing *E. coli* in an overall yield of 65% (Tobimatsu et al., 1997) (Protocol 9.2). It is extracted from the crude membrane fraction with 1% Brij 35, followed by DEAE-cellulose column chromatography. This method is based on its low solubility. DD consists of equimolar amounts of three subunits with M_rs of 60,000 (α), 30,000 (β), and 19,000 (γ), which correspond to the products of *pddABC* genes, respectively. The molecular weight estimated by nondenaturing PAGE on a gradient gel was 220,000. The subunit structure of DD is thus $\alpha_2\beta_2\gamma_2$. Components F and S were identified as the β subunit and the $\alpha_2\gamma_2$ complex, respectively, by sodium dodecyl sulfate-polyacrylamide gel electrophoresis (SDS-PAGE) analysis (Tobimatsu et al., 1997).

Soluble DDs are produced by *E. coli* carrying expression plasmids for DDs with the full-length α subunit, the full-length β subunit or β subunit lacking the residues Gluβ2-Lysβ20, and the full-length γ subunit or γ subunit lacking the residues Asnγ2-Argγ16 (Tobimatsu et al., 2005). These soluble DDs can be purified by a conventional procedure, i.e., $(NH_4)_2SO_4$ fractionation and column chromatography on Sepharose CL 6B and DEAE-cellulose in this order. His_6-tagged DD is also produced by *E. coli* carrying another expression plasmid encoding DD with the full-length α subunit and the β and γ subunits lacking the residues Gluβ2-Aspβ45 and Asnγ2-Serγ37, respectively, with a His_6-tag followed by a HRV3C protease cleavage site before Glyβ46 (Shibata et al., 2018). The cell extract prepared from overexpressing *E. coli* is applied to a cOmplete His-Tag Purification Resin column (Roche Molecular Systems, Inc.). After washing the column, glutathione *S*-transferase-tagged HRV3C protease is added, followed by overnight digestion at 4 °C. The recovered protein is further purified by size-exclusion chromatography on a HiLoad 26/60 Superdex 200 pg column (GE Healthcare). This method was employed for crystallization of DD.

4.2 Glycerol dehydratase

Recombinant wild-type GD of *K. pneumonia* ATCC 25955 was purified from overexpressing *E. coli* by a conventional method (Yamanishi et al., 2002) (Protocol 9.3), i.e., by $(NH_4)_2SO_4$ fractionation and column chromatography on Sepharose 6B, hydroxyapatite, and Sephadex G-200, in a yield of 63%. Upon nondenaturing PAGE in the presence of substrate, a major upper band and a minor lower band were seen whose molecular weights were estimated to be 220,000 and 200,000, respectively, by nondenaturing PAGE on a Multigel 2/15 gradient gel (Cosmo Bio Co., Ltd.). The major band is active GD upon activity staining (Protocol 9.1). GD consists of equimolar amounts of three subunits with M_rs of 61,000 (α), 22,000 (β) and 16,000 (γ), which correspond to the products of *gldABC* genes, respectively. The subunit structure of GD is thus $\alpha_2\beta_2\gamma_2$. The minor lower band is most likely $\alpha_2\gamma_2$. Upon DEAE-cellulose chromatography in the absence of substrate, GD is separated into components A and B that are identified by SDS-PAGE as the β subunit and the $\alpha_2\gamma_2$ complex, respectively. The addition of component A to the purified GD preparation converted the lower band to the upper one in nondenaturing PAGE with a concomitant increase of enzyme activity.

His_6-tagged GD is purified from overexpressing *E. coli* by nickel-affinity chromatography on a Ni-NTA agarose (Qiagen GmbH) column, and the His_6-tag is removed by digestion with thrombin, followed by passing through a Ni-NTA agarose column (Yamanishi et al., 2002) (Protocol 9.4). Purified GD is sometimes contaminated with component B due to its partial dissociation into components A and B. The highest specific activity is obtained by supplying the enzyme with excess component A, followed by passing through a Sepharose 6B column. This method was employed for the crystallization of GD. Component A is separately purified from *E. coli* overproducing His_6-tagged β subunit, as described for GD.

4.3 Hybrid and chimeric enzymes of diol dehydratase and glycerol dehydratase

We constructed the expression plasmids for hybrid enzymes between *K. oxytoca* DD and *K. pneumonia* GD in all the combinations of subunits by gene engineering techniques (Sakai et al., 2007). Although all the hybrid enzymes were produced in *E. coli* at high levels, only two hybrid enzymes, i.e., $\alpha_G\beta_D\gamma_D$ and $\alpha_G\beta_D\gamma_G$ showed high activity, where suffixes D and G represent DD and GD, respectively. It was shown with purified hybrid enzymes that the susceptibility to glycerol inactivation and monovalent cation specificity are primarily determined by the origin of their α subunits.

DD, a low-solubility enzyme, can be converted to a soluble enzyme by eliminating short N-terminal sequence from either β or γ subunit or both (Tobimatsu et al., 2005). On the contrary, we prepared a low-solubility GD by construction of an expression plasmid encoding a chimeric enzyme which carries the N-terminal 60 residues of β_D plus the C-terminal 167 residues of β_G and the N-terminal 33 residues of γ_D plus the C-terminal 140 residues of γ_G (Tobimatsu, Nishiki, Morimoto, Miyata, & Toraya, 2009). Almost all the activity of this chimeric enzyme was recovered in the precipitant fraction when overexpressing *E. coli* cells were disrupted by sonication.

4.4 Ethanolamine ammonia-lyase

Recombinant wild-type EAL of *E. coli* was purified to homogeneity from overexpressing *E. coli* in an overall yield of 31% (Akita et al., 2010) (Protocol 9.5). This method is based on its low solubility. EAL is extracted from the crude membrane fraction with 1% Brij 35, followed by DEAE-cellulose column chromatography. It should be noted that *E. coli* JM109 carrying pUSI2ENd(EAL)2 produces EAL in unbalanced subunit ratio ($\alpha \gg \beta$), whereas *E. coli* carrying pUSI2ENd(EAL) produces it in well balanced ratio ($\alpha/\beta \sim 1$). Therefore, *E. coli* carrying pUSI2ENd(EAL) is preferable. EAL consists of equimolar amounts of two subunits with M_rs of 49,000 (α) and 32,000 (β), which correspond to the products of *eutBC* genes, respectively. Its molecular weight determined by gel filtration on a calibrated Superose 6 column was approximately 470,000. The subunit structure of EAL is thus $\alpha_6\beta_6$.

Since the short N-terminal truncation of the β subunit increases the solubility of EAL without affecting the catalytic properties, soluble EALs were produced by *E. coli* carrying expression plasmids encoding EALs with the N-terminal His_6-tagged α subunit and the β subunit lacking the residues Aspβ2-Cysβ29 and with the full-length α subunit and the C-terminal His_6-tagged β subunit lacking the residues Aspβ2-Cysβ29 (Akita et al., 2010). These truncated EALs can be purified from overexpressing *E. coli* by chromatography on a Ni-NTA agarose column and a Superose 6 column attached to an FPLC system (GE Healthcare, UK) (Protocol 9.6). A truncated EAL is resistant to aggregation and oxidative inactivation even at high concentrations. Another EAL with the full-length α subunit and the His_6-tagged β subunit lacking the residues Lysβ4-Cysβ43 was also purified from overexpressing *E. coli* by a Ni-NTA agarose column chromatography for X-ray structure analysis (Shibata et al., 2010).

5. Enzyme assays and catalytic properties

5.1 Enzyme assays

DD, GD, and EAL are assayed by the same procedures, because reaction products are aldehydes in each case. Aldehyde products formed per unit time at 30 or 37 °C are converted to the azine derivatives by reaction with 3-methyl-2-benzothiazolinone hydrazone (MBTH) (Toraya, Ushio, Fukui, & Hogenkamp, 1977) (Protocol 9.7) or to 2,4-dinitrophenylhydrazones by reaction with 2,4-dinitrophenylhydrazine (DNPH) (Abeles, 1966). The former are measured spectrophotometrically at 305 nm, whereas the latter at 540 nm after color development in pyridine with alcoholic KOH. The MBTH method is superior to the DNPH method in sensitivity, brevity, and color stability.

In a coupled assay, aldehyde products formed are continuously reduced to the corresponding alcohols in the presence of excess 1,4-dihydronicotinamide adenine dinucleotide (NADH) and alcohol dehydrogenase (Toraya, Krodel, Mildvan, & Abeles, 1979) (Protocol 9.8). The decrease in the absorbance of NADH at 340 nm per unit time is a measure of the enzyme activity. Yeast alcohol dehydrogenase is used for the products propionaldehyde and acetaldehyde, whereas horse-liver alcohol dehydrogenase or more preferably *K. pneumoniae* 1,3-propanediol dehydrogenase (DhaT) (Johnson & Lin, 1987) is used for the product β-hydroxypropionaldehyde formed from glycerol (Yamanishi et al., 2012). The coupled assay method is superior to the MBTH method for kinetic analysis, while the MBTH method is more convenient and easier for routine assays.

5.2 Substrate specificity, mechanism-based inactivation, and inactivation in the absence of substrate

Substrate specificity of DD is shown in Table 1. The dehydration of 1,2-propanediol proceeds almost linearly with time, whereas glycerol serves as both a good substrate and a potent inactivator for DD (Bachovchin et al., 1977; Toraya et al., 1976) and GD (Poznanskaya, Yakusheva, & Yakovlev, 1977; Schneider & Pawelkiewicz, 1966). Other substrates also cause inactivation, but to a lesser degree. EAL also undergoes mechanism-based inactivation gradually by the natural substrate ethanolamine (Bradbeer, 1965b) and quickly by the substrate analog 2-amino-1-propanol (Carty, Babior, & Abeles, 1974). The inactivation rate constant k_{inact} can be obtained from a change in the slope of a tangent to the time course of the reaction determined by the NADH-alcohol dehydrogenase coupled method (Toraya, Krodel, et al., 1979), or alternatively by non-linear

Table 1 Substrate specificity of DD and mechanism-based inactivation by substrate or its analogs[a].

Substrate or its analog	k_{cat} [b] (s^{-1})	K_m (mM)	K_i (mM)	k_{inact} [c] (min^{-1})	k_{cat}/k_{inact} $\times 10^{-4}$	Reference
1,2-Propanediol	337, 330	0.18, 0.04		0.014, 0.024	144, 83	*d,e,f*
Glycerol	371, 173	1.54, 1.2		1.15	1.9, 0.90	*d,g*
1,2-Ethanediol	286	1.08		0.098[h]	18	*d*
1,2-Butanediol	7.2	2.5	1.97	0.066	0.80	*d,i,j*
2,3-Butanediol	15.1	0.64		0.068[h]	1.3	*d*
3-Butene-1,2-diol			0.03	1.08	<0.0001	*j*
3-Butyne-1,2-diol			0.02	1.02	<0.0001	*j*
Thioglycerol			0.1	1.14	<0.0001	*j*
Styrene glycol			38.4			*i*
cis-Cyclopentanediol			0.017			*k*
trans-Cyclopentanediol			0.66			*k*
cis-Cyclohexanediol			8.6			*k*
trans-Cyclohexanediol			0.34			*k*

[a]Determined at 37 °C. For other substrate analogs, see Bachovchin, Eagar, Moore, and Richards (1977).
[b]Obtained from the maximum velocity (V_{max}).
[c]Calculated from a change in the slope of a tangent to the time course curves of the reaction which is determined by the alcohol dehydrogenase-NADH coupled method.
[d]Toraya et al. (1976).
[e]Toraya, Krodel, et al. (1979).
[f]Kawata et al. (2006).
[g]Yamanishi et al. (2012).
[h]Bachovchin, Eagar, Moore, and Richards (1977).
[i]Toraya and Fukui (1972).
[j]Toraya et al. (2008).
[k]Yamanishi, M., Adachi, M., Oshima, M. and Toraya, T. (unpublished results).

curve-fitting (Yamanishi et al., 2012) as described briefly in the footnote to Table 2. The k_{cat}/k_{inact} ratio is an important indication for the tendency of mechanism-based inactivation, because it shows the average numbers of catalytic turnovers per enzyme molecule before inactivation.

For the mechanistic study of DD, we used 3-butene-1,2-diol, 3-butyne-1,2-diol, and thioglycerol as substrate analogs (Toraya et al., 2008). They serve as pseudosubstrates ($k_{cat}/k_{inact} < 1$). In addition, DD, GD, and EAL undergo rapid mechanism-based inactivation by various compounds, such as substrate analogs, pseudosubstrates, product analogs, etc. (for reviews, see Toraya, 2000, 2003, 2014 and references therein). For mechanism-based inactivation by coenzyme analogs, see Section 6.4. Generally, stabilization or loss of intermediate radicals by side reactions results in the damage of coenzyme, and damaged cofactor(s) remain tightly bound to the active site and are not displaceable by intact AdoCbl, which leads to the inactivation of holoenzymes (Toraya et al., 2008). Apoenzymes themselves are not inactivated or damaged during the process, because active apoenzyme is recovered by an acid $(NH_4)_2SO_4$ treatment. We discovered reactivating factors for inactivated DD, GD, and EAL (see chapter "Reactivating chaperones for coenzyme B_{12}-dependent diol and glycerol dehydratases and ethanolamine ammonia-lyase" by Toraya et al.).

Holoenzymes of DD, GD, and EAL undergo rapid inactivation in the absence of substrate as well (Kaplan & Stadtman, 1968; Lee & Abeles, 1963; Poznanskaya et al., 1977). In the case of DD, the inactivation occurs by the reaction of activated coenzyme's Co-C bond with O_2, forming tightly bound hydroxocobalamin (OH-Cbl) (Wagner, Lee, Frey, & Abeles, 1966) and 5′-peroxyadenosine that is further decomposed to adenosine 5′-aldehyde, adenine, and adenosine (Schwartz & Frey, 2007).

5.3 Involvements of metal ions

Requirements of DD and GD for monovalent cations (Lee & Abeles, 1963; Pawelkiewicz & Zagalak, 1965) were investigated using K^+-free apoenzymes that are easily prepared by Sephadex G-25 gel filtration. DD is fully activated by K^+ or other monovalent cations with similar ionic radii (NH_4^+, Rb^+) (Toraya, Sugimoto, Tamao, Shimizu, & Fukui, 1971). Na^+ and Cs^+ are 57% and 44%, respectively, as active as K^+, while Li^+ is inactive. The selectivity pattern of GD is substantially different, i.e., K^+, NH_4^+, and Rb^+ are most active, whereas Cs^+ only slightly active, and Na^+ and Li^+ totally inactive (Toraya & Fukui, 1977). Monovalent cations have been shown to play essential roles in the cobalamin binding (Toraya et al., 1971) and in the homolysis of the coenzyme's Co-C bond (Schwartz & Frey, 2007).

DD is strongly inhibited by the incubation of apoenzyme with ethylenediaminetetraacetate (EDTA) or ethylene glycol bis(2-aminoethylether)-*N*,*N*,*N*′,*N*′-tetraacetate (EGTA) in the absence of substrate (Toraya, Honda, & Mori, 2010). The metal analyses of both recombinant enzyme and purified native enzyme from *K. oxytoca* revealed that DD contains ~2 mol of tightly bound calcium per mol of enzyme. The calcium-deprived, EDTA-free apoDD obtained by the EDTA treatment, followed by ultrafiltration (Protocol 9.9), requires Ca^{2+} for catalytic activity and stability. We thus concluded that DD is a K^+-requiring calcium-metalloenzyme (Toraya et al., 2010).

6. Mechanistic studies

6.1 How reaction intermediates are identified: Optical and EPR spectroscopy

Which form of hydrogen migrates from substrate to product via the intermediate 5′-deoxyadenosine? An important hint is obtained by measuring the UV-visible spectra of cobalamin intermediates in the "reacting" holoenzymes. Upon incubation with DD (Toraya, Krodel, et al., 1979) or EAL (Bandarian & Reed, 2000) in the presence of substrates, AdoCbl was rapidly converted to cob(II)alamin (Figs. 1B(5) and 3A) , although the spectra of holoenzymes in the absence of substrate were essentially identical to that of free AdoCbl (Toraya et al., 1977; Wagner et al., 1966). This indicated that the Co-C bond of the coenzyme is homolytically cleaved during catalysis and suggests that a hydrogen migrates as a hydrogen atom ($H^{\bullet}$). However, it should be noted that suicidally inactivated holoenzymes also show the spectrum of cob(II)alamin. To distinguish the cob(II)alamin intermediate in the steady-state from that in the inactivated complex, electron paramagnetic resonance (EPR) measurements are needed.

The X-band EPR spectra of reacting DD (Finlay, Valinsky, Mildvan, & Abeles, 1973; Toraya, Krodel, et al., 1979; Valinsky, Abeles, & Fee, 1974), GD (Cockle, Hill, Williams, Davies, & Foster, 1972), and EAL (Babior, Moss, Orme-Johnson, & Beinert, 1974) provide useful information regarding cob(II)alamin-organic radical pairs. The spectrum obtained with DD showed a high-field doublet signal with g values of 1.95 and 2.04 and a low-field broad signal with a g value of 2.2–2.4 (a prominent signal, $g = 2.21$) (Fig. 3B1 and C1) (Protocol 9.10). The former was assigned to a single organic radical and the latter to the low-spin Co(II) of cob(II)alamin.

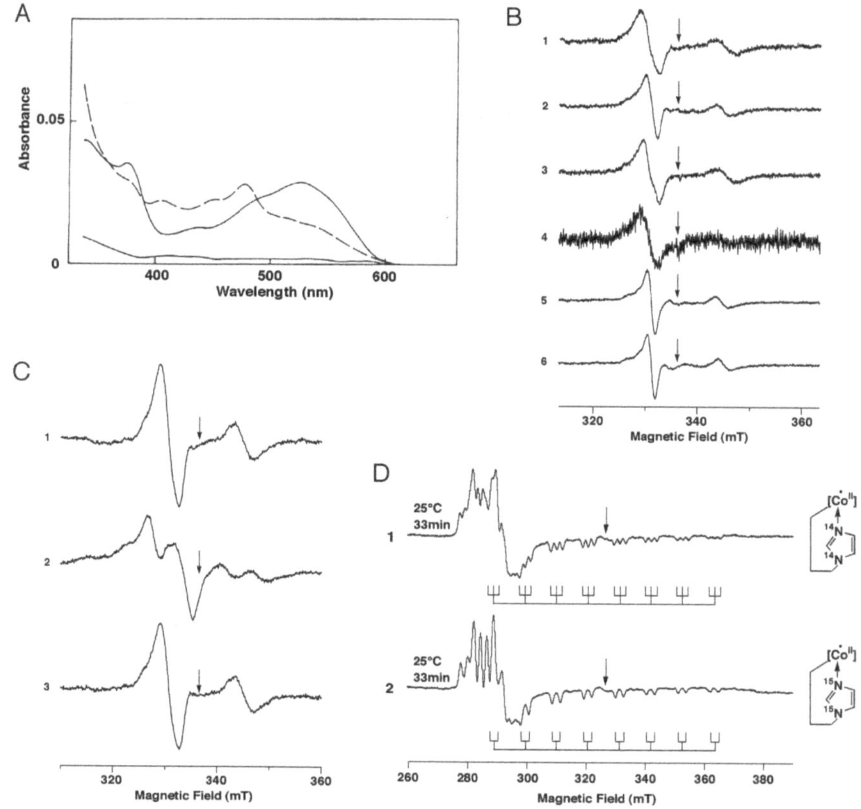

Fig. 3 Optical and EPR spectroscopy for the identification of a radical intermediate in the DD reaction and for the determination of the cobalamin-binding mode of DD. (A) Optical spectra of reacting holoDD (30 °C, 10 min, *broken line*), free AdoCbl (*solid line*), and apoenzyme (*bottom spectrum*). (B) The doublet signals of reacting holoDD (4 °C, 1 min) in X-band EPR spectra (Protocol 9.10) obtained with (1) unlabeled, (2) 1,1-dideuterated, (3) 2-deuterated, (4) 3,3,3-trideuterated, (5) 1,1,2-trideuterated, and (6) 1,1,2,3,3,3-hexadeuterated 1,2-propanediols. (C) The same as (B) obtained with (1) unlabeled, (2) 1-^{13}C-labeled, and (3) 2-^{13}C-labeled 1,2-propanediols. (D) EPR spectra of the DD-bound, (1) unlabeled and (2) [*imidazole*-^{15}N]-labeled cob(II)inamide imidazolylpropyl phosphate (25 °C, 33 min). Microwave power: 5 mW (B), 10 mW (C), or 20 mW (D). The arrows correspond to $g = 2.0$.

These signals have been interpreted to arise from a weak interaction of organic radical with low-spin Co(II) from a distance equal or greater than 10 Å (Boas, Hicks, Pilbrow, & Smith, 1978; Buettner & Coffman, 1977; Schepler, Dunham, Sands, Fee, & Abeles, 1975). The formation of such radical pairs indicates the homolysis of the coenzyme's Co–C bond during catalysis and hence suggests that these reactions proceed by a radical mechanism, in which a hydrogen migrates as $H^{\bullet}$.

An organic radical intermediate in the DD reaction can be characterized by EPR with isotopically labeled substrates. Only with substrates deuterated on C1, was a significant line-width narrowing of the doublet signal observed (Fig. 3B) (Yamanishi, Ide, Murakami, & Toraya, 2005). Furthermore, a distinct change in the hyperfine splitting was seen with [1-^{13}C]-1, 2-propanediol, but not with the [2-^{13}C]-counterpart (Fig. 3C). From these results, the organic radical intermediate in the DD reaction was identified as the 1,2-propanediol-1-yl radical, a C1-centered substrate-derived radical. This discovery consolidated the evidence for a radical mechanism in DD catalysis.

Organic radical intermediates formed by EAL from 2-amino-1-propanol and ethanolamine were also identified as C1-centered substrate radicals (Babior, Moss, et al., 1974; Bender, Poyner, & Reed, 2008). Thus, it was concluded that these AdoCbl-dependent reactions proceed by a radical mechanism, as illustrated in the minimal mechanism for DD and EAL (Fig. 1H). When other substrates, such as 1,2-ethanediol, glycerol, and *meso*-2,3-butanediol, were used in the DD reaction, there was a negative correlation between the magnitude of their exchange interaction between Co(II) and an organic radical in terms of *J*-value and their tendency toward the mechanism-based inactivation (k_{inact}) (Toraya et al., 1997). Hence, cob(II)alamin may not be simply a spectator but a stabilizer for organic radical intermediates.

The EPR spectra of DD completely inactivated with coenzyme analogs or substrate analogs resemble that of Co(II) corrinoid without organic-radical coupling partners (Toraya et al., 2008; Yamanishi, Yamada, Ishida, Yamauchi, & Toraya, 1998; Yamanishi, Yamada, Muguruma, et al., 1998). EPR is useful for determining the mode of cobalamin binding by enzymes. Since the coenzyme analog adenosylcobinamide 3-imidazolylpropyl phosphate (AdoCbiP-ImPr) induces mechanism-based inactivation (Toraya & Ishida, 1991), Co(II) corrinoid can be easily introduced to the active site of DD, GD, or EAL. When EPR spectra were measured after the mechanism-based inactivation with this analog, octet hyperfine lines showed superhyperfine splitting into triplets with an unlabeled analog due to the interaction of unpaired electron of Co(II) with ^{14}N nucleus ($I = 1$), whereas they showed the splitting into doublets with the [*imidazole*-^{15}N]-labeled analog due to the interaction of unpaired electron with ^{15}N nucleus ($I = 1/2$) (Fig. 3D).

These observations offered evidence for the so-called base-on mode of cobalamin binding by DD (Yamanishi, Yamada, Ishida et al., 1998; Yamanishi, Yamada, Muguruma, et al., 1998), GD (Yamanishi et al., 2002), and EAL (Akita et al., 2010), i.e., with 5,6-dimethylbenzimidazole (DBI) coordinating to the cobalt atom in the lower axial position. This is consistent with their X-ray structures determined later.

6.2 How catalytic radical is formed: Structure-function studies with coenzyme analogs

The adenosyl radical functions as the central catalyst in AdoCbl-dependent rearrangements. It is formed by the homolysis of coenzyme's Co-C bond and initiates catalysis. To investigate the mechanism of Co-C bond activation (labilization) by apoenzymes, we have synthesized a lot of AdoCbl analogs (Protocol 9.11) in which one of the structural components is substituted by a closely related group and examined for their coenzymic activity and binding affinity (K_m or K_i) for DD (Table 2) and EAL (data not shown) (for syntheses, see original papers). The status of the Co-C bond was also monitored spectroscopically with both active and inactive analogs.

Regarding AdoCbl analogs with a modified adenosyl group (adenine or β-ribose moiety), only small modifications are permitted without complete loss of activity (Table 2). The importance of the adenine nitrogen atoms for the manifestation of catalytic function and for the activation of Co-C bond decreases in the order N7 > 6-NH_2 > N3 > N1 (Toraya et al., 1977, 1986; Ushio et al., 1984). The rigid β-D-ribofuranose ring is essential for activity, and the importance of functional groups on it decreases in the order β-D-ribofuranose ≫ 3′-OH > 2′-OH > -O- (Ichikawa & Toraya, 1988). 2-Adeninylethylcobalamin (AdeEtCbl) (Toraya et al., 1987) and 2′,3′-secoAdoCbl dialdehyde (Ichikawa & Toraya, 1988) undergoes the Co-C bond heterolysis forming cob(III)alamin and homolysis, respectively, by the interaction with apoenzyme, although they are inactive analogs. Medium chain ω-(adeninyl)alkylcobalamins are strong competitive inhibitors (Toraya et al., 1977). Adenosylmethylcobalamin shows a trace of coenzymic activity in DD and EAL, whereas adenosylethylcobalamin does not (Fukuoka et al., 2005).

Isomers of AdoCbl analogs in which one of the three amide groups of the propionamide side chain of the corrin ring is converted to $-COO^-$, $-COOCH_3$, or $-CONHCH_3$ vary considerably in coenzymic activity for DD (Table 2) (Toraya & Abeles, 1980; Toraya, Krodel, et al., 1979).

Table 2 Relative activity and binding affinity of the coenzyme analogs that we studied with DD[a].

Moiety modified	Analog[b]	Relative activity (%)	K_m or K_i (μM)	k_{cat}/k_{inact}[c] $\times 10^{-4}$	Reference
	AdoCbl	(100)	0.80	144	*d*
Adenine	1-deazaAdoCbl	56	2.7	142	*e*
	3-deazaAdoCbl	46	2.5	20	*e*
	3-isoAdoCbl	36	1.4		*d*
	N^6-MeAdoCbl	17	4.2		*f*
	Nebularyl-Cbl	9	1.7		*d*
	7-deazaAdoCbl	Inactive	1.1		*e*
	N^6-EtAdoCbl	Inactive	5.4		*f*
	N^6,N^6-Me$_2$AdoCbl	Inactive	3.5		*e*
	1-MeAdoCbl	Inactive	5.3		*f*
	Cytidyl-Cbl	Inactive	6.1		*d*
	BI-ribosyl-Cbl	Inactive	3.3		*d*
	Inosyl-Cbl	Inactive	~22		*d*
	GuoCbl	Inactive	~30		*e*
	1,N^6-ethenoAdoCbl	Inactive	~35		*f*
	3,N^4-ethenocytidyl-Cbl	Inactive	1.4		*f*
β-Ribose	*ara*AdoCbl	61	1.4		*d*
	Aristeromycyl-Cbl	42	1.1		*d*
	2′-dAdoCbl	31	1.3	5.5	*g*
	3′-dAdoCbl	19	2.3	7.4	*g*
	2′,3′-secoAdoCbl	Inactive	4.0		*g*
	2′,3′-secoAdoCbl dialdehyde	Inactive	5.8		*g*
	L-AdoCbl	Inactive	1.2		*d*
	AdoMeCbl	0.27	1.7	0.3	*h*
	AdoEtCbl	Inactive	0.58, 1.5		*a,h*
	AdePeCbl	Inactive	0.27		*d*
	AdeBuCbl	Inactive	0.47		*d*
	AdeEtCbl	Inactive	2.6		*i*

Table 2 Relative activity and binding affinity of the coenzyme analogs that we studied with DD[a].—cont'd

Moiety modified	Analog[b]	Relative activity (%)	K_m or K_i (μM)	k_{cat}/k_{inact}[c] $\times 10^{-4}$	Reference
Peripheral side chain	AdoepiCbl	14	13		*j*
	AdoCbl(*d*-OH)	86	2.3		*k*
	AdoCbl(*b*-OH)	11	4.5	1.28	*k*
	AdoCbl(*e*-OH)	66	8.0		*k*
	AdoCbl(*d*-OCH_3)	14	3.1		*k*
	AdoCbl(*b*-OCH_3)	7	1.9	1.32	*k*
	AdoCbl(*e*-OCH_3)	14			*k*
	AdoCbl(*d*-$NHCH_3$)	16	2.1	5.06	*k*
	AdoCbl(*b*-$NHCH_3$)	41	0.94	10.0	*k*
	AdoCbl(*e*-$NHCH_3$)	43	1.5		*k*
Base	Ado(BI)Cba	95	0.96	75	*l*
	Ado(Im)Cba	59	1.04	8.9	*l*
α-Ribose	AdoCbiP-DBIPr	59	0.82	76	*m*
	AdoCbiP-DBIBu	50	11.2		*m*
	AdoCbiP-DBIHx	Inactive	38		*m*
	AdoCbiP-DBIEt	Inactive	52		*m*
Base, α-ribose	AdoCbiP-BIPr	40	2.7	9.6	*l*
	AdoCbiP-PyPr	25	0.29	1.4	*n*
	AdoCbiP-ImPr	8	0.99	0.33	*m*
	AdoCbl(N^3-Me)	Inactive			*o*
	AdoCbiP-Me	Inactive	2.5	<0.0018	*p*
	AdoCbi	Inactive	20–30		*o*

[a]Determined at 37 °C.

[b]Abbreviations used for analogs: Cbl, cobalamin; epiCbl, 13-epicobalamin; Cbl(*b*-, *d*-, or *e*-OH), cobalamin *b*-, *d*-, or *e*-carboxylate; Cbl(*b*-, *d*-, or *e*-OCH_3), cobalamin *b*-, *d*-, or *e*-methyl ester; Cbl(*b*-, *d*-, or *e*-$NHCH_3$), cobalamin *b*-, *d*-, or *e*-methylamide; Cbl(N^3-Me), cobalamin with N^3-methylated DBI; (BI)Cba, benzimidazolylcobamide; (Im)Cba, imidazolylcobamide; Cbi, cobinamide; CbiP, cobinamide phosphate; Ado, adenosyl; Me, methyl; Et, ethyl; Pr, propyl; Bu, butyl; Hx, hexyl; *ara*Ado, arabinoadenosyl; BI-ribosyl, benzimidazoleribosyl; dAdo, deoxyadenosyl; AdoMe, adenosylmethyl; AdoEt, adenosylethyl; AdePe, adeninylpentyl; AdeBu, adeninylbutyl; AdeEt, adeninylethyl; DBI, 5,6-dimethylbenzimidazolyl; BI, benzimidazolyl; Py, pyridyl; Im, imidazolyl; AdoCbiP-Me, AdoCbi methyl phosphate.

Continued

Table 2 Relative activity and binding affinity of the coenzyme analogs that we studied with DD[a].—cont'd

[c] k_{cat} was obtained from the maximum velocity (V_{max}). k_{inact} was calculated from a change in the slope of a tangent to the time course curves of the reaction which is determined by the alcohol dehydrogenase-NADH coupled method. Alternatively, the time course was traced digitally, and k_{cat} was determined by fitting the data to a straight line ($a + bx$). k_{inact} was determined by fitting the data to a decaying exponential [($y_0 + A\exp(-Bx)$)] using IGOR PRO (WaveMetrics Inc.).
[d] Toraya et al. (1977).
[e] Toraya et al. (1986).
[f] Ushio, Fukui, and Toraya (1984).
[g] Ichikawa and Toraya (1988).
[h] Fukuoka, Nakanishi, Hannak, Kräutler, and Toraya (2005).
[i] Toraya et al. (1987).
[j] Toraya, Shirakashi, Fukui, and Hogenkamp (1975).
[k] Toraya, Krodel, et al. (1979).
[l] Fukuoka et al. (2002).
[m] Toraya and Ishida (1991).
[n] Toraya, Miyoshi, Mori, and Wada (1994).
[o] Ishida et al. (1993).
[p] Ishida and Toraya (1993).

With the *d*-site modifications, relative activity decreases in the order $-COO^- \gg -CONHCH_3$, $-COOCH_3$, suggesting inhibition by steric effects, while at the *b*-site the activity decreases in the order $-CONHCH_3 \gg -COO^- > -COOCH_3$, suggesting the importance of hydrogen-bond donation from coenzyme to apoenzyme. The specific tight interactions of the corrin ring side chains with apoenzyme are likely essential for catalysis by facilitating the Co-C bond cleavage and by contributing to the stabilization of radical intermediates. Cleavage of the Co-C bond of sterically hindered alkylcobalamins bearing neither adenine nor functional groups, such as neopentylcobalamin and isobutylcobalamin, is accelerated by binding to apoenzyme (Toraya & Ishida, 1988).

AdoCbl analogs in which the α-D-ribose of the nucleotide moiety is substituted by trimethylene and tetramethylene groups are partially active coenzymes for DD (Table 2), indicating that neither α-D-ribofuranose ring nor the functional groups on it are essential for catalysis (Toraya & Ishida, 1991). The loop size corresponding to ribose is optimal for catalysis and for tight binding to apoenzyme. Therefore, the D-ribose moiety is important as a spacer to keep the DBI base in the proper position. The coenzyme activity of the trimethylene analogs in which DBI is replaced by artificial bases increases with bulkiness of the base (Fukuoka et al., 2002; Toraya & Ishida, 1991; Toraya et al., 1994), suggesting that a bulky base or longer Co-C bond facilitates the radical formation and stabilization. The fact that imidazolylcobamide coenzyme (Ado(Im)Cba) is an active coenzyme but undergoes inactivation with high probability also supports this idea. The *Co*-coordinating compact loop of nucleotide moiety (Ishida et al., 1993) and the phosphate group (Ishida & Toraya, 1993) are indispensable for tight binding to apoenzyme.

6.3 Plausible mechanism of the Co-C bond activation

AdoCbl is a cofactor that preserves a highly-reactive primary-carbon radical in masked form in living organisms. The apoenzyme controls the reactivity of the coenzyme by activating its organometallic bond only when necessary. Since the Co-C bond is sufficiently weak (homolytic bond dissociation energy, ~30 kcal/mol) (Halpern, Kim, & Leung, 1984; Hay & Finke, 1986), only a modest labilization by interaction with apoprotein is required for its homolysis. The rate acceleration of its Co-C bond cleavage by DD and EAL in the presence of substrate is estimated to be $\sim 10^{12\pm1}$-fold (Bandarian & Reed, 2000: Hay & Finke, 1987; Valinsky et al., 1974). This acceleration corresponding to $\Delta\Delta G^{\ddagger}$ of 16–19 kcal/mol at 37 °C could be achieved by utilizing the coenzyme- and substrate-binding energies.

The structure-function studies suggest a model of Co-C bond activation—that is, AdoCbl would be fixed at least two sites, i.e., the cobalamin-binding site and the adenine-binding site, and a change of their relative positions brings about the coenzyme Co-C bond activation. The cobalamin moiety is tightly bound to apoenzyme by specific interactions through peripheral amide side chains of the corrin ring (Toraya, Krodel, et al., 1979) and the phosphate group of the nucleotide moiety (Ishida & Toraya, 1993) (for structural evidence for cobalamin-binding site, see Section 7.2). The cobalamin binding to this site, of itself, labilizes the Co-C bond of the coenzyme to some extent (Toraya & Ishida, 1988).

The presence of adenosyl group-binding site in DD was biochemically demonstrated. When apoenzyme was incubated first with [U-^{14}C] 5'-deoxyadenosine and then with cob(II)alamin, followed by gel filtration on Bio-Gel P-6, irreversible 1:1:1 ternary complex was formed between enzyme, cob(II)alamin, and the nucleoside (Toraya, 1985). If the order of addition was reversed, no 5′-deoxyadenosine was bound to the enzyme·cob(II)alamin complex. The adenosyl group-binding site in DD is thus spatially limited and highly specific for adenine nucleosides. From the K_D value for 5′-deoxyadenosine, ΔG° upon binding of the coenzyme adenosyl group to DD at 37 °C was estimated to be approximately −6.6 kcal/mol (Toraya et al., 1986). Later, the presence of adenine-binding pockets in DD, GD, and EAL has been shown by X-ray analyses. Strong attraction of the coenzyme's adenine moiety of the adenosyl group to this site could produce a kind of angular strain or tensile force that leads to the labilization of the Co-C bond (Toraya et al., 1987).

Substrate binding assists or induces the Co-C bond homolysis. The bond becomes activated upon coenzyme binding to apoenzyme, but in the absence of substrate only a small fraction of the coenzyme is in the dissociated form. Upon the addition of substrate, most of the coenzyme undergoes Co-C bond homolysis (Toraya, Krodel, et al., 1979) at a kinetically competent rate (Valinsky et al., 1974). Substrate-induced conformational changes of the enzyme or substrate binding itself brings about further strain which shifts the equilibrium so that a major fraction is in the dissociated form.

6.4 How radicals are prevented from side reactions: Suicide coenzymes and a pseudocoenzyme

The time courses of DD reaction with AdoCbl and its DBIyl trimethylene analog (AdoCbiP-DBIPr) as coenzymes are essentially linear for at least 20 min (k_{cat}/k_{inact}, ~1,440,000 and 700,000, respectively) (Toraya & Ishida, 1991), whereas the reactions with "suicide coenzymes," such as imidazolyl and pyridyl counterparts, are accompanied by inactivation during catalysis with high probabilities (k_{cat}/k_{inact}, ~3,300 and 14,000, respectively) (Table 2) (Fukuoka et al., 2002; Toraya & Ishida, 1991; Toraya et al., 1994). Adenosylcobinamide methyl phosphate (AdoCbiP-Me) is a "pseudocoenzyme" and causes rapid irreversible inactivation in the presence of substrate, although it is inactive as coenzyme (k_{cat}/k_{inact}, ~0) (Ishida & Toraya, 1993). Upon incubation of AdoCbiP-ImPr or the pyridyl trimethylene analog (AdoCbiP-PyPr) with apoenzyme in the presence of substrate, EPR spectra showed that cob(II)alamin-like species accumulates as inactivation proceeds, but organic radicals quickly disappear (Yamanishi, Yamada, Ishida, et al., 1998). When AdoCbiP-Me was incubated with apoenzyme in the presence of substrate, a cob(II)inamide-like signal was observed as well, but no signals of organic radicals were seen (Yamanishi, Yamada, Ishida, et al., 1998). These results indicated that the quenching of organic radical intermediates results in inactivation during catalysis. In the inactivation by AdoCbiP-ImPr, AdoCbiP-Me, and AdoCbl *b*-monocarboxylate analog (AdoCbl(*b*-OH)) or in the inactivation of mutant enzymes with AdoCbl, the only product derived from the adenosyl group were identified as 5′-deoxyadenosine by thin-layer chromatography (TLC) with a labeled coenzyme (Protocols 9.12 and 9.13)

(Ishida & Toraya, 1993; Toraya & Abeles, 1980; Toraya & Ishida, 1991) or by HPLC with unlabeled coenzyme (Protocol 9.14) (Kinoshita et al., 2008; Ogura, Kunita, Mori, Tobimatsu, & Toraya, 2008). Inactivated complexes can be resolved by an acid $(NH_4)_2SO_4$ treatment into the intact apoenzyme and aquacobalamin (aqCbl) (Fig. 1B(3)), indicating that the apoenzyme itself is not damaged in the inactivation processes. Inactivation by these analogs takes place only in the presence of substrate and is thus considered to be mechanism-based—that is, at least one of the radical intermediates common to the normal catalytic cycle is unstable and quenched by undesirable side reactions, which makes it impossible to regenerate the coenzyme and thus leads to enzyme inactivation by tightly bound damaged cofactor. The *Co*-coordinating bulky base or longer Co-C bond may be important for preventing radical intermediates from side reactions.

7. X-ray structure analysis

7.1 Crystallization

Since the wild-type DD is a low-solubility enzyme, recombinant DD was first crystalized in the presence of detergent lauryldimethylamine oxide (LDAO) as a complex with cyanocobalamin (CN-Cbl, Fig. 1B(2)) with (Masuda, Shibata, Morimoto, Toraya, & Yasuoka, 2000; Shibata et al., 1999) (Protocol 9.15) and without (Shibata, Masuda, Morimoto, Yasuoka, & Toraya, 2002) substrate 1,2-propanediol. Crystals were grown by sandwich-drop vapor diffusion method at 4 °C against a reservoir containing 15%(w/v) polyethylene glycol (PEG) 6000, 0.23 M $(NH_4)_2SO_4$, 0.02 M Tris·HCl buffer (pH 8.0), and 0.2% LDAO. Crystals of (*R*)- or (*S*)-1,2-propanediol-bound (Shibata et al., 2003) and glycerol-bound (Yamanishi et al., 2012) DD·CN-Cbl complexes were obtained by essentially the same method. Crystals of the complex of DD with the inactive coenzyme analog adeninylpentylcobalamin (AdePeCbl) (Fig. 1B (4), $n = 5$) were also grown under the same conditions in the presence of both AdePeCbl and substrate (Masuda et al., 2000). For the crystallization of the complex of DD with AdoCbl itself, a soluble recombinant DD consisting of the full-length α subunit and the β and γ subunits lacking the residues Gluβ2-Aspβ45 and Asnγ2-Serγ37, respectively, was purified and dialyzed against oxygen-free buffer in an anaerobic chamber because of the sensitivity of holoenzyme to O_2 (Shibata et al., 2018) (Protocol 9.16). After the decomposition of contaminating substrate by the addition of a trace amount of coenzyme, holoenzyme was formed by incubating the substrate-free apoenzyme with AdoCbl and crystallized under anaerobic conditions. The best crystals were obtained with 10.7% (w/v) PEG 3000, 8.6% (w/v) PEG 2000 monomethyl ether, and 0.05 M Tris·HCl buffer (pH 7.6) at 20 °C. For crystals in the presence of substrate, crystals were soaked in the mother liquid containing 5% (v/v) 1,2-propanediol for 6 h.

Recombinant GD is converted to the enzyme·CN-Cbl·1,2-propanediol complex, and the complex was crystallized by the same methods as those for wild-type DD, except that the detergent LDAO was not included (Yamanishi et al., 2002).

Since EAL is also a low-solubility enzyme, a soluble recombinant EAL consisting of the full-length α subunit and the β subunits lacking the residues Lysβ4-Cysβ43 was purified, and the EAL·CN-Cbl· ethanolamine complex was crystallized by the sitting-drop vapor diffusion method (Shibata et al., 2010). The optimized well solution was 6.0–7.0% (w/v) PEG 6000, 24–26% (v/v) glycerol, 5.0% (v/v) 2-propanol, and 0.1 M HEPES-NaOH (pH 7.0). For the substrate-free form and the EAL·AdePeCbl·ethanolamine complex (Protocol 9.17) crystals, the optimized well solution was 6.0–7.0% (w/v) PEG 4000, 24–26% (v/v) glycerol, 1.0% (v/v) 2-methyl-2, 4-pentanediol, and 0.1 M imidazole·HCl (pH 6.3) (Shibata et al., 2010). To obtain the crystals of the EAL·CN-Cbl·2-amino-1-propanol complex, substrate-free crystals were harvested and soaked for 30 min in the mother liquid supplemented with 10 mM racemic 2-amino-1-propanol (pH 8.0) as well as each component of the substrate-free final buffer. The ternary complexes with each enantiomer were obtained by soaking for 30 min in the mother liquid supplemented with 10 mM (*R*)- or (*S*)-2-amino-1-propanol (pH 8.0) as well as each component of the substrate-free final buffer (Shibata, Higuchi, & Toraya, 2011). For crystallization of the complex of EAL with AdoCbl itself, we re-screened new crystallization conditions (Protocol 9.18) (Shibata et al., 2018). Diffractive crystals were obtained from 12.5% (w/v) PEG 1500 and 0.05 M sodium malonate/imidazole/borate buffer (pH 9.0) and from 10% (w/v) PEG 6000, 0.1 M NH_4Cl, and 0.05 M Tris·HCl buffer (pH 8.0) at 4 °C. For crystals in the presence of substrate, crystals were soaked in the mother liquid containing 0.1 M 2-amino-1-propanol (pH 8.0) for 10 min.

Methods of diffraction data collection (Protocol 9.19) and structure determination (Protocol 9.20) are described in Section 9.

7.2 X-ray structures and structure-based mechanisms

The X-ray structures of DD (Masuda et al., 2000; Shibata et al., 1999), GD (Yamanishi et al., 2002), and EAL (Shibata et al., 2010) were first solved as complexes with the inactive coenzyme analog CN-Cbl, because their complexes with AdoCbl are rapidly inactivated in the absence of substrate. The structures of DD and GD are similar, and their overall structures are dimers of heterotrimers, i.e., $(\alpha\beta\gamma)_2$ (Fig. 4A). In contrast, EAL exists as a trimer of the $(\alpha\beta)_2$ dimer, namely $[(\alpha\beta)_2]_3$ (Fig. 4D). In all three enzymes, cobalamin is bound between the α and β subunits in the base-on mode, i.e., with DBI coordinated to the cobalt atom (Fig. 4B and E), as predicted by EPR spectroscopy, and this is the first crystallographic indication of the base-on mode of cobalamin binding. The active sites (Fig. 4C and F) are in the C-terminal side of the $(\beta/\alpha)_8$ barrel, so-called *t*riose phosphate *i*so*m*erase (TIM) barrel, of α subunits, and the β subunits cover the lower part of the cobalamin. A salt bridge is formed between the ε-NH_2 group of Lysβ135 and the phosphate group of cobalamin. Such architectures explain that spatial isolation prevents reactive radical intermediates from side reactions with O_2 or other molecules. The substrate-coordinated metal ion observed in the X-ray structure of DD·CN-Cbl complex was reassigned as calcium (Fig. 4B and C) by our biochemical (Toraya et al., 2010) and theoretical (Kamachi, Takahata, Toraya, & Yoshizawa, 2009) investigations, whereas another metal ion near the adenine ring in the X-ray structure of DD·AdePeCbl complex was assigned as K^+ (Masuda et al., 2000) (Fig. 4B). Ca^{2+} would assist the OH group migration through its stronger Lewis acidity (Kamachi, Doitomi, Takahata, Toraya, & Yoshizawa, 2011). An unusually long Co-base bond seems to favor homolysis of the Co-C bond and thus favor radical catalysis.

The X-ray structures of the complexes of DD and EAL with the adenine-containing inactive analog AdePeCbl revealed the enzyme-coenzyme interactions at the adenine moiety (Fig. 4B and E) (Masuda et al., 2000; Shibata et al., 2010). The adenine moiety is anchored in the adenine-binding pocket in parallel to the pyrrole ring C of the corrin ring. In DD, for example, it is accommodated in the pocket by hydrogen-bonding of N3 with the side-chain OH group of Serα224 and of 6-NH_2, N1, and N7 with main-chain amide groups of nearby amino acid residues and a water molecule. The presence of such pockets is as expected (see Section 6.3) and provide a molecular basis of the enzyme's strict specificity for the adenosyl group of the coenzyme. Modeling studies of DD and EAL imply that tight enzyme-coenzyme interactions at both cobalamin moiety and the adenine ring of adenosyl group induce marked angular strains (C5′-Co-N22 and Co-C5′-C4′ bond angles) and tensile force (Co-C5′ bond length) which inevitably lead to the fission of coenzyme's Co-C bond (steric strain model of the Co-C bond cleavage) (Fig. 4G, (1)→(2)). The adenosyl radical in the "catalytic position" is modeled by

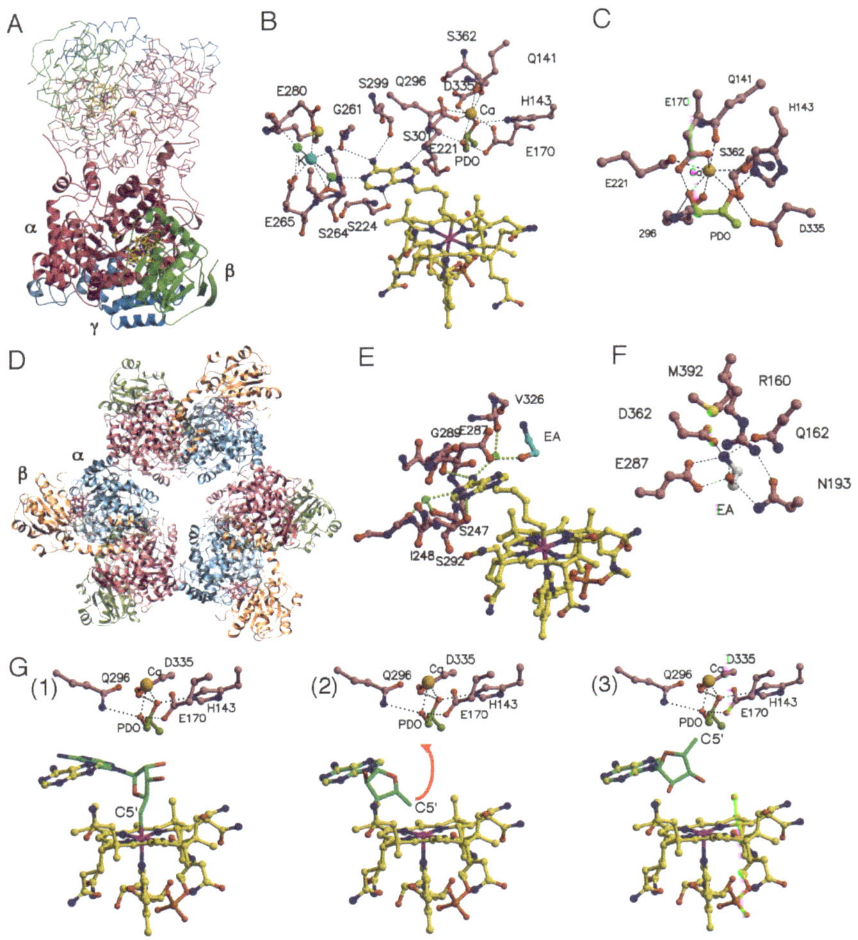

Fig. 4 X-ray structures of DD and EAL. (A) Overall structure $(\alpha\beta\gamma)_2$ of DD. (B) Adenine-binding pocket of DD. (C) Active-site residues of DD interacting with the substrate 1,2-propanediol (PDO) and Ca^{2+}. (D) Overall structure $[(\alpha\beta)_2]_3$ of EAL. (E) Adenine-binding pocket of EAL. (F) Active-site residues of EAL interacting with the substrate ethanolamine (EA). (G) Steric strain model of the Co-C bond cleavage and ribosyl rotation model of the radical propagation in DD. In this modeling study, superimposition of AdoCbl over the DD-bound AdePeCbl at both the cobalamin moiety and the adenine ring induces marked steric strains ((1)→(2)), and subsequent ribosyl rotation (*red arrow*) around the glycosidic linkage so that the C5′ is closest to C1 of PDO explains how the radical center accesses substrates ((2)→(3)). Drawn with Molscript (Kraulis, 1991)/Raster3D (Merritt & Bacon, 1997) (A–C and E–G) and with Chimera (Pettersen et al., 2004) (D).

superimposing its adenine ring on the adenine-binding pocket, followed by ribosyl rotation around the *N*-glycosidic bond so that C5′ is accessible to substrate/product (ribosyl rotation model of the radical propagation) (Fig. 4G, (2)→(3)). Similar models can be built with EAL as well.

The mechanisms of substrate-induced Co-C bond homolysis in DD and EAL were also solved from the X-ray structures of substrate-free forms of DD·CN-Cbl and EAL·CN-Cbl complexes (Shibata et al., 2002, 2010). In DD, as compared with the structure of its substrate-bound form, the β subunit is tilted by ~3° with respect to the α subunit in the substrate-free form, and cobalamin is also tilted so that pyrrole rings A and D are significantly lifted up toward the substrate-binding site. This substrate-induced conformational change increases the strain of the Co-C bond and thus contributes to its additional labilization and cleavage. In EAL, such subunit tilting is not observed, but Gluα287 (Fig. 4F) is fixed at the substrate-binding site when substrate binds and has a direct contact with the 2′-OH group of the ribose moiety of modeled AdoCbl, which may contribute to the substrate-induced Co-C bond dissociation through the stabilization of post-homolysis state.

Enzymes are generally stereospecific for chiral substrates, but DD and EAL are exceptional and act on both enantiomers of 1,2-propanediol and 2-amino-1-propanol, respectively (Diziol, Haas, Rétey, Graves, & Babior, 1980; Rétey, Umani-Ronchi, & Arigoni, 1966; Rétey, Umani-Ronchi, Seibl, & Arigoni, 1966; Zagalak, Frey, Karabatsos, & Abeles, 1966). To rationalize their apparent lack of stereospecificity and to establish the enantiomer-specific stereochemical courses of the reactions, we analyzed the X-ray structures of each enantiomer-bound forms of DD·CN-Cbl and EAL·CN-Cbl complexes. In DD, (*R*)- and (*S*)-enantiomers are bound in a symmetrical manner, with respect to the plane including Ca^{2+} and O1 and O2 atoms of substrate without changing the hydrogen-bonding interactions with active-site residues (Fig. 5B) (Shibata et al., 2003). C5′ of the adenosyl radical in the catalytic conformation lies on this symmetrical plane and abstracts the nearest hydrogen atoms, i.e., the *pro-R* and *pro-S* hydrogens from (*R*)- and (*S*)-enantiomers, respectively (Fig. 5A). The OH group of the substrate radical migrates from C2 to C1 by an energetically favorable suprafacial shift, resulting in the inversion at C1, with clockwise and counterclockwise rotations of the C1-C2 bond of (*R*)- and (*S*)-enantiomers, respectively, if viewed from Ca^{2+}. A modeling study of 1,1-diol intermediates indicated that the new radical center at C2 becomes close to the CH_3 group of 5′-deoxyadenosine and back-abstracts a hydrogen atom from it with the inversion at C2. Stereospecific dehydration of the 1,1-diols is rationalized by assuming that Aspα335 and Gluα170

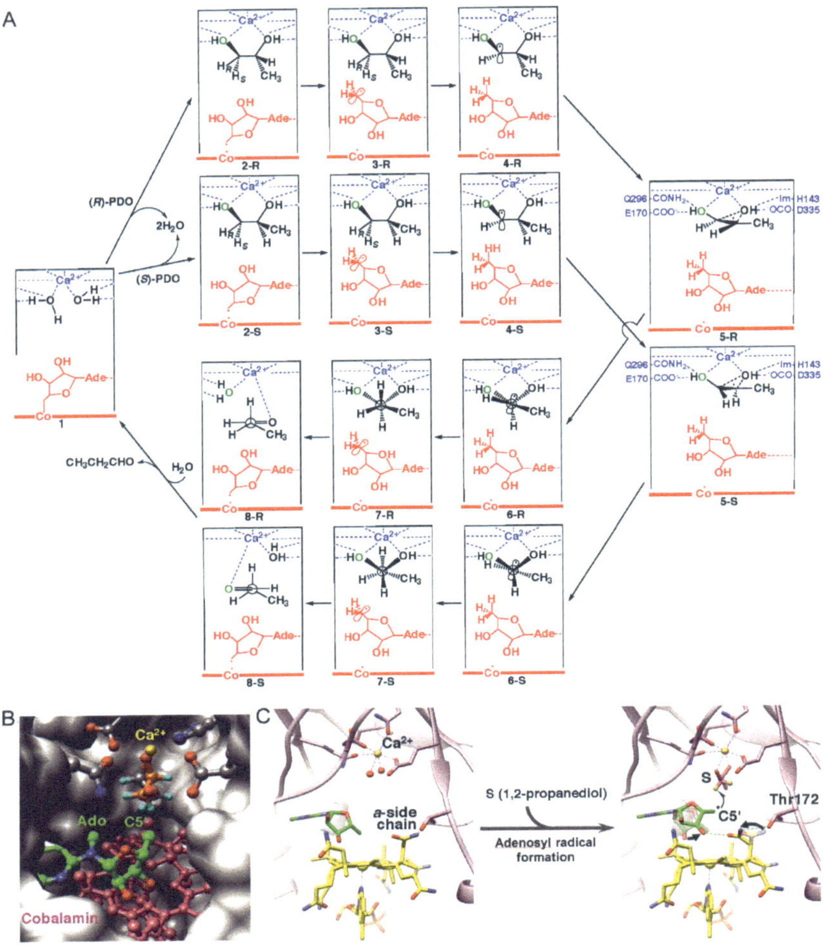

Fig. 5 (A) Overall refined mechanism and stereochemical courses of the DD reaction with enantiomeric substrates. PDO, 1,2-propanediol; Im, imidazole group. (B) (*R*)- (*orange*) and (*S*)- (*light gray*) enantiomers bind to DD in a symmetrical manner, with respect to the plane including Ca^{2+} (*yellow sphere*) and O1 and O2 atoms of substrate. C5′ of adenosyl radical (*green*) in the catalytic conformation comes on this symmetrical plane and abstracts the nearest hydrogen atoms (*sky blue*). (C) Plausible model for the synchronized conformational changes of peripheral *a*-side chain of the corrin ring with the radical shuttling motion of adenosyl radical between original and catalytic positions. In the absence of substrate, the *a*-side chain is away from the ribosyl moiety of adenosyl radical, but in its presence, it takes an inward conformation by hydrogen bonding with Thrα172 and contributes to the formation and positional stabilization of adenosyl radical by hydrogen bonding with its 3′-OH group. This side chain may serve as a gate for opening and closing the channel for a plausible substrate/product pathway. Drawn with Chimera (Pettersen et al., 2004) (B and C).

function as bases in the dehydration of (*R*)- and (*S*)-enantiomers, respectively. The structure-based refined mechanism and stereochemical courses of DD reaction are depicted in Fig. 5A.

In EAL, the adenosyl radical abstracts the *pro-S* hydrogen atom on C1, i.e., the nearest one, from both enantiomeric substrates (Shibata et al., 2011) (Fig. 6A). The NH_2 group migrates from C2 to C1 by a suprafacial shift with inversion at C1 for both enantiomeric substrate radicals, although the absolute configurations of 1-amino-1-propanols are not yet known. (*R*)-2-Amino-1-propanol is deaminated with the inversion at C2, whereas the rotameric intermediate from the (*S*)-enantiomer would undergo flipping to the rotamer from the (*R*)-enantiomer and then back-abstract a hydrogen atom, resulting in the retention at C2. This preference of the active site for the rotamer from the (*R*)-enantiomer may be explained by the steric repulsion of the (*S*)-enantiomer-derived product radical at C3 with the Pheα329 and Leuα402 residues. The structure-based refined mechanism and stereochemical courses of EAL reaction are illustrated in Fig. 6A. It was thus concluded that DD and EAL are not non-stereospecific, but rather that each enantiomer actually reacts stereospecifically.

For information regarding the enzyme-coenzyme interaction at the β-ribose moiety or its vicinity, we then analyzed the X-ray structures of DD and EAL with AdoCbl itself in the presence and absence of substrates (Shibata et al., 2018). In the presence of substrates, the structures of both enzymes revealed that the peripheral *a*-acetamide side chain of the corrin ring directly interacts with the 3′-OH group of the adenosyl group to maintain the group in the catalytic position, and that the direction of this side chain itself is determined by hydrogen bonding with Thrα172 in DD and Proα194 in EAL (Figs. 5C and 6C). In the absence of substrate, the structure of resting holoenzyme has been obtained for the first time with EAL in which the adenosyl group is in the original position (Fig. 6B). In contrast, the structure of holoDD without substrate was the same as that with substrate, suggesting that holoDD without substrate has undergone the Co-C bond homolysis probably by reaction with the contaminating substrate. Thus, the *a*-side chain likely swings between the original and catalytic positions in a synchronized manner with the radical shuttling between the coenzyme and substrate/product—that is, the *a*-side chain controls the radical reactions by promoting the formation and positionally stabilizing the adenosyl radical (Figs. 5C and 6C). This side chain may also serve as a gate for opening and closing the channel for a plausible substrate/product pathway.

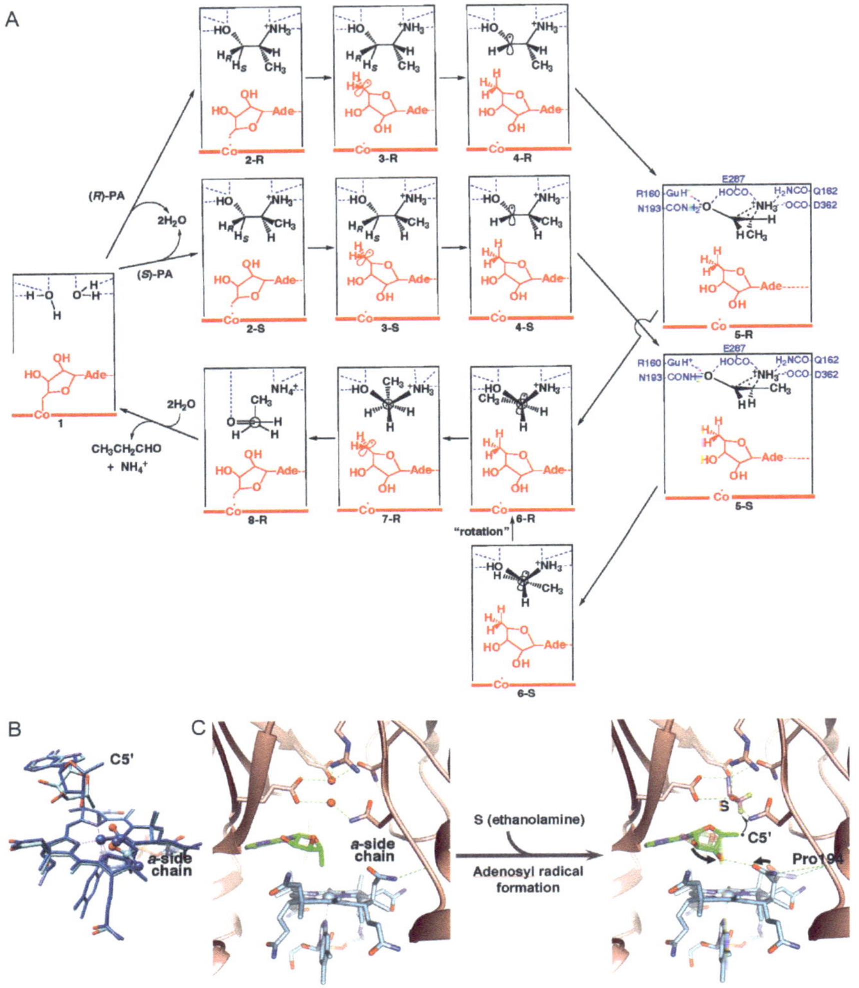

Fig. 6 (A) Overall refined mechanism and stereochemical courses of the EAL reaction with enantiomeric substrates. PA, 2-amino-1-propanol; GuH^+, guanidinium group. (B) Conformational changes of adenosyl group and *a*-side chain depending on the presence and absence of substrate. The AdoCbl molecules of EAL·AdoCbl complexes with (*blue*) and without (*sky blue*) 2-amino-1-propanol are superimposed on their corrin rings. (C) The model for the synchronized conformational changes of *a*-side chain and adenosyl radical between the original and catalytic positions in EAL. The same model as Fig. 5C is applicable to Fig. 6C, except that Proα194 in EAL plays the same role as Thrα172 in DD. Drawn with Chimera (Pettersen et al., 2004) (B and C).

8. Functional analysis of amino acid residues

8.1 Chemical modification studies

We demonstrated with SH-modifying reagents that DD has two reactive cysteine residues - one is essential for catalytic activity and located at or in close proximity to the coenzyme binding site, while the other not essential (Kuno, Toraya, & Fukui, 1981a). Modification with 2,3-butanedione and phenylglyoxal suggested an essential guanidinium group of arginine is involved in the coenzyme binding (Kuno, Toraya, & Fukui, 1980). Pyridoxal 5′-phosphate modified one essential lysine residue that is located in component F and involved in the subunit interaction (Kuno, Toraya, & Fukui, 1981b). Dye-sensitized photooxidation and ethoxycarbonylation showed that imidazole group(s) located in component F are essential for subunit association (Kuno, Fukui, & Toraya, 1990). However, functional analyses by chemical modifications are generally time-consuming and applicable only to a limited number of amino acid residues.

8.2 Mutational studies of active-site residues: Establishment of refined mechanisms

Site-directed mutagenesis offers a powerful tool to investigate the functions of active-site residues. Desired mutations are introduced to enzymes using a QuikChange™ Site-Directed Mutagenesis kit (Stratagene, CA) (Protocol 9.21). The importance of a specific active-site residue in DD was first estimated by so-called alanine scanning, i.e., mutant enzymes in which one of the residues interacting with substrate and/or Ca^{2+} (Fig. 4C) is mutated to Ala are prepared, and their activity determined (Kawata et al., 2006). The Eα170A and Dα335A mutants were totally inactive, and the Hα143A mutant showed only a trace of activity, indicating that Gluα170, Aspα335, and Hisα143 are catalytic residues. Qα141A, Qα296A, and Sα362A showed partial activity. In order to analyze the functions of active-site residues, other mutant DDs (Hα143L, Hα143Q, Hα143E, Hα143K, Eα170D, Eα170H, Eα170Q, and Dα335N) were also prepared, and their catalytic properties investigated in details (Kawata et al., 2006; Kinoshita et al., 2008). From the results obtained, the roles of active-site residues can be summarized as follows: (1) Glnα296 is important for substrate binding and Glnα296 and Glnα141 for preventing the enzyme from mechanism-based inactivation. (2) The presence and the position of $-COO^-$ group in the residue α170 are essential for catalysis as well as for the continuous progress of catalytic cycles.

(3) The roles of Gluα170 and Aspα335 are to participate in the binding of substrate and intermediates and keep them properly oriented and to function as a base in the dehydration of the 1,1-diol intermediate. (4) In addition, Gluα170 seems to stabilize the transition state for the OH group migration from C2 to C1 through temporary deprotonation of the spectator OH group on C1. (5) The hydrogen-bonding interaction between the migrating OH group on C2 of substrate and the side chain of Hisα143 is important not only for catalysis but also for protecting radical intermediates from side reactions.

The deuterium kinetic isotope effects for overall reaction indicated a hydrogen abstraction step is partially rate-determining with the Hα143A mutant, but fully rate-determining with the wild-type and Hα143Q mutant. The inactivation of Hα143A during catalysis was mechanism-based: cob(II) alamin without an organic-radical coupling-partner accumulated, 5′-deoxyadenosine was formed from the adenosyl group, and the apoenzyme itself was not damaged. In addition, a computational study suggested that substrate-coordinated Ca^{2+} lowers the activation barrier for the OH group migration by serving as a Lewis acid (Kamachi et al., 2011). By taking the roles of active-site residues and metal ions into consideration, a refined overall mechanism depicted in Fig. 5A for DD has been established. The validity of the radical DD reaction along this pathway has been verified by theoretical computations (Doitomi, Kamachi, Toraya, & Yoshizawa, 2016; Kamachi et al., 2011; Kamachi, Toraya, & Yoshizawa, 2004, 2007).

In EAL, alanine scanning showed that Rα160A, Qα162A, Nα193A, Eα287A, and Dα362A are essentially inactive, indicating that Argα160, Glnα162, Asnα193, Gluα287, and Aspα362 (Fig. 4F) are all important for catalysis and for the high-affinity binding of substrates (Mori et al., 2014). To analyze the roles of the active-site residues, other mutant EALs (Rα160K, Qα162E, Qα162H, Qα162K, Nα193D, Eα287D, Eα287H, Eα287Q, Dα362E, and Dα362N) were also prepared, and their catalytic properties investigated in details. From the results obtained, the roles of active-site residues can be summarized as follows: (1) The $-COO^-$ group at the residue α287 is absolutely required for activity and the Co-C bond cleavage. There is a spatially optimal position for it, suggesting that Gluα287 contributes to the Co-C bond homolysis, stabilizes the transition state for NH_2 group migration from C2 to C1 through partial deprotonation of the spectator OH group, and functions as a base in the elimination of ammonia. (2) A positive charge and/or the hydrogen bond at the residue α160 and the hydrogen bonds at the residues α162 and α193 with substrate are important for catalysis and for preventing radical intermediate from side reactions. (3) Argα160 would stabilize the trigonal transition state in the NH_2 group migration by electrostatic catalysis and hydrogen bonding with spectator OH group. (4) Asnα193 contributes to keep an appropriate position and direction of the guanidinium group of Argα160 as well. (5) Hydrogen-bond acceptors are necessary at the residue α162, but hydrogen-bond donors are rather harmful. Glnα162 might stabilize the trigonal transition state by accepting a hydrogen bond from migrating NH_3^+. (6) The activity is sensitive to the position of $-COO^-$ group at the residue α362. Aspα362 would assist the Co-C bond homolysis indirectly and stabilize the trigonal transition state by accepting a hydrogen bond from migrating NH_3^+ and electrostatic interaction. By taking the roles of active-site residues into consideration, a refined overall mechanism illustrated in Fig. 6A for EAL has been established.

8.3 Mutational studies of AdoCbl-binding residues

Since the presence of adenine-binding pocket and cobalamin-binding site was structurally confirmed, the importance of adenine anchoring and ion pairing were assessed with DD by mutating the Serα224 and Lysβ135 residues. Sα224A, Sα224N, and Sα224C mutants were 3–19% as active as that of the wild-type with k_{cat}/k_{inact} less than 2% that of the wild-type (Ogura et al., 2008; Shibata et al., 2018). Sα224N and Sα224A holoenzymes showed low susceptibility to O_2 in the absence of substrate. Thus, it was concluded that Serα224 is not absolutely required for catalysis but important for the Co-C bond activation and preventing the enzyme from inactivation. The inactivation of Sα224A during catalysis was shown to be mechanism-based. In contrast, Kβ135A, Kβ135Q, and Kβ135R retained 58–76% of activity with k_{cat}/k_{inact} 25–75% that of the wild-type (Ogura et al., 2008). It was shown that the presence of positive charge at the residue β135 increases the affinity for cobalamins, whereas the introduction of a negative charge there (Kβ135E) markedly decreases the activity (2%) and increases the probability of inactivation (k_{cat}/k_{inact}, 7.2%), and lowers the affinity for AdoCbl or CN-Cbl.

The X-ray structures of DD and EAL revealed that the adenosyl group is maintained in the catalytic position by hydrogen bonds between 3′-OH of its β-ribose moiety and the peripheral *a*-acetamide side chain of the corrin ring and between this side chain and Thrα172 in DD and Proα194 in EAL.

The Tα172A mutant of DD was essentially inactive with k_{cat}/k_{inact} of 0.4% that of the wild-type enzyme, while Tα172S showed 10% activity with k_{cat}/k_{inact} of 5.1% (Shibata et al., 2018). Thus, this mutational study also confirmed that the *a*-side chain of the corrin ring precisely controls the radical reactions by promoting the formation and positionally stabilizing the adenosyl group through the interaction with Thrα172 in DD.

8.4 Rational redesign of diol dehydratase for glycerol-inactivation resistance and wide specificity

For industrial use of DD, we attempted to redesign it to be resistant to mechanism-based inactivation by glycerol and to widen substrate specificity (Yamanishi et al., 2012). The X-ray structure of glycerol-bound DD·CN-Cbl complex indicated that glycerol is bound to the active site preferentially in the same conformation as that of (*S*)-1,2-propanediol (Fig. 4B and C), i.e., the G_s conformation, in which the *pro-S*-CH_2OH group is oriented to the hydrogen-abstracting site and the 3′-OH group is hydrogen-bonded to Serα301, but not to nearby Glnα336. Since it was reported that glycerol bound in this conformation primarily contributes to the inactivation reaction (Bachovchin et al., 1977), we prepared Sα301A, Qα336A, and Sα301A/Qα336A mutants. k_{cat}/k_{inact} showed that Sα301A and Qα336A are substantially more resistant to glycerol inactivation than the wild-type enzyme, and the degree of preference for (*S*)-1,2-propanediol decreased on these mutations (Yamanishi et al., 2012). It is intriguing that the substrate specificities toward longer-chain 1,2-diols significantly increased on the Sα301A/Qα336A double mutation, probably because these amino-acid substitutions yielded more space for accommodating a longer alkyl group on C3.

9. Protocol

9.1. Activity staining for DD and GD

1.1. Carry out PAGE of cell-free extract under nondenaturing conditions, as described by Davis (1964) in the presence of 0.1 M 1,2-propanediol.

1.2. After PAGE, immerse the gel in a solution consisting of 0.035 M potassium phosphate buffer (pH 8.0), 0.1 M 1,2-propanediol, 0.05 M KCl, and 15 μM AdoCbl, and incubate it at 37 °C for 10–15 min in the dark.

1.3. Transfer the gel immediately to a solution of 0.4% (w/v) DNPH in 2 M HCl, and allow it to stand for 15 min.

1.4. Remove excess DNPH by immersing the gel in 2 M HCl solution with several changes. The band of DD is located by the precipitate of 2,4-dinitrophenylhydrazone of propionaldehyde against a black background.

Note: Since AdoCbl is light-sensitive, all the operations should be done in dim light. This is the case for all the experiments in which organocobalamins are used, even though not indicated so.

9.2. Purification of recombinant wild-type DD

2.1. Grow *E. coli* JM 109 carrying pUSI2E(DD) aerobically at 37 °C in LB medium containing 0.1% 1,2-propanediol and 50 μg/mL ampicillin, and add 1 mM IPTG for induction at an OD_{600} of ~0.9. After cultivation for further 6 h, harvest cells, and wash them twice with 0.05 M potassium phosphate buffer (pH 8.0) (Buffer A) containing 2% 1,2-propanediol.

2.2. All the operations are carried out at 0–4 °C. Suspend cells (4.5 g, wet weight) in 16 mL of Buffer A containing 2% 1,2-propanediol, 2 mM EDTA, 2 mM phenylmethanesulfonyl fluoride (PMSF), and disrupt by sonication for 10 min at the output of 240 W with Kaijo Corp. TA-5287 ultrasonic cell destruction system (Japan). DD activity in the homogenate is accurately assayed after dilution in Brij 35-containing buffer.

2.3. After centrifugation at 17,400 × *g* for 30 min, most of enzyme activity is recovered in the crude membrane fraction. Wash the crude membrane three times with 24 mL of Buffer A containing 2% 1,2-propanediol, 2 mM EDTA, and 1 mM PMSF by repeated sonication and centrifugation at 17,400 × *g* for 30 min.

2.4. Wash the crude membrane further three times with 24 mL of the same buffer supplemented with 0.2% Brij 35, and suspend it in 20 mL of 0.01 M potassium phosphate buffer (pH 8.0) containing 1% Brij 35, 2% 1,2-propanediol, 2 mM EDTA, and 1 mM PMSF. Sonicate the suspension to effect extraction of the enzyme. The extracted enzyme obtained by centrifugation at 100,000 × *g* for 30 min is almost homogeneous, as judged by SDS-PAGE.

2.5. Dilute the extract with 20 mL of ice-cold 1% Brij 35 containing 1 mM PMSF and 2% 1,2-propanediol, and load it onto a DEAE-cellulose column (bed volume 75 mL) which has been equilibrated with 0.01 M potassium phosphate buffer (pH 8.0) containing 2% 1,2-propanediol.

2.6. Wash the column successively with 800 mL of 0.01 M potassium phosphate buffer (pH 8.0) containing 1% Brij 35, 1 mM PMSF, and 2% 1,2-propanediol, with 500 mL of 0.01 M potassium phosphate buffer (pH 8.0) containing 1 mM PMSF and 2% 1,2-propanediol, and with 100 mL of 0.01 M potassium phosphate buffer (pH 8.0) containing 2% 1,2-propanediol.

2.7. Elute the enzyme with about 100 mL of Buffer A containing 0.15 M KCl, 2% 1,2-propanediol, and 0.02 M sucrose monocaprate. The enzyme obtained is homogeneous, and ~fivefold purification is achieved in an overall yield of 65%. Specific activity is about 106 units/mg of protein.

9.3. Purification of recombinant wild-type GD by a conventional method

3.1. Grow *E. coli* JM109 carrying pUSI2E(GD) aerobically at 37 °C in LB medium containing 0.1% 1,2-propanediol and 50 μg/mL ampicillin, and add 1 mM IPTG for induction at an OD_{600} of ~0.9. After cultivation for further 5 h, harvest cells by centrifugation and wash them twice with Buffer A.

3.2. All the operations are carried out at 0–4 °C. Suspend cells in Buffer A containing 2 mM PMSF, disrupt them by sonication for 10 min, and remove cell debris by centrifugation.

3.3. To the supernatant, add finely powdered $(NH_4)_2SO_4$ to a concentration of 50% saturation at 4 °C with stirring. After 60 min, centrifuge the suspension, and re-dissolve the precipitate in Buffer A.

3.4. Load the solution onto Sepharose 6B column which has been equilibrated with Buffer A containing 2% 1,2-propanediol, and develop the column with the same buffer. Combine the enzyme-containing peak fractions, and dialyze them for 12 h against 40 volumes of 1.5 mM potassium phosphate buffer (pH 8.0) containing 0.2% 1,2-propanediol with a buffer exchange.

3.5. Load the sample onto a hydroxyapatite column which has been equilibrated with 2 mM potassium phosphate buffer (pH 8.0) containing 0.2% 1,2-propanediol. After washing the column with 5 mM potassium phosphate buffer (pH 8.0) containing 0.2% 1,2-propanediol, elute the enzyme with 13 mM potassium phosphate buffer (pH 8.0) containing 2% 1,2-propanediol.

3.6. Concentrate the eluate, and load it onto a Sephadex G-200 column which has been equilibrated with 0.02 M potassium phosphate buffer (pH 8.0) containing 2% 1,2-propanediol. Develop the column with the same buffer, and combine the enzyme-containing peak fractions. Specific activity is about 65 units/mg.

9.4. Purification of recombinant wild-type GD by nickel affinity chromatography

4.1. Grow *E. coli* JM109 carrying pUSI2E($\alpha_G\beta_G H_6\gamma_G$) aerobically at 30 °C in Terrific broth containing 0.1% 1,2-propanediol and 50 μg/mL ampicillin, and add 1 mM IPTG for induction at an OD_{600} of ~ 0.9. After cultivation for further 7 h, harvest cells by centrifugation, and wash them twice with Buffer A.

4.2. All the operations are carried out at 0–4 °C. Suspend cells in Buffer A containing 2 mM PMSF, disrupt them by sonication, and remove cell debris by centrifugation.

4.3. Mix the extract containing His_6-tagged GD with an equal volume of Buffer A containing 0.02 M imidazole and 0.6 M KCl, and load it onto a Ni-NTA agarose (Qiagen GmbH, Germany) column which has been equilibrated with Buffer A containing 0.01 M imidazole and 0.3 M KCl.

4.4. After washing the column with Buffer A containing 0.01 M imidazole and 0.3 M KCl, elute the enzyme with Buffer A containing 0.05–0.1 M imidazole and 0.3 M KCl.

4.5. After dialysis against 40 volumes of 0.05 M Tris·HCl buffer (pH 8.0) containing 2% 1,2-propanediol, 0.15 M KCl and 2.5 mM $CaCl_2$, digest His_6-tagged GD with thrombin at 25 °C for 2 h, and run it through the Ni-NTA agarose column to remove the His_6-tag peptide.

4.6. Purify the β subunit from *E. coli* BL21 (DE3) carrying pET19b(H6β_G), as described above for His_6-tagged GD.

4.7. Because a part of the enzyme loses its β subunit, concentrate the enzyme solution, and supply with purified β subunit by incubation at 30 °C for 30 min, followed by Sepharose 6B gel filtration to remove unbound excess β subunit. The enzyme is purified fivefold in a yield of 18%.

9.5. Purification of recombinant wild-type EAL

5.1. Grow *E. coli* JM109 carrying pUSI2ENd(EAL) aerobically at 30 or 37 °C in LB medium containing 50 μg/mL ampicillin. At an OD_{600} of ~0.9, add 1 mM IPTG for induction. After cultivation for further 5–6 h, harvest cells by centrifugation, and wash them twice with Buffer A.

5.2. All the operations are carried out at 0–4 °C. Suspend cells (15.4 g, wet weight) in 31 mL of Buffer A containing 0.01 M ethanolamine, 2 mM EDTA, and 2 mM PMSF, and disrupt them by sonication for 10 min at 240 W after adjusting pH to 8.2.

5.3. After centrifugation at 23,000 × *g* for 30 min, wash the crude membrane twice with 50 mL of the same buffer and three times with 31 mL of Buffer A containing 0.01 M ethanolamine, 1 mM PMSF, and 0.2% Brij 35.

5.4. Extract EAL from the precipitate with 31 mL of the same buffer by increasing the Brij 35 concentration to 1%. Sonicate the suspension and collect the supernatant by centrifugation at 23,000 × *g* for 20 min. Repeat this procedure twice, and centrifuge the combined supernatant at 100,000 × *g* for 20 min. The extracted EAL is almost homogeneous.

5.5. Dilute the enzyme solution into four volumes of ice-cold 0.01 M ethanolamine containing 1 mM PMSF, 1% Brij 35, and 5 mM 2-mercaptoethanol, and apply it to a DEAE-cellulose column (bed volume, 50 mL) which has been equilibrated with 0.01 M potassium phosphate buffer (pH 8.0) containing 0.01 M ethanolamine, 1% Brij 35, and 5 mM 2-mercaptoethanol. Wash the column successively with 250 mL of the same buffer and 250 mL of the same buffer without Brij 35.

5.6. Elute EAL with Buffer A containing 0.08 M KCl, 0.01 M ethanolamine, 1 mM PMSF, 0.02 M sucrose monocaprate, and 5 mM 2-mercaptoethanol, and combine EAL-containing fractions.

9.6. Purification of His_6-tagged EALs by nickel affinity chromatography

6.1. All the operations are carried out at 0–4 °C. Suspend *E. coli* cells (2–3.2 g, wet weight) expressing each His_6-tagged EAL in two volumes of Buffer A containing 0.01 M ethanolamine and 1 mM PMSF, and disrupt them by sonication. After centrifugation at 23,000 × *g* for 30 min, add 1/2 volume of Buffer A containing 0.3 M KCl, 0.01 M ethanolamine, 0.01 M imidazole, and 1 mM PMSF to the supernatant.

6.2. Load the mixture onto a Ni-NTA agarose column (bed volume, 4 mL) which has been equilibrated with the same buffer. Pass the effluent through the column twice.

6.3. After washing the column successively with 40 mL of the same buffer and then 40–80 mL of the same buffer containing 0.02–0.05 M imidazole, elute His_6-tagged EAL with 40–70 mL of the same buffer containing 0.1 M imidazole.

6.4. Combine EAL-containing fractions, dialyze them against Buffer A containing 0.01 M ethanolamine and 5 mM 2-mercaptoethanol, and load them onto a Superose 6 column (10/300GL) attached to an FPLC system (GE Healthcare, UK) that has been equilibrated with the same buffer. Develop the column with the same buffer at a flow rate of 0.4 mL/min, and combine the EAL-containing fractions.

9.7. DD assay by a MBTH method

7.1. (See *Note* to Protocol 9.1) A usual assay mixture contains an appropriate amount of DD, 0.2 M 1,2-propanediol, 0.05 M KCl, 0.035 M potassium phosphate buffer (pH 8.0), and 15 μM AdoCbl, in a total volume of 1.0 mL. Carry out minus enzyme or minus AdoCbl experiment as a control.

7.2. After incubation at 37 °C for 10 min, cool the mixture in an ice-water bath, and terminate the enzyme reaction by adding 1 mL of 0.1 M potassium citrate buffer (pH 3.6). The mixture to which this citrate buffer has been added before incubation is taken as a blank test.

7.3. Add 0.5 mL of 0.1% MBTH·HCl, and incubate the mixture at 37 °C for further 15 min.

7.4. Add 1 mL of water, and measure the absorbance at 305 nm using a blank as a reference. The apparent molar extinction coefficient at 305 nm for the colored product formed from propionaldehyde is $16.4 \times 10^3\ M^{-1}cm^{-1}$. One unit is defined as the amount of enzyme activity catalyzing the formation of 1 μmol of propionaldehyde per min under the standard assay conditions.

Note: 1. For routine assays, keep the amount of enzyme to be assayed between 0.003 and 0.03 unit. In experiments in which more than 0.03 unit of enzyme is used, measure the absorbance after dilution.

2. The amount of propionaldehyde produced under these conditions varies linearly with at least 0.001–0.09 unit of enzyme.

3. GD and EAL can be similarly assayed by the MBTH method.

9.8. DD assay by a coupled assay method

8.1. (See *Note* to Protocol 9.1) A usual assay mixture is composed of an appropriate amount of DD, 0.2 M 1,2-propanediol, 50 μg of yeast alcohol dehydrogenase, 0.2 mM NADH, 0.04 M potassium phosphate buffer (pH 8.0), and 15 μM AdoCbl, in a total volume of 1.0 mL.

8.2. The reaction is carried out at 37 °C in a cuvette with a 1.0-cm light path. Start the reaction by adding coenzyme or DD to the remainder of the mixture which has been preincubated at 37 °C for 5–10 min.

8.3. Decrease in the absorbance of NADH at 340 nm per unit time is a measure of the DD activity. One unit is defined as the amount of enzyme activity catalyzing the formation of 1 μmol of propionaldehyde per min under the standard conditions.

Note: 1. For routine assays, keep the amount of enzyme to be assayed between 0.001 and 0.05 unit.

2. GD and EAL can be similarly assayed by an alcohol dehydrogenase-NADH coupled method.

3. For the assay of glycerol-dehydrating activity, use 100 μg of horse-liver alcohol dehydrogenase or *K. pneumoniae* 1,3-propanediol dehydrogenase (DhaT) instead of yeast enzyme.

9.9. Preparation of calcium-deprived apoDD and its activation by Ca^{2+}

9.1. Incubate substrate-free apoenzyme at 37 °C for 10 min with 0.2 mM EDTA (pH 8.0), followed by ultrafiltration through a YM-10 Diaflo membrane (Millipore). The calcium content of the resulting enzyme decreases to almost zero.

9.2. Calcium-deprived, EDTA-free apoenzyme thus obtained can be reconstituted into calcium-containing enzyme by the preincubation at 30 °C for 10 min with various concentrations of $CaCl_2$ in the absence of substrate.

9.3. Measure DD activity after activation by added Ca^{2+}. The activity is recovered in a concentration-dependent manner (apparent K_m for Ca^{2+}, 0.88 mM).

Note: In the presence of substrate, calcium is not removed at all by the EDTA treatment, followed by ultrafiltration.

9.10. EPR measurements of reacting holoDD

10.1. (See *Note* to Protocol 9.1) Mix 100–150 units (5–6.9 nmol) of substrate-free apoenzyme in 0.6 mL of Buffer A containing

20 mM sucrose monocaprate at 0 °C with 50 nmol of AdoCbl in 0.05 mL in a quartz EPR tube (o.d. 5 mm) stoppered with a rubber septum.

10.2. Replace the air in the tube with argon (purity, 99.999%) by repeated evacuation and flushing with argon three times.

10.3. Form holoenzyme by incubation of the mixture at 25 °C for 3 min, and then cool in an ice-water bath.

10.4. Start the reaction by adding 50–100 mmol 1,2-propanediol in 0.05 mL by a syringe through the septum. After mixing, allow the mixture to stand at 4 °C for 1 min, and then freeze it in an isopentane-bath that has been cooled to its melting point (approximately −160 °C). Keep the freeze-quenched samples in a liquid nitrogen bath until measurements.

10.5. Transfer the sample to the EPR cavity, and cool with a cold nitrogen gas flow controlled by an Eurotherm B-VT 2000 temperature controller.

10.6. Take EPR spectra at −130 °C on a Bruker EPS-380E EPR spectrometer modified with a Gunn diode X-band microwave unit. Other conditions for measurement are: EPR microwave frequency, 9.5 GHz (X-band); modulation amplitude, 0.1–1 mT; modulation frequency, 100 kHz; microwave power, 1–10 mW. Each spectrum is the average of four scans.

10.7. After the first measurement of EPR spectrum, incubate the mixture at 25 °C for 3 min, and freeze it again. Do the second EPR measurement at −130 °C.

10.8. After the second measurement, incubate the mixture at 25 °C for additional 30 min, and freeze it again. Do the third EPR measurement at −130 °C.

Note: At a microwave power of 10 mW, the intensity of the doublet signal due to organic radical is almost saturated, while that of the broad signal due to cob(II)alamin is not.

9.11. Conversion of cobalamin analogs to the coenzyme form

11.1. (See *Note* to Protocol 9.1) Cobalamin analogs in which the corrin ring, its peripheral amide side chain, or the lower nucleotide loop is modified can be converted to their adenosyl forms and examined for coenzymic activity. Reduce 7–10 μmol of the cyano form of such analog in 3 mL of water with 25 mg of $NaBH_4$ or with 0.3 g of zinc powder in 10% NH_4Cl.

11.2. Add 25 mg of 5'-iodo-5′-deoxyadenosine in 3 mL of dimethylformamide to the system through a serum cap in the dark.

11.3. After 20–30 min, desalt corrinoid products by phenol extraction, followed by back-extraction from phenol into water by adding diethyl ether. Collect the water layer, wash it with diethyl ether to remove phenol, and then evaporate it to dryness *in vacuo*.

11.4. Purify the adenosyl form of analog by ion-exchange cellulose or silica gel column chromatography and reversed-phase high-performance liquid chromatography (HPLC) using an octadecylsilica column. The choice of columns depends on the properties of desired product.

9.12. Identification of adenosyl group-derived product(s) with labeled coenzyme

12.1. (See *Note* to Protocol 9.1) The adenosyl group-derived product(s) formed in the mechanism-based inactivation of DD can be identified with a radiolabeled coenzyme or its analog. For example, incubate apoenzyme (10 units) at 37 °C for 0.5–2 h with 1.0 nmol of a labeled suicide coenzyme or pseudocoenzyme, such as [8-^{14}C]AdoCbl(*b*-CH) (specific activity, 493 Bq/nmol) in 0.1 mL of 0.04 M potassium phosphate buffer (pH 8.0) containing 1 M 1,2-propanediol.

12.2. Add ethanol to a final concentration of 80% (v/v) and 2 nmol of cold 5′-deoxyadenosine as a carrier, and heat the mixture at 90 °C for 3 min to effect denaturation of the enzyme.

12.3. After concentration to 0.1–0.2 mL *in vacuo*, chromatograph the sample on a Whatman 3MM paper with 1-butanol-water (178:28 by volume) using AdoCbl(*b*-OH) and 5′-deoxyadenosine as markers.

12.4. Detect radioactive areas by radioactive scanning of the chromatogram. Two radioactive bands coincide with the positions of AdoCbl(*b*-OH) and 5′-deoxyadenosine markers.

12.5. Cut out both radioactive areas, and elute each area with water. The radioactivity that co-migrates with 5′-deoxyadenosine is quantitatively equal to the amount of enzyme inactivated.

12.6. Compare R_f value(s) of the radioactive nucleoside(s) with those of markers by TLC on silica gel with ethyl acetate-methanol (6:1 by volume) and water-saturated 2-butanol as well as by paper chromatography with 1 M ammonium acetate-90% ethanol (3:7 by volume). Use 5′-deoxyadenosine, 5′,8-cyclic adenosine, 4′,5′-anhydroadenosine, adenosine 5′-aldehyde, adenosine, and adenine as markers. In all the cases, virtually all the radioactivity co-migrated with 5′-deoxyadenosine, indicating that this is the only product derived from the adenosyl group of the coenzyme analog.

Note: This method is superior in sensitivity to that described in Protocol 9.14 and thus preferable even when enough amounts of high-purity enzyme is not available.

9.13. Radioactive synthesis of a labeled coenzyme

13.1. [8-^{14}C]Adenosine (New England Nuclear) is converted to 5′-chloro-5′-deoxyadenosine by chlorination with the hexamethylphosphoramide (HMPA)-$SOCl_2$ reagent (Kikugawa & Ichino, 1971). Add a solution of 30 μL of $SOCl_2$ in 0.1 mL of HMPA to labeled adenosine (0.2 mg, 0.37 MBq).

Safety considerations: $SOCl_2$ causes skin burns and eye damage. HMPA causes respiratory tract, eye, and skin irritation. To avoid contact, wear protective gloves and clothing, eye and face protection, and use only in a well-ventilated area.

13.2. After standing at room temperature for 20 h with exclusion of moisture, add 2 mL of cold water to the reaction mixture to destroy the excess reagent.

13.3. Apply the resulting aqueous solution to a column (bed volume 1.8 mL) of Dowex 50 × 8 (H^+).

13.4. Wash the column with water, and then develop it with 1 M ammonium hydroxide. Combine the nucleoside fractions, and evaporate it to dryness *in vacuo*. As judged from silica gel TLC with water-saturated 2-butanol, the yield of 5′-chloro-5′-[8-^{14}C]deoxyadenosine is 80–90% theoretical.

13.5. Add 25 mg of NH_4Cl and 15 mg of zinc powder to 2.3 μmol of CN-Cbl or its analog, such as CN-Cbl(*b*-OH), in 0.25 mL of water. Stir the suspension vigorously for 40 min under argon atmosphere.

13.6. (See *Note* to Protocol 9.1) Add 5′-chloro-5′-[8-^{14}C] deoxyadenosine obtained above in 0.3 mL of 60% aqueous ethanol to the system through a serum cap.

13.7. After 1.5 h, add 2 mL of ethanol and filter off the zinc powder remaining. Evaporate the filtrate to dryness under reduced pressure, and take up the residue in a small amount of water.

13.8. Separate the corrinoids by paper chromatography on a Whatman 3MM filter paper with water-saturated 2-butanol, followed by paper electrophoresis in 0.5 M acetic acid (pH 2.7) at a voltage gradient of approximately 10 V/cm.

13.9. The labeled coenzyme or its analog, such as [8-^{14}C] AdoCbl(*b*-OH), thus purified is homogeneous by criteria of paper chromatography on a Whatman 3MM paper with water-saturated 2-butanol containing 1% acetic acid and TLC on a silica gel plate with 2-butanol-acetic acid-water (127:1:50 by volume). The yield is approximately 15% theoretical.

Note: If radioactive synthesis does not work, lower the specific radioactivity of labeled adenosine to 1/40, i.e., 0.37 MBq/8 mg of [8-^{14}C]adenosine.

9.14. Identification of adenosyl group-derived product(s) with unlabeled coenzyme

14.1. (See *Note* to Protocol 9.1) The adenosyl group-derived product(s) formed in the mechanism-based inactivation of DD can be identified even with unlabeled coenzyme or its analog. For example, incubate 1.0 mg (4.6 nmol) of the apoenzyme of Hα143A mutant at 37 °C for 30 min in the dark with 15 μM AdoCbl in the presence of 0.1 M 1,2-propanediol.

14.2. Add ethanol to a final concentration of 80%, and heat the mixture at 90 °C for 10 min to effect denaturation of the enzyme, followed by centrifugation.

14.3. Evaporate the supernatant to dryness and take up the residue in 0.5 mL of 15% methanol containing 1% acetic acid.

14.4. Analyze the nucleoside product(s) from the adenosyl group by HPLC using a Cosmosil C_{18} column (0.46 × 15 cm) (Nacalai Tesque, Kyoto, Japan) with 15% methanol containing in 1% acetic acid as a mobile phase. Use the same nucleoside markers as those used in Protocol 9.12. From the retention time, the only nucleoside product derived from the adenosyl group is identified as 5′-deoxyadenosine.

14.5. Quantify the product from its peak height by developing the calibrated column at a flow rate of 0.4 mL/min.

Note: If HPLC is not available, or if enough amounts of high-purity enzyme are not available, identification with radiolabeled coenzyme or its analog is recommended (Protocols 9.12 and 9.13).

9.15. Crystallization of DD·CN-Cbl and DD·AdePeCbl complexes in the presence of substrate

15.1. (See *Note* to Protocol 9.1) Add CN-Cbl or AdePeCbl powder to the substrate-bound apoenzyme to a final concentration of 2 mM, and dissolve the powder by gentle agitation using a rotator at 4 °C. Then, incubate the mixture again at 30 °C for additional 30 min.

15.2. Remove excess cobalamin by dialysis against 0.025 M potassium phosphate buffer (pH 8.0) containing 0.2 M KCl and 1 mM dithiothreitol (DTT) to a final cobalamin concentration of 20 μM.

15.3. Prepare precipitant solution consisting of 0.02 M Tris·HCl buffer (pH 8.0), 0.24 M $(NH_4)_2SO_4$, 15% (w/v) PEG 6000, and 0.20% LDAO, and put the precipitant solution into each of the reservoir wells of a commercially available 24-well crystallization plate.

15.4. Mix 10 μL of sample solution and 10 μL of precipitant solution on the protein well of the plate.

15.5. Keep the plates at 4 °C for several months, and check the formation of crystals.

9.16. Crystallization of DD·AdoCbl complex using soluble DD under anaerobic conditions

16.1. All the operations are carried out in a polyvinyl chloride anaerobic chamber (Coy Laboratory Products). Dialyze aerobically-purified enzyme twice against oxygen-free buffer containing 0.02 M potassium phosphate buffer (pH 8.0) containing 0.2 M KCl and 1 mM DTT at 4 °C.

16.2. (See *Note* to Protocol 9.1) To decompose a trace amount of remaining substrate, add a trace amount of AdoCbl to the dialyzed sample solution to a final molar ratio of 1:100 to that of apoenzyme, and incubate the solution at 30 °C for 30 min in the dark.

16.3. Add AdoCbl powder to the substrate-free apoenzyme to a final concentration of 1 mM, and incubate the mixture again at 30 °C for additional 30 min.

16.4. Remove excess AdoCbl by dialysis against 0.025 M potassium phosphate buffer (pH 8.0) containing 0.2 M KCl and 1 mM DTT to a final AdoCbl concentration of 20 μM.

16.5. Prepare precipitant solution consisting of 10.7% (w/v) PEG 3000, 8.6% (w/v) PEG 2000 monomethyl ether, and 0.05 M Tris·HCl buffer (pH 7.6) outside of the anaerobic chamber, and put the precipitant solution into each of the reservoir wells of commercially available 24-well crystallization plates.

16.6. Transfer the plates immediately into the anaerobic chamber, and leave the plates unsealed.

16.7. Seal the plates, and incubate the plates for 1 h in the anaerobic chamber. Then, open seals from the plates for 1 min, and reseal them again. Repeat these processes several times.

16.8. Mix 2 μL of holoenzyme solution and 2 μL of precipitant solution of the reservoir wells on the protein wells of the plates.

16.9. Store the plates at 20 °C in the anaerobic chamber or pack them in oxygen-tolerant bags containing AGELESS EYE oxygen indicators and AGELESS oxygen absorbers (Mitsubishi Gas Chemical, Inc., Japan), and incubate them at 20 °C outside of the chamber. Check the oxygen indicator to ensure that the oxygen concentration is kept less than 0.1% during crystal growth. Crystals appear within a week.

9.17. Crystallization of substrate-free form of EAL·CN-Cbl complex and soaking with substrates

17.1. Dialyze purified enzyme against buffer containing 0.02 M potassium phosphate buffer (pH 8.0) containing 0.2 M KCl and 1 mM DTT at 4 °C.

17.2. To decompose a trace amount of remaining substrate, add a trace amount of AdoCbl to the dialyzed sample solution to a final molar ratio of 1:100 to that of apoenzyme, and incubate the solution at 30 °C for 30 min in the dark.

17.3. Add CN-Cbl powder to the substrate-free apoenzyme to a final concentration of 2 mM, and dissolve the powder by gentle agitation using a rotator at 4 °C. Then, incubate the mixture again at 30 °C for additional 30 min.

17.4. Remove excess CN-Cbl by dialysis against 0.01 M Tris·HCl buffer (pH 8.0) containing 0.2 M KCl and 1 mM DTT to a final CN-Cbl concentration of 20 μM.

17.5. Prepare precipitant solution consisting of 6.0–7.0% (w/v) PEG 4000, 24–26% (v/v) glycerol, 1.0% (v/v) 2-methyl-2, 4-pentanediol, and 0.1 M imidazole·HCl buffer (pH 6.3), and put the precipitant solution into each of the reservoir wells of a commercially available 96-well crystallization plate.

17.6. Mix 1 μL of sample solution and 1 μL of precipitant solution on the protein well of the plate.

17.7. Keep the plates at 4 °C for several weeks, and check the formation of crystals.

17.8. Harvest substrate-free crystals, and soak them at 4 °C for 30 min in the mother liquor supplemented with 10 mM ethanolamine (pH 8.0) or 10 mM 2-amino-1-propanol (pH 8.0) as well as each component of the substrate-free final buffer.

9.18. Crystallization of EAL·AdoCbl complex using soluble EAL under anaerobic conditions

18.1. (See *Note* to Protocol 9.1) Prepare holoEAL from purified His_6-tagged soluble EAL in an anaerobic chamber according to Protocol 9.16, except for using 0.01 M Tris·HCl buffer (pH 8.0) instead of 0.02 M potassium phosphate buffer (pH 8.0) for dialysis.

18.2. Prepare precipitant solution consisting of 12.5% (w/v) PEG 1500, 0.05 M sodium malonate, imidazole, and boric acid buffer (pH 9.0) or 10% (w/v) PEG 6000, 0.1 M NH_4Cl, and 0.05 M Tris·HCl buffer (pH 8.0), and setup crystallization plates as Protocol 9.16 in an anaerobic chamber.

18.3. Store the crystallization plates at 4 °C under the anaerobic condition. Crystals appear within a week.

9.19. X-ray diffraction data collection

19.1. (See *Note* to Protocol 9.1) For the native data set, soak the crystals briefly in a cryoprotectant solution containing 15% (v/v) PEG 400 for DD or 20% (v/v) glycerol for EAL and the components of the above crystallization conditions. Before collecting diffraction data, flash-cool the crystal in a 100 K stream of cryogenic nitrogen gas.

19.2. Solve crystal structures of DD and EAL by the molecular replacement (MR) method by using the published structures of the enzymes. However, the initial structure of the enzyme can be determined by the multiple isomorphous replacement (MIR) method using ethylmercurythiosalicylate (EMTS) and ethylmercury chloride derivative crystals. For preparation of heavy-atom derivatives, soak crystals in the crystallization solution containing harvest buffer containing 0.1 mM EMTS or 0.1 mM ethylmercury chloride for 1 day.

Safety considerations: EMTS and ethylmercury chloride are very toxic. Wear protective gloves, eye or face protection, and protective clothing to avoid contact with skin and eyes.

19.3. Collect diffraction data sets of the frozen crystals at 100 K using a synchrotron beamline for protein crystallography. DD crystals usually diffract up to resolutions range between 1.7 and 2.0 Å. The best conditions for the diffraction experiment depend on the beamlines, but the followings are typical parameters for DD crystals: wavelength, 1.0000 Å; crystal to detector distance, 150–250 mm; oscillation range per image, 0.1–0.5°; Total oscillation range, 360°.

19.4. The collected diffraction data are indexed, integrated and scaled using the HKL2000 program package (Otwinowski & Minor, 1997) or the XDS program (Kabsch, 2010).

9.20. Structure determination

20.1. For structure determination by the MR method, any published structures available in the Protein Data Bank can be used for the initial search model. The initial models can be easily obtained using any MR programs. Also, initial phase determination of DD is performed by the MIR method using the data sets of the EMTS and ethylmercury chloride derivatives. We highly recommend using the latest program packages (CCP4 (Collaborative Computational Project, 1994) or PHENIX (Liebschner et al., 2019)) for crystal structure determination. Both programs can perform heavy-atom detection, initial phase determination, phase improvement, model building, and model improvement automatically.

20.2. For the initial structure determined by the MR method, simply replace or modify the altered parts (point mutations, cofactors, and substrates) using a protein modeling program such as COOT (Emsley, Lohkamp, Scott, & Cowtan, 2010) if the electron density map unambiguously displays the correct structure. In the case of structures determined by the MIR method, manual chain trace and model building may be needed.

20.3. Refine the modified model using a refinement program such as phenix.refine in PHENIX and REFMAC (Murshudov et al., 2011).

9.21. Site-directed mutagenesis

21.1. Site-directed mutageneses are generated using a QuikChange™ Site-Directed Mutagenesis kit (Stratagene, CA). First, design the mutagenic sense primer for a mutant according to the manufacturer's instructions. Suggested guidelines are: (1) primers length should be 25–45 in bases, (2) with a melting temperature (T_m) of $\geq 78\,°C$. $T_m = 81.5 + 0.41$ (%GC) − (675/N) − %mismatch (N is the primer length in bases). (3) the desired mutation should be in the middle of the primer with ~10–15 bases of correct sequence on both sides. (4) the primers optimally should have a minimum GC content of 40% and should terminate in one or more C or G bases. For example, 5′-ccgtcccagcaggcgaaagtcaccaacgtc aaagataac-3′ is designed as the mutagenic sense primer for the Hα143K mutant. Use the oligonucleotide having the complementary sequence in the opposite direction as the respective antisense primer.

21.2. Synthesize nicked mutant strand by temperature cycling using *PfuTurbo* DNA polymerase with mutagenic sense and antisense primers and pUSI2ENd(DD) as a template (~16 cycles).

21.3. Digest methylated, non-mutated parental DNA template with *Dpn*I restriction enzyme. Mutated plasmid containing nicked circular strands is not digested.

21.4. Transform the circular, nicked dsDNA into XL1-Blue supercompetent cells, in which the nicks in the mutated plasmid are repaired.

21.5. Confirm the insert of the mutant expression plasmid by nucleotide sequencing to have intended mutations but not to have unintended mutations.

10. Conclusion

Function resides in structure. The functions of enzymes must be understood on the bases of their structures. Here we have outlined our combined biochemical and structural approaches to studying AdoCbl-dependent dehydratases and a deaminase. Our major research interests/questions were: (1) how do the enzymes form catalytic radicals at the active site? (2) how do the enzymes utilize the high reactivity of radicals for catalysis by suppressing undesirable side reactions? (3) How do living organisms address inactivation when the enzymes are inactivated by undesirable side reactions?

Regarding questions (1) and (2), we have obtained convincing solutions to date, as described above. Computational studies also support our conclusion. Concerning question (3), we have discovered reactivating chaperones that are essential for activity-maintaining systems (for details, see chapter "Reactivating chaperones for coenzyme B_{12}-dependent diol and glycerol dehydratases and ethanolamine ammonia-lyase" by Toraya et al.). In this chapter, specific methods for investigating eliminases are also presented in Section 9, which may hopefully be useful for other investigators to start working with these fascinating enzymes.

Acknowledgments

This work was supported by Grants-in-Aid for Scientific Research from Japan Society for the Promotion of Science and the Ministry of Education, Culture, Sports, Science and Technology and other research grants. We are grateful to all the people who worked with us at Okayama University, Kyoto University, Brandeis University, and University of Hyogo.

References

Abeles, R. H. (1966). Dioldehydrase. *Methods in Enzymology, 9*, 686–689.

Abeles, R. H., Brownstein, A. M., & Randles, C. H. (1960). β-Hydroxypropionaldehyde, an intermediate in the formation of 1,3-propanediol by *Aerobacter aerogenes*. *Biochimica et Biophysica Acta, 41*, 530–531.

Abeles, R. H., & Dolphin, D. (1976). The vitamin B_{12} coenzyme. *Accounts of Chemical Research, 9*, 114–120.

Abeles, R. H., & Zagalak, B. (1966). The nature of the hydrogen transfer in the dimethylbenzimidazolycobamide coenzyme-catalyzed conversion of 1,2-propanediol to propionaldehyde. *Journal of Biological Chemistry, 241*, 1245–1246.

Akita, K., Hieda, N., Baba, N., Kawaguchi, S., Sakamoto, H., Nakanishi, Y., et al. (2010). Purification and some properties of wild-type and N-terminal-truncated ethanolamine ammonia-lyase of *Escherichia coli*. *Journal of Biochemistry, 147*, 83–93.

Babior, B. M. (1975). Mechanism of cobalamin-dependent rearrangements. *Accounts of Chemical Research, 8*, 376–384.

Babior, B. M., Carty, T. J., & Abeles, R. H. (1974). The mechanism of action of ethanolamine ammonia-lyase, a B_{12}-dependent enzyme: The reversible formation of 5'-deoxyadenosine from adenosylcobalamin during the catalytic process. *Journal of Biological Chemistry, 249*, 1689–1695.

Babior, B. M., Moss, T. H., Orme-Johnson, W. H., & Beinert, H. (1974). The mechanism of action of ethanolamine ammonia-lyase, a B_{12}-dependent enzyme. The participation of paramagnetic species in the catalytic deamination of 2-aminopropanol. *Journal of Biological Chemistry, 249*, 4537–4544.

Bachovchin, W. W., Eagar, R. G., Jr., Moore, K. W., & Richards, J. H. (1977). Mechanism of action of adenosylcobalamin: glycerol and other substrate analogues as substrates and inactivators for propanediol dehydratase- -kinetics, stereospecificity, and mechanism. *Biochemistry, 16*, 1082–1092.

Bandarian, V., & Reed, G. H. (2000). Isotope effects in the transient phases of the reaction catalyzed by ethanolamine ammonia-lyase: Determination of the number of exchangeable hydrogens in the enzyme-cofactor complex. *Biochemistry, 39*, 12069–12075.

Banerjee, R. (Ed.). (1999). *Chemistry and Biochemistry of B_{12}*. New York: John Wiley & Sons.
Barker, H. A., Weissbach, H., & Smyth, R. D. (1958). A coenzyme containing pseudovitamin B_{12}. *Proceedings of the National Academy of Sciences of the United States of America, 44*, 1093–1097.
Bender, G., Poyner, R. R., & Reed, G. H. (2008). Identification of the substrate radical intermediate derived from ethanolamine during catalysis by ethanolamine ammonia-lyase. *Biochemistry, 47*, 11360–11366.
Blackwell, C. M., Scarlett, F. A., & Turner, J. M. (1977). Microbial metabolism of amino alcohols. Control of formation and stability of partially purified ethanolamine ammonia-lyase in *Escherichia coli*. *Journal of General Microbiology, 98*, 133–139.
Blackwell, C. M., & Turner, J. M. (1978). Microbial metabolism of amino alcohols. Purification and properties of coenzyme B_{12}-dependent ethanolamine ammonia-lyase of *Escherichia coli*. *Biochemical Journal, 175*, 555–563.
Boas, J. F., Hicks, P. R., Pilbrow, J. R., & Smith, T. D. (1978). Interpretation of electron spin resonance spectra due to some B_{12}-dependent enzyme reactions. *Journal of the Chemical Society, Faraday Transactions II, 74*, 417–431.
Bobik, T. A., Xu, Y., Jeter, R. M., Otto, K. E., & Roth, J. R. (1997). Propanediol utilization genes (*pdu*) of *Salmonella typhimurium*: Three genes for the propanediol dehydratase. *Journal of Bacteriology, 179*, 6633–6639.
Bradbeer, C. (1965a). The clostridial fermentations of choline and ethanolamine: I. Preparation and properties of cell-free extracts. *Journal of Biological Chemistry, 240*, 4669–4674.
Bradbeer, C. (1965b). The clostridial fermentations of choline and ethanolamine: II. Requirement for a cobamide coenzyme by an ethanolamine deaminase. *Journal of Biological Chemistry, 240*, 4675–4681.
Buettner, G. R., & Coffman, R. E. (1977). EPR determination of the Co(II)-free radical magnetic geometry of the "doublet" species arising in a coenzyme B_{12} enzyme reaction. *Biochimica et Biophysica Acta, 480*, 495–505.
Carty, T. J., Babior, B. M., & Abeles, R. H. (1974). The mechanism of action of ethanolamine ammonia-lyase, a B_{12}-dependent enzyme. Evidence for two intermediates in the catalytic process. *Journal of Biological Chemistry, 249*, 1683–1688.
Chang, G. W., & Chang, J. T. (1975). Evidence for the B_{12}-dependent enzyme ethanolamine deaminase in *Salmonella*. *Nature, 254*, 150–151.
Cockle, S. A., Hill, H. A. O., Williams, R. J. P., Davies, S. P., & Foster, M. A. (1972). The detection of intermediates during the conversion of propane-1,2-diol to propionaldehyde by glyceroldehydrase, a coenzyme B_{12} dependent reaction. *Journal of the American Chemical Society, 94*, 275–277.
Collaborative Computational Project, Number 4. (1994). The *CCP4* suite: programs for protein crystallography. *Acta Crystallographica Section D, 50*, 760–763.
Daniel, R., Bobik, T. A., & Gottschalk, G. (1998). Biochemistry of coenzyme B_{12}-dependent glycerol and diol dehydratases and organization of the encoding genes. *FEMS Microbiology Reviews, 22*, 553–566.
Davis, B. J. (1964). Disc electrophoresis– II. Method and application to human serum proteins. *Annals of the New York Academy of Science, 121*, 404–427.
Diziol, P., Haas, H., Rétey, J., Graves, S. W., & Babior, B. M. (1980). The substrate-dependent steric course of the ethanolamine ammonia-lyase reaction. *European Journal of Biochemistry, 106*, 211–224.
Doitomi, K., Kamachi, T., Toraya, T., & Yoshizawa, K. (2016). Computational mutation study of the roles of catalytic residues in coenzyme B_{12}-dependent diol dehydratase. *Bulletin of Chemical Society of Japan, 89*, 955–964.
Dolphin, D. (Ed.). (1982). *Vol. 2. B_{12}*. New York: John Wiley & Sons.
Emsley, P., Lohkamp, B., Scott, W. G., & Cowtan, K. (2010). Features and development of Coot. *Acta Crystallographica Section D, 66*, 486–501.

Essenberg, M. K., Frey, P. A., & Abeles, R. H. (1971). Studies on the mechanism of hydrogen transfer in the coenzyme B_{12} dependent dioldehydrase reaction II. *Journal of the American Chemical Society, 93*, 1242–1251.

Faust, L. P., & Babior, B. M. (1992). Overexpression, purification, and some properties of the AdoCbl-dependent ethanolamine ammonia-lyase from *Salmonella typhimurium*. *Archives of Biochemistry and Biophysics, 294*, 50–54.

Faust, L. P., Connor, J. A., Roof, D. M., Hoch, J. A., & Babior, B. M. (1990). Cloning, sequencing, and expression of the genes encoding the adenosylcobalamin-dependent ethanolamine ammonia-lyase of *Salmonella typhimurium*. *Journal of Biological Chemistry, 265*, 12462–12466.

Finlay, T. H., Valinsky, J., Mildvan, A. S., & Abeles, R. H. (1973). Electron spin resonance studies with dioldehydrase. Evidence for radical intermediates in reactions catalyzed by coenzyme B_{12}. *Journal of Biological Chemistry, 248*, 1285–1290.

Forage, R. G., & Foster, M. A. (1979). Resolution of the coenzyme B_{12} dependent dehydratases of *Klebsiella* sp. and *Citrobacter freundii*. *Biochimica et Biophysica Acta, 569*, 249–258.

Forage, R. G., & Foster, M. A. (1982). Glycerol fermentation in *Klebsiella pneumoniae*: Functions of the coenzyme B_{12}-dependent glycerol and diol dehydratases. *Journal of Bacteriology, 149*, 413–419.

Forage, R. G., & Lin, E. C. C. (1982). *dha* system mediating aerobic and anaerobic dissimilation of glycerol in *Klebsiella pneumoniae* NCIB 418. *Journal of Bacteriology, 151*, 591–599.

Frey, P. A., & Abeles, R. H. (1966). The role of the B_{12} coenzyme in the conversion of 1,2-propanediol to propionaldehyde. *Journal of Biological Chemistry, 241*, 2732–2733.

Frey, P. A., Essenberg, M. K., & Abeles, R. H. (1967). Studies on the mechanism of hydrogen transfer in the cobamide coenzyme-dependent dioldehydrase reaction. *Journal of Biological Chemistry, 242*, 5369–5377.

Frey, P. A., Essenberg, M. K., Abeles, R. H., & Kerwar, S. S. (1970). Comments on a proposed mechanism of action of B_{12} coenzyme. *Journal of the American Chemical Society, 92*, 4488–4489.

Frey, P. A., Kerwar, S. S., & Abeles, R. H. (1967). The participation of the two non-equivalent C-5′ hydrogens of B_{12}-coenzyme in the catalytic process. *Biochemical and Biophysical Research Communications, 29*, 873–874.

Fukuoka, M., Nakanishi, Y., Hannak, R. B., Kräutler, B., & Toraya, T. (2005). Homoadenosylcobalamins as probes for exploring the active sites of coenzyme B_{12}-dependent diol dehydratase and ethanolamine ammonia-lyase. *FEBS Journal, 272*, 4787–4796.

Fukuoka, M., Yamada, S., Miyoshi, S., Yamashita, K., Yamanishi, M., Zou, X., et al. (2002). Functions of the D-ribosyl moiety and the lower axial ligand of the nucleotide loop of coenzyme B_{12} in diol dehydratase and ethanolamine ammonia-lyase reactions. *Journal of Biochemistry, 132*, 935–943.

Halpern, J., Kim, S.-H., & Leung, T. W. (1984). Cobalt-carbon bond dissociation energy of coenzyme B_{12}. *Journal of the American Chemical Society, 106*, 8317–8319.

Hartmanis, M. G. N., & Stadtman, T. C. (1986). Diol metabolism and diol dehydratase in *Clostridium glycolicum*. *Archives of Biochemistry and Biophysics, 245*, 144–152.

Hay, B. P., & Finke, R. G. (1986). Thermolysis of the Co-C bond of adenosylcobalamin. 2. Products, kinetics, and Co-C bond dissociation energy in aqueous solution. *Journal of the American Chemical Society, 108*, 4820–4829.

Hay, B. P., & Finke, R. G. (1987). Thermolysis of the Co-C bond in adenosylcorrins. 3. Quantification of the axial base effect in adenosylcobalamin by the synthesis and thermolysis of axial base-free adenosylcobinamide. Insights into the energetics of enzyme-assisted cobalt-carbon bond homolysis. *Journal of the American Chemical Society, 109*, 8012–8018.

Hosoi, N., Morimoto, K., Ozaki, C., Kitamoto, Y., & Ichikawa, Y. (1978). Enzyme activities involved in the metabolism of 1,2-propanediol by *Propionibacterium freudenreichii*. *Journal of Fermentation Technology, 56*, 566–572.

Ichikawa, M., & Toraya, T. (1988). Roles of the β-D-ribofuranose ring and the functional groups of the D-ribose moiety of adenosylcobalamin in the diol dehydratase reaction. *Biochimica et Biophysica Acta, 952*, 191–200.

Ishida, A., Ichikawa, M., Kobayashi, K., Hitomi, T., Kojima, S., & Toraya, T. (1993). Importance of the nucleotide loop moiety coordinated to the cobalt atom of adenosylcobalamin for coenzymic function in the diol dehydrase reaction. *Journal of Nutritional Science and Vitaminology, 39*, 115–125.

Ishida, A., & Toraya, T. (1993). Adenosylcobinamide methyl phosphate as a pseudocoenzyme for diol dehydrase. *Biochemistry, 32*, 1535–1540.

Jeter, R. M. (1990). Cobalamin-dependent 1,2-propanediol utilization by *Salmonella typhimurium*. *Journal of General Microbiology, 136*, 887–896.

Johnson, E. A., & Lin, E. C. C. (1987). *Klebsiella pneumoniae* 1,3-propanediol:NAD^+ oxidoreductase. *Journal of Bacteriology, 169*, 2050–2054.

Kabsch, W. (2010). *XDS*. *Acta Crystallographica Section D, 66*, 125–132.

Kamachi, T., Doitomi, K., Takahata, M., Toraya, T., & Yoshizawa, K. (2011). Catalytic roles of the metal ion in the substrate-binding site of coenzyme B_{12}-dependent diol dehydratase. *Inorganic Chemistry, 50*, 2944–2952.

Kamachi, T., Takahata, M., Toraya, T., & Yoshizawa, K. (2009). What is the identity of the metal ions in the active sites of coenzyme B_{12}-dependent diol dehydratase? A computational mutation analysis. *Journal of Physical Chemistry B, 113*, 8435–8438.

Kamachi, T., Toraya, T., & Yoshizawa, K. (2004). Catalytic roles of active-site amino acid residues of coenzyme B_{12}-dependent diol dehydratase: Protonation state of histidine and pull effect of glutamate. *Journal of American Chemical Society, 126*, 16207–16216.

Kamachi, T., Toraya, T., & Yoshizawa, K. (2007). Computational mutation analysis of hydrogen abstraction and radical rearrangement steps in the catalysis of coenzyme B_{12}-dependent diol dehydratase. *Chemistry, 13*, 7864–7873.

Kaplan, B. H., & Stadtman, E. R. (1968). Ethanolamine deaminase, a cobamide coenzyme-dependent enzyme: I. Purification, assay, and properties of the enzyme. *Journal of Biological Chemistry, 243*, 1787–1793.

Kawata, M., Kinoshita, K., Takahashi, S., Ogura, K., Komoto, N., Yamanishi, M., et al. (2006). Survey of catalytic residues and essential roles of glutamate-α170 and aspartate-α335 in coenzyme B_{12}-dependent diol dehydratase. *Journal of Biological Chemistry, 281*, 18327–18334.

Kikugawa, K., & Ichino, M. (1971). Direct halogenation of sugar moiety of nucleosides. *Tetrahedron Letters, 12*, 87–90.

Kinoshita, K., Kawata, M., Ogura, K., Yamasaki, A., Watanabe, T., Komoto, N., et al. (2008). Histidine-α143 assists 1,2-hydroxyl group migration and protects radical intermediates in coenzyme B_{12}-dependent diol dehydratase. *Biochemistry, 47*, 3162–3173.

Kofoid, E., Rappleye, C., Stojiljkovic, I., & Roth, J. (1999). The 17-gene ethanolamine (*eut*) operon of *Salmonella typhimurium* encodes five homologues of carboxysome shell proteins. *Journal of Bacteriology, 181*, 5317–5329.

Kohara, Y., Akiyama, K., & Isono, K. (1987). The physical map of the whole *E. coli* chromosome: Application of a new strategy for rapid analysis and sorting of a large genomic library. *Cell, 50*, 495–508.

Kraulis, P. J. (1991). *MOLSCRIPT*: a program to produce both detailed and schematic plots of protein structures. *Journal of Applied Crystallography, 24*, 946–950.

Kuno, S., Fukui, S., & Toraya, T. (1990). Essential histidine residues in coenzyme B_{12}-dependent diol dehydrase: Dye-sensitized photooxidation and ethoxycarbonylation. *Archives of Biochemistry and Biophysics, 277*, 211–217.

Kuno, S., Toraya, T., & Fukui, S. (1980). Coenzyme B_{12}-dependent diol dehydrase: chemical modification with 2,3-butanedione and phenylglyoxal. *Archives of Biochemistry and Biophysics*, *205*, 240–245.

Kuno, S., Toraya, T., & Fukui, S. (1981a). Reactive sulfhydryl groups of coenzyme B_{12}-dependent diol dehydrase: differential modification of essential and nonessential ones. *Archives of Biochemistry and Biophysics*, *210*, 474–480.

Kuno, S., Toraya, T., & Fukui, S. (1981b). Chemical modification of coenzyme B_{12}-dependent diol dehydrase with pyridoxal 5'-phosphate: lysyl residue essential for interaction between two components of the enzyme. *Archives of Biochemistry and Biophysics*, *211*, 722–730.

LaMattina, J. W., Keul, N. D., Reitzer, P., Kapoor, S., Galzerani, F., Koch, D. J., et al. (2016). 1,2-Propanediol dehydration in *Roseburia inulinivorans*: STRUCTURAL BASIS FOR SUBSTRATE AND ENANTIOMER SELECTIVITY. *Journal of Biological Chemistry*, *291*, 15515–15526.

Leal, N. A., Havemann, G. D., & Bobik, T. A. (2003). PduP is a coenzyme-a-acylating propionaldehyde dehydrogenase associated with the polyhedral bodies involved in B_{12}-dependent 1,2-propanediol degradation by *Salmonella enterica* serovar Typhimurium LT2. *Archives of Microbiology*, *180*, 353–361.

Lee, H. A., Jr., & Abeles, R. H. (1963). Purification and properties of dioldehydrase, an enzyme requiring a cobamide coenzyme. *Journal of Biological Chemistry*, *238*, 2367–2373.

Levin, B. J., & Balskus, E. P. (2018). Characterization of 1,2-propanediol dehydratases reveals distinct mechanisms for B_{12}-dependent and glycyl radical enzymes. *Biochemistry*, *57*, 3222–3226.

Liebschner, D., Afonine, P. V., Baker, M. L., Bunkóczi, G., Chen, V. B., Croll, T. I., et al. (2019). Macromolecular structure determination using X-rays, neutrons and electrons: recent developments in *Phenix*. *Acta Crystallographica Section D*, *75*, 861–877.

Magnusson, O. T., & Frey, P. A. (2002). Interactions of diol dehydrase and 3', 4'-anhydroadenosylcobalamin: Suicide inactivation by electron transfer. *Biochemistry*, *41*, 1695–1702.

Masuda, J., Shibata, N., Morimoto, Y., Toraya, T., & Yasuoka, N. (2000). How a protein generates a catalytic radical from coenzyme B_{12}: X-ray structure of a diol-dehydratase-adeninylpentylcobalamin complex. *Structure*, *8*, 775–788.

McGee, D. E., Carroll, S. S., Bond, M. W., & Richards, J. H. (1982). Diol dehydratase: N-terminal amino acid sequences and subunit stoichiometry. *Biochemical and Biophysical Research Communications*, *108*, 547–551.

McGee, D. E., & Richards, J. H. (1981). Purification and subunit characterization of propanediol dehydratase, a membrane-associated enzyme. *Biochemistry*, *20*, 4293–4298.

Merritt, E. A., & Bacon, D. J. (1997). Raster3D: photorealistic molecular graphics. *Methods in Enzymology*, *277*, 505–524.

Mickelson, M. N., & Werkman, C. H. (1940). The dissimilation of glycerol by coli-aerogenes intermediates. *Journal of Bacteriology*, *39*, 709–715.

Mori, K., Oiwa, T., Kawaguchi, S., Kondo, K., Takahashi, Y., & Toraya, T. (2014). Catalytic roles of substrate-binding residues in coenzyme B_{12}-dependent ethanolamine ammonia-lyase. *Biochemistry*, *53*, 2661–2671.

Murshudov, G. N., Skubák, P., Lebedev, A. A., Pannu, N. S., Steiner, R. A., Nicholls, R. A., et al. (2011). *REFMAC5* for the refinement of macromolecular crystal structures. *Acta Crystallographica Section D*, *67*, 355–367.

Obradors, N., Badía, J., Baldomà, L., & Aguilar, J. (1988). Anaerobic metabolism of the L-rhamnose fermentation product 1,2-propanediol in *Salmonella typhimurium*. *Journal of Bacteriology*, *170*, 2159–2162.

O'Brien, J. R., Raynaud, C., Croux, C., Girbal, L., Soucaille, P., & Lanzilotta, W. N. (2004). Insight into the mechanism of the B_{12}-independent glycerol dehydratase from *Clostridium butyricum*: preliminary biochemical and structural characterization. *Biochemistry*, *43*, 4635–4645.

Ogura, K., Kunita, S., Mori, K., Tobimatsu, T., & Toraya, T. (2008). Roles of adenine anchoring and ion pairing at the coenzyme B_{12}-binding site in diol dehydratase catalysis. *FEBS Journal, 275*, 6204–6216.

Otwinowski, Z., & Minor, W. (1997). Processing of X-ray diffraction data collected in oscillation mode. *Methods in Enzymology, 276*, 307–326.

Pawelkiewicz, J., & Zagalak, B. (1965). Enzymic conversion of glycerol into β-hydroxypropionaldehyde in a cell-free extract from *Aerobacter aerogenes*. *Acta Biochimica Polonica, 12*, 207–218.

Pettersen, E. F., Goddard, T. D., Huang, C. C., Couch, G. S., Greenblatt, D. M., Meng, E. C., et al. (2004). UCSF Chimera—A visualization system for exploratory research and analysis. *Journal of Computational Chemistry, 25*, 1605–1612.

Poppe, L., & Rétey, J. (1997). Kinetic investigations with inhibitors that mimic the posthomolysis intermediate in the reactions of coenzyme-B_{12}-dependent glycerol dehydratase and diol dehydratase. *European Journal of Biochemistry, 245*, 398–401.

Poznanskaja, A. A., Tanizawa, K., Soda, K., Toraya, T., & Fukui, S. (1979). Coenzyme B_{12}-dependent diol dehydrase: Purification, subunit heterogeneity, and reversible association. *Archives of Biochemistry and Biophysics, 194*, 379–386.

Poznanskaya, A. A., Yakusheva, M. I., & Yakovlev, V. A. (1977). Study of the mechanism of action of adenosylcobalamin-dependent glycerol dehydratase from *Aerobacter aerogenes*. II. The inactivation kinetics of glycerol dehydratase complexes with adenosylcobalamin and its analogs. *Biochimica et Biophysica Acta, 484*, 236–243.

Raynaud, C., Sarçabal, P., Meynial-Salles, I., Croux, C., & Soucaille, P. (2003). Molecular characterization of the 1,3-propanediol (1,3-PD) operon of *Clostridium butyricum*. *Proceedings of the National Academy of Sciences of the United States of America, 100*, 5010–5015.

Rétey, J., Umani-Ronchi, A., & Arigoni, D. (1966). On the stereochemistry of the propanediol dehydrase reaction. *Experientia, 22*, 72–73.

Rétey, J., Umani-Ronchi, A., Seibl, J., & Arigoni, D. (1966). On the mechanism of the propanediol dehydrase reaction. *Experientia, 22*, 502–503.

Roof, D. M., & Roth, J. R. (1988). Ethanolamine utilization in *Salmonella typhimurium*. *Journal of Bacteriology, 170*, 3855–3863.

Sakai, T., Yamasaki, A., Toyofuku, S., Nishiki, T., Yunoki, M., Komoto, N., et al. (2007). Construction and characterization of hybrid dehydratases between adenosylcobalamin-dependent diol and glycerol dehydratases. *Journal of Nutritional Science and Vitaminology, 53*, 102–108.

Sauvageot, N., Pichereau, V., Louarme, L., Hartke, A., Auffray, Y., & Laplace, J. (2002). Purification, characterization and subunits identification of the diol dehydratase of *Lactobacillus collinoides*. *European Journal of Biochemistry, 269*, 5731–5737.

Scarlett, F. A., & Turner, J. M. (1976). Microbial metabolism of amino alcohols. Ethanolamine catabolism mediated by coenzyme B_{12}-dependent ethanolamine ammonia-lyase in *Escherichia coli* and *Klebsiella aerogenes*. *Journal of General Microbiology, 95*, 173–176.

Schepler, K. L., Dunham, W. R., Sands, R. H., Fee, J. A., & Abeles, R. H. (1975). A physical explanation of the EPR spectrum observed during catalysis by enzymes utilizing coenzyme B_{12}. *Biochimica et Biophysica Acta, 397*, 510–518.

Schneider, Z., Larsen, E. G., Jacobson, G., Johnson, B. C., & Pawelkiewicz, J. (1970). Purification and properties of glycerol dehydrase. *Journal of Biological Chemistry, 245*, 3388–3396.

Schneider, Z., & Pawelkiewicz, J. (1966). The properties of glycerol dehydratase isolated from *Aerobacter aerogenes*, and the properties of the apoenzyme subunits. *Acta Biochimica Polonica, 13*, 311–328.

Schwartz, P. A., & Frey, P. A. (2007). Dioldehydrase: An essential role for potassium ion in the homolytic cleavage of the cobalt-carbon bond in adenosylcobalamin. *Biochemistry, 46*, 7293–7301.

Seyfried, M., Daniel, R., & Gottschalk, G. (1996). Cloning, sequencing, and overexpression of the genes encoding coenzyme B_{12}-dependent glycerol dehydratase of *Citrobacter freundii*. *Journal of Bacteriology*, *178*, 5793–5796.
Sheppard, D. E., & Roth, J. R. (1994). A rationale for autoinduction of a transcriptional activator: Ethanolamine ammonia-lyase (EutBC) and the operon activator (EutR) compete for adenosyl-cobalamin in *Salmonella typhimurium*. *Journal of Bacteriology* *176*, 1287–1296.
Shibata, N., Higuchi, Y., & Toraya, T. (2011). How coenzyme B_{12}-dependent ethanolamine ammonia-lyase deals with both enantiomers of 2-amino-1-propanol as substrates: Structure-based rationalization. *Biochemistry*, *50*, 591–598.
Shibata, N., Masuda, J., Morimoto, Y., Yasuoka, N., & Toraya, T. (2002). Substrate-induced conformational change of a coenzyme B_{12}-dependent enzyme: Crystal structure of the substrate-free form of diol dehydratase. *Biochemistry*, *41*, 12607–12617.
Shibata, N., Masuda, J., Tobimatsu, T., Toraya, T., Suto, K., Morimoto, Y., et al. (1999). A new mode of B_{12} binding and the direct participation of a potassium ion in enzyme catalysis: X-ray structure of diol dehydratase. *Structure*, *7*, 997–1008.
Shibata, N., Nakanishi, Y., Fukuoka, M., Yamanishi, M., Yasuoka, N., & Toraya, T. (2003). Structural rationalization for the lack of stereospecificity in coenzyme B_{12}-dependent diol dehydratase. *Journal of Biological Chemistry*, *278*, 22717–22725.
Shibata, N., Sueyoshi, Y., Higuchi, Y., & Toraya, T. (2018). Direct participation of a peripheral side chain of a corrin ring in coenzyme B_{12} catalysis. *Angewandte Chemie International Edition*, *57*, 7830–7835.
Shibata, N., Tamagaki, H., Hieda, N., Akita, K., Komori, H., Shomura, Y., et al. (2010). Crystal structures of ethanolamine ammonia-lyase complexed with coenzyme B_{12} analogs and substrates. *Journal of Biological Chemistry*, *285*, 26484–26493.
Smiley, K. L., & Sobolov, M. (1962). A cobamide-requiring glycerol dehydrase from an acrolein-forming lactobacillus. *Archives of Biochemistry and Biophysics*, *97* 538–543.
Sobolov, M., & Smiley, K. L. (1960). Metabolism of glycerol by an acrolein-forming Lactobacillus. *Journal of Bacteriology*, *79*, 261–266.
Stroinski, A., Pawelkiewicz, J., & Johnson, B. C. (1974). Allosteric interactions in glycerol dehydratase: Purification of enzyme and effects of positive and negative cooperativity for glycerol. *Archives of Biochemistry and Biophysics*, *162*, 321–330.
Tanizawa, K., Nakajima, N., Toraya, T., Tanaka, H., & Soda, K. (1987). Re-investigation of the protein structure of coenzyme B_{12}-dependent diol dehydrase. *Zeitschrift für Naturforschung C*, *42*, 353–359.
Tobimatsu, T., Azuma, M., Hayashi, S., Nishimoto, K., & Toraya, T. (1998). Molecular cloning, sequencing and characterization of the genes for adenosylcobalamin-dependent diol dehydratase of *Klebsiella pneumoniae*. *Bioscience, Biotechnology, and Biochemistry*, *62*, 1774–1777.
Tobimatsu, T., Azuma, M., Matsubara, H., Takatori, H., Niida, T., Nishimoto, K., et al. (1996). Cloning, sequencing, and high level expression of the genes encoding adenosylcobalamin-dependent glycerol dehydrase of *Klebsiella pneumoniae*. *Journal of Biological Chemistry*, *271*, 22352–22357.
Tobimatsu, T., Hara, T., Sakaguchi, M., Kishimoto, Y., Wada, Y., Isoda, M., et al. (1995). Molecular cloning, sequencing, and expression of the genes encoding adenosylcobalamin-dependent diol dehydrase of *Klebsiella oxytoca*. *Journal of Biological Chemistry*, *270*, 7142–7148.
Tobimatsu, T., Kawata, M., & Toraya, T. (2005). The N-terminal regions of β and γ subunits lower the solubility of adenosylcobalamin-dependent diol dehydratase. *Bioscience, Biotechnology, and Biochemistry*, *69*, 455–462.
Tobimatsu, T., Nishiki, T., Morimoto, M., Miyata, R., & Toraya, T. (2009). Low-solubility glycerol dehydratase, a chimeric enzyme of coenzyme B_{12}-dependent glycerol and diol dehydratases. *Archives of Microbiology*, *191*, 199–206.
Tobimatsu, T., Sakai, T., Hashida, Y., Mizoguchi, N., Miyoshi, S., & Toraya T. (1997). Heterologous expression, purification, and properties of diol dehydratase, an adenosylcobalamin-dependent enzyme of *Klebsiella oxytoca*. *Archives of Biochemistry and Biophysics*, *347*, 132–140.

Toraya, T. (1985). The binding site for the adenosyl group of coenzyme B_{12} in diol dehydrase. *Archives of Biochemistry and Biophysics, 242*, 470–477.
Toraya, T. (1999). Diol dehydrase and glycerol dehydrase. In R. Banerjee (Ed.), *Chemistry and Biochemistry of B_{12}* (pp. 783–809). New York: John Wiley & Sons.
Toraya, T. (2000). Radical catalysis of B_{12} enzymes: structure, mechanism, inactivation, and reactivation of diol and glycerol dehydratases. *Cellular and Molecular Life Sciences, 57*, 106–127.
Toraya, T. (2003). Radical catalysis in coenzyme B_{12}-dependent isomerization (eliminating) reactions. *Chemical Reviews, 103*, 2095–2127.
Toraya, T. (2014). Cobalamin-dependent dehydratases and a deaminase: radical catalysis and reactivating chaperones. *Archives of Biochemistry and Biophysics, 544*, 40–57.
Toraya, T., & Abeles, R. H. (1980). Inactivation of dioldehydrase in the presence of a coenzyme B_{12} analog. *Archives of Biochemistry and Biophysics, 203*, 174–180.
Toraya, T., & Fukui, S. (1972). Coenzyme B_{12}-dependent propanediol dehydratase systems. Ternary complex between apoenzyme, coenzyme, and substrate analog. *Biochimica et Biophysica Acta, 284*, 536–548.
Toraya, T., & Fukui, S. (1977). Immunochemical evidence for the difference between coenzyme-B_{12}-dependent diol dehydratase and glycerol dehydratase. *European Journal of Biochemistry, 76*, 285–289.
Toraya, T., Honda, S., & Fukui, S. (1979). Fermentation of 1,2-propanediol with 1,2-ethanediol by some genera of *Enterobacteriaceae*, involving coenzyme B_{12}-dependent diol dehydratase. *Journal of Bacteriology, 139*, 39–47.
Toraya, T., Honda, S., Kuno, S., & Fukui, S. (1978). Coenzyme B_{12}-dependent diol dehydratase: Regulation of apoenzyme synthesis in *Klebsiella pneumoniae* (*Aerobacter aerogenes*) ATCC 8724. *Journal of Bacteriology, 135*, 726–729.
Toraya, T., Honda, S., & Mori, K. (2010). Coenzyme B_{12}-dependent diol dehydratase is a potassium ion-requiring calcium metalloenzyme: Evidence that the substrate-coordinated metal ion is calcium. *Biochemistry, 49*, 7210–7217.
Toraya, T., & Ishida, A. (1988). Acceleration of cleavage of the carbon-cobalt bond of sterically hindered alkylcobalamins by binding to apoprotein of diol dehydrase. *Biochemistry, 27*, 7677–7681.
Toraya, T., & Ishida, A. (1991). Roles of the D-ribose and 5,6-dimethylbenzimidazole moieties of the nucleotide loop of adenosylcobalamin in manifestation of coenzymic function in the diol dehydrase reaction. *Journal of Biological Chemistry, 266*, 5430–5437.
Toraya, T., Krodel, E., Mildvan, A. S., & Abeles, R. H. (1979). Role of peripheral side chains of vitamin B_{12} coenzymes in the reaction catalyzed by dioldehydrase. *Biochemistry, 18*, 417–426.
Toraya, T., Kuno, S., & Fukui, S. (1980). Distribution of coenzyme B_{12}-dependent diol dehydratase and glycerol dehydratase in selected genera of *Enterobacteriaceae* and *Propionibacteriaceae*. *Journal of Bacteriology, 141*, 1439–1442.
Toraya, T., Matsumoto, T., Ichikawa, M., Itoh, T., Sugawara, T., & Mizuno, Y. (1986). The synthesis of adenine-modified analogs of adenosylcobalamin and their coenzymic function in the reaction catalyzed by diol dehydrase. *Journal of Biological Chemistry, 261*, 9289–9293.
Toraya, T., Miyoshi, S., Mori, M., & Wada, K. (1994). The synthesis of a pyridyl analog of adenosylcobalamin and its coenzymic function in the diol dehydratase reaction. *Biochimica et Biophysica Acta, 1204*, 169–174.
Toraya, T., Shirakashi, T., Fukui, S., & Hogenkamp, H. P. C. (1975). Coenzyme action of adenosyl-13-epicobalamin in the diol dehydrase system. *Biochemistry, 14*, 3949–3952.
Toraya, T., Shirakashi, T., Kosuga, T., & Fukui, S. (1976). Substrate specificity of coenzyme B_{12}-dependent diol dehydrase: Glycerol as both a good substrate and a potent inactivator. *Biochemical and Biophysical Research Communications, 69*, 475–480.
Toraya, T., Sugimoto, Y., Tamao, Y., Shimizu, S., & Fukui, S. (1971). Propanediol dehydratase system. Role of monovalent cations in binding of vitamin B_{12} coenzyme or its analogs to apoenzyme. *Biochemistry, 10*, 3475–3484.

Toraya, T., Tamura, N., Watanabe, T., Yamanishi, M., Hieda, N., & Mori, K. (2008). Mechanism-based inactivation of coenzyme B_{12}-dependent diol dehydratase by 3-unsaturated 1,2-diols and thioglycerol. *Journal of Biochemistry, 144*, 437–446.

Toraya, T., Uesaka, M., & Fukui, S. (1974). Coenzyme B_{12} dependent diol dehydrase system. Dissociation of the enzyme into two different protein components and some properties of the components. *Biochemistry, 13*, 3895–3899.

Toraya, T., Ushio, K., Fukui, S., & Hogenkamp, H. P. C. (1977). Studies on the mechanism of the adenosylcobalamin-dependent diol dehydrase reaction by the use of analogs of the coenzyme. *Journal of Biological Chemistry, 252*, 963–970.

Toraya, T., Watanabe, N., Ichikawa, M., Matsumoto, T., Ushio, K., & Fukui, S. (1987). Activation and cleavage of the carbon-cobalt bond of adeninylethylcobalamin by diol dehydrase. *Journal of Biological Chemistry, 262*, 8544–98550.

Toraya, T., Yamanishi, M., Muguruma, H., Ushio, K., Yamauchi, J., & Kawamura, T. (1997). An electron paramagnetic resonance study on the mechanism-based inactivation of adenosylcobalamin-dependent diol dehydrase by glycerol and other substrates. *Biochimica et Biophysica Acta, 1337*, 11–16.

Ushio, K., Fukui, S., & Toraya, T. (1984). Coenzymic function of 1- or N^6-substituted analogs of adenosylcobalamin in the diol dehydratase reaction. *Biochimica et Biophysica Acta, 788*, 318–326.

Valinsky, J. E., Abeles, R. H., & Fee, J. A. (1974). Electron spin resonance studies on diol dehydrase. III. Rapid kinetic studies on the rate of formation of radicals in the reaction with propanediol. *Journal of the American Chemical Society, 96*, 4709–4710.

Wagner, O. W., Lee, H. A., Jr., Frey, P. A., & Abeles, R. H. (1966). Studies on the mechanism of action of cobamide coenzymes. Chemical properties of the enzyme-coenzyme complex. *Journal of Biological Chemistry, 241*, 1751–1762.

Wallis, O. C., Johnson, A. W., & Lappert, M. F. (1979). Studies on the subunit structure of the adenosylcobalamin-dependent enzyme ethanolamine ammonia-lyase. *FEBS Letters, 97*, 196–199.

Yamanishi, M., Ide, H., Murakami, Y., & Toraya, T. (2005). Identification of the 1,2-propanediol-1-yl radical as an intermediate in adenosylcobalamin-dependent diol dehydratase reaction. *Biochemistry, 44*, 2113–2118.

Yamanishi, M., Kinoshita, K., Fukuoka, M., Saito, T., Tanokuchi, A., Ikeda, Y., et al. (2012). Redesign of coenzyme B_{12} dependent diol dehydratase to be resistant to the mechanism-based inactivation by glycerol and act on longer chain 1,2-diols. *FEBS Journal, 279*, 793–804.

Yamanishi, M., Yamada, S., Ishida, A., Yamauchi, J., & Toraya, T. (1998). EPR spectroscopic evidence for the mechanism-based inactivation of adenosylcobalamin-dependent diol dehydratase by coenzyme analogs. *Journal of Biochemistry, 124*, 598–601.

Yamanishi, M., Yamada, S., Muguruma, H., Murakami, Y., Tobimatsu, T., Ishida, A., et al. (1998). Evidence for axial coordination of 5,6-dimethylbenzimidazole to the cobalt atom of adenosylcobalamin bound to diol dehydratase. *Biochemistry, 37*, 4799–4803.

Yamanishi, M., Yunoki, M., Tobimatsu, T., Sato, H., Matsui, J., Dokiya, A., et al. (2002). The crystal structure of coenzyme B_{12}-dependent glycerol dehydratase in complex with cobalamin and propane-1,2-diol. *European Journal of Biochemistry, 269*, 4484–4494.

Zagalak, B., Frey, P. A., Karabatsos, G. L., & Abeles, R. H. (1966). The stereochemistry of the conversion of D and L 1,2-propanediols to propionaldehyde. *Journal of Biological Chemistry, 241*, 3028–3035.

CHAPTER TEN

Reactivating chaperones for coenzyme B_{12}-dependent diol and glycerol dehydratases and ethanolamine ammonia-lyase

Tetsuo Toraya[a,*], Takamasa Tobimatsu[a], Naoki Shibata[b], and Koichi Mori[a]

[a]Graduate School of Natural Science and Technology, Okayama University, Tsushima-naka, Kita-ku, Okayama, Japan
[b]Graduate School of Life Science, University of Hyogo, 3-2-1 Koto, Kamigori-cho, Ako-gun, Hyogo, Japan
*Corresponding author: e-mail address: toraya@cc.okayama-u.ac.jp

Contents

Methods in Enzymology, Volume 668
ISSN 0076-6879
https://doi.org/10.1016/bs.mie.2021.11.028

Abstract

Adenosylcobalamin (AdoCbl) or coenzyme B_{12}-dependent enzymes tend to undergo mechanism-based inactivation during catalysis or inactivation in the absence of substrate. Such inactivation may be inevitable because they use a highly reactive radical for catalysis, and side reactions of radical intermediates result in the damage of the coenzyme. How do living organisms address such inactivation when enzymes are inactivated by undesirable side reactions? We discovered reactivating factors for radical B_{12} eliminases. They function as releasing factors for damaged cofactor(s) from enzymes and thus mediate their exchange for intact AdoCbl. Since multiple turnovers and chaperone functions were demonstrated, they were renamed "reactivases" or "reactivating chaperones." They play an essential role in coenzyme recycling as part of the activity-maintaining systems for B_{12} enzymes. In this chapter, we describe our investigations on reactivating chaperones, including their discovery, gene cloning, preparation, characterization, activity assays, and mechanistic studies, that have been conducted using a wide range of biochemical and structural methods that we have developed.

1. Introduction

Glycerol is fermented by a significant number of bacteria with accumulating 1,3-propanediol (Mickelson & Werkman, 1940; Sobolov & Smiley, 1960). Adenosylcobalamin (AdoCbl) (Fig. 1A)-dependent glycerol dehydratase (GD) plays an essential role in this type of fermentation by producing the electron acceptor β-hydroxypropionaldehyde (Abeles, Brownstein, & Randles, 1960; Toraya, Kuno, & Fukui, 1980) (see Fig. 2E in chapter "Coenzyme B_{12}-dependent eliminases: Diol and glycerol dehydratases and ethanolamine ammonia-lyase" by Toraya et al.). Diol dehydratase (DD) and GD are AdoCbl-dependent isofunctional enzymes (Toraya & Fukui, 1977), both of which catalyze the conversion of 1,2-propanediol, glycerol, and 1,2-ethanediol to the corresponding aldehydes (Fig. 1C) (Lee & Abeles, 1963; Schneider & Pawelkiewicz, 1966; Toraya, Shirakashi, Kosuga, & Fukui, 1976). Some genera of *Enterobacteriaceae*, such as *Klebsiella* and *Citrobacter*, grown anaerobically on glycerol and on 1,2-propanediol produce AdoCbl-dependent GD and DD, respectively (Forage & Foster, 1982; Toraya et al., 1980; Toraya & Fukui, 1977). In contrast, *Klebsiella oxytoca* ATCC 8724 produces only DD, although it grows well on glycerol (Toraya, Honda, Kuno, & Fukui, 1978). This indicates that GD is functionally substituted by DD that is induced at a low level when bacteria are grown in a glycerol medium. In this sense, glycerol is a natural substrate of DD as well. Despite these metabolic roles, the holoenzymes of both GD and DD (holoGD and holoDD) undergo rapid, mechanism-based inactivation by the growth substrate glycerol (Bachovchin, Eagar, Moore, & Richards, 1977; Poznanskaya, Yakusheva, & Yakovlev, 1977; Schneider & Pawelkiewicz, 1966; Toraya et al., 1976), the result of damage to the enzyme-bound coenzyme.

Ethanolamine ammonia-lyase (EAL) catalyzes AdoCbl-dependent deamination of ethanolamine to acetaldehyde (Fig. 1D) (Bradbeer, 1965b). The enzyme is involved in the fermentation of this substrate (Bradbeer, 1965a) or the utilization of this substrate as a sole nitrogen source (Chang & Chang, 1975) (see Fig. 2D in chapter "Coenzyme B_{12}-dependent eliminases: Diol and glycerol dehydratases and ethanolamine ammonia-lyase" by Toraya et al.). The holoenzyme of EAL (holoEAL) also undergoes

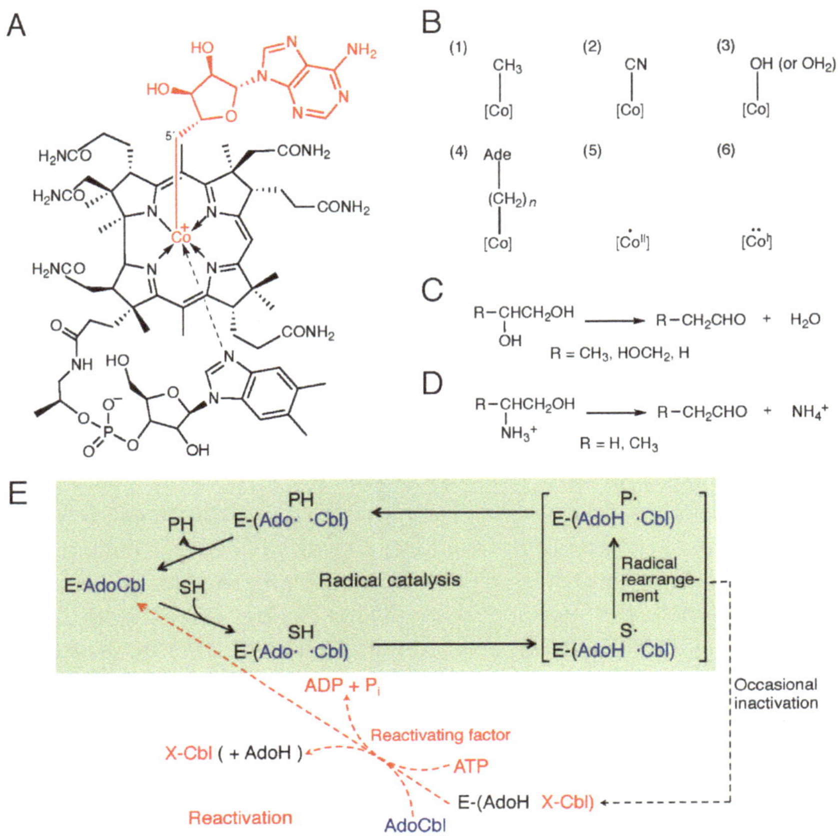

Fig. 1 AdoCbl-dependent dehydratases and a deaminase. (A) Coenzyme B_{12} (AdoCbl). (B) Some analogs of vitamin B_{12}. (1) MeCbl. (2) CN-Cbl. (3) OH-Cbl (or aqCbl). (4) AdePeCbl (n = 5), AdeBuCbl (n = 4), and AdeEtCbl (n = 2). (5) Cob(II)alamin (B_{12r}). (6) Cob(I)alamin (B_{12s}). (C) Reactions catalyzed by DD and GD. (D) Reactions catalyzed by EAL. (E) Mechanism-based inactivation by natural substrates and reactivation by reactivating factors. E, apoenzyme; SH, substrate; S•, substrate radical; PH, product; P•, product radical; Ado•, adenosyl radical; AdoH, 5′-deoxyadenosine; Cbl, cobalamin; •Cbl, cob(II)alamin; X-Cbl, damaged cofactor(s).

gradual mechanism-based inactivation by the natural substrate, ethanolamine, during catalysis (Bradbeer, 1965b; Mori, Bando, Hieda, & Toraya, 2004) as well as inactivation in the absence of substrate (Kaplan & Stadtman, 1968).

One of our major research interests was: how do living organisms address the inevitable inactivation of radical B_{12} enzymes that are indispensable for the dissimilation of growth substrates by themselves? Our discovery of reactivating factors in living organisms has provided convincing solutions to this question. This chapter focuses on the major biochemical and structural methodologies we have developed in our studies of these protein factors over more than 35 years, which are arranged along our research story from metabolic studies to our current structure-function studies. The methodologies for studying the AdoCbl-dependent eliminases DD, GD, and EAL are described in another chapter in this volume (see chapter "Coenzyme B_{12}-dependent eliminases: Diol and glycerol dehydratases and ethanolamine ammonia-lyase" by Toraya et al.).

2. Mechanism-based inactivation by natural substrates and *in situ* reactivation

2.1 Diol dehydratase (DD) and glycerol dehydratase (GD)

As described above, glycerol is a natural substrate not only for GD but also for DD. Despite their metabolic roles, both enzymes undergo mechanism-based inactivation by glycerol (Bachovchin, Eagar, Moore, & Richards, 1977; Poznanskaya et al., 1977; Schneider & Pawelkiewicz, 1966; Toraya et al., 1976), accompanying the irreversible cleavage of the coenzyme's Co—C bond (Fig. 1E). In the inactivation, AdoCbl is converted to 5′-deoxyadenosine (Bachovchin, Moore, & Richards, 1978) and an unidentified cobalamin that shows an alkylcobalamin-like spectrum in DD (Fig. 2B) (Bachovchin, Eagar, Moore, & Richards, 1977; Mori & Toraya, 1999), but the latter is gradually degraded to hydroxocobalamin (OH-Cbl) (Fig. 1B(3)) during dialysis (Yamanishi et al., 2012). The holoenzymes of both dehydratases undergo inactivation by O_2 in the absence of substrate as well (Lee & Abeles, 1963; Poznanskaya et al., 1977; Wagner, Lee, Frey, & Abeles, 1966). Damaged cofactor(s) of inactivated holoenzymes are not displaceable by free, intact AdoCbl, thus resulting in the irreversible inactivation of the enzymes.

For a long time, why these enzymes necessary for glycerol fermentation are suicidally inactivated by the growth substrate glycerol itself remained enigmatic. Does the inactivation of these enzymes by natural substrates take place in the cell or not? Does reactivation occur in the cell? We examined these possibilities by using bacterial cells permeabilized by toluene treatment, so-called *in situ* systems, and found with permeabilized *Klebsiella pneumoniae* ATCC 25955 and *K. oxytoca* ATCC 8724 cells that GD and DD did not apparently undergo inactivation in the presence of AdoCbl, ATP, and Mg^{2+} (or Mn^{2+}) (Protocol 9.1), although inactivation by glycerol occurs *in situ* as well (Honda, Toraya, & Fukui, 1980; Ushio, Honda, Toraya, & Fukui, 1982). Rapid reactivation of glycerol-inactivated enzymes took place *in situ* upon the addition of ATP and Mg^{2+} (or Mn^{2+}) to the completely inactivated system in the presence of AdoCbl (Fig. 1E) (Honda et al., 1980) (Protocol 9.2). These findings solved the above-mentioned enigma or apparent inconsistency. It was shown that the relative affinity of GD or DD for aquacobalamin (aqCbl) (Fig. 1B(3)), cyanocobalamin (CN-Cbl) (Fig. 1B(2)), or methylcobalamin (MeCbl) (Fig. 1B(1)) is lowered by ATP and Mn^{2+}, whereas that for adeninylbutylcobalamin (AdeBuCbl) (Fig. 1B(4), $n=4$) or adeninylethylcobalamin (AdeEtCbl) (Fig. 1B(4), $n=2$) is not (Ushio et al., 1982). Since no reducing agents were required for the *in situ* reactivation, the Co-adenosylation mechanism, *i.e.*, regeneration of AdoCbl from the enzyme-bound damaged cofactor, was less likely. The absolute requirement for free AdoCbl in the reactivation suggested that the B_{12}-exchange mechanism, *i.e.*, displacement of the enzyme-bound damaged cofactor by intact AdoCbl, is more likely (Honda et al., 1980; Ushio et al., 1982). Since reactivation was detectable only *in situ* but not *in vitro*, it remained unclear for a long time whether a specific proteinaceous factor was involved in the reactivation or not.

2.2 Ethanolamine ammonia-lyase (EAL)

EAL also undergoes mechanism-based inactivation gradually by the natural substrate ethanolamine (Bradbeer, 1965b; Mori et al., 2004) and quickly by the substrate analog 2-amino-1-propanol (Carty, Babior, & Abeles, 1974) during catalysis (Fig. 1E) as well as inactivation in the absence of substrate (Kaplan & Stadtman, 1968). In either case, the irreversible cleavage of the coenzyme's Co—C bond takes place, and the damaged cofactor remains tightly bound to the enzyme and causes inactivation (Fig. 1E). When *Escherichia coli* JM109 is grown in the presence of both ethanolamine and CN-Cbl, EAL is induced to a high level (Mori et al., 2004; Scarlett & Turner, 1976). EAL in permeabilized *E. coli* cells also underwent gradual and complete inactivation by ethanolamine during catalysis within 20 min. Such inactivation did not take place at all when ATP and Mg^{2+} were supplemented in addition to AdoCbl (Mori et al., 2004). This indicates that the inactivated holoenzyme in permeabilized cells undergoes rapid *in situ* reactivation (Fig. 1E) (Protocols 9.1 and 9.2). It should be noted that, unlike the cases of GD and DD, the reactivation of inactivated EAL is observed not only *in situ* but also *in vitro*, *i.e.*, in homogenates (Mori et al., 2004).

3. Discovery and identification of the gene(s) encoding reactivating factors

3.1 Diol dehydratase-reactivating factor (DD-R) genes

Since the induction by glycerol of some factor(s) necessary for the *in situ* reactivation was indirectly suggested, we had an intuition that some protein(s) might be encoded by nearby gene(s) for GD or DD. When DD genes had been cloned for the first time in our laboratory, we searched for gene(s) encoding a putative reactivating factor for DD by genetic engineering techniques. We found that glycerol-inactivated holoDD in permeabilized *E. coli* cells carrying a plasmid containing the DD genes (*pddABC*) and two open reading frames (ORFs) in the 3′-flanking region undergoes rapid reactivation in the presence of free AdoCbl, ATP, and Mg^{2+}, whereas that in the cells carrying a plasmid containing DD genes alone or DD genes and a ORF in the 5′-flanking region does not (Mori, Tobimatsu, Hara, & Toraya, 1997; Mori, Tobimatsu, & Toraya, 1997). The inactive complexes of DD with CN-Cbl, aqCbl, and pentylcobalamin (PeCbl) in the cells also underwent rapid activation, but the complex with adeninylpentylcobalamin (AdePeCbl) (Fig. 1B(4), $n = 5$) did not (Mori, Tobimatsu, & Toraya, 1997). These results showed that certain protein(s) encoded by ORFs in the 3′-flanking region of DD genes are responsible for the *in situ* (re)activation. These two ORFs in the 3′-flanking region were identified as the genes encoding a reactivating factor for inactivated DD (Mori, Tobimatsu, Hara, & Toraya, 1997). They were designated *ddrAB* genes as they encode a "*DD-re*activating factor (DD-R)." It turned out that they correspond to *pduGH* genes of *Salmonella typhimurium* (see Fig. 2A in chapter "Coenzyme B_{12}-dependent eliminases: Diol and glycerol dehydratases and ethanolamine ammonia-lyase" by Toraya et al.).

The *ddrAB* genes encode polypeptides consisting of 610 and 125 amino acid residues with predicted molecular weights of 64,266 and 13,620, respectively (Mori, Tobimatsu, Hara, & Toraya, 1997). We constructed the expression plasmids pCXV(5b-6) and pCXV(6/5b) in which *ddrAB* genes are placed in the natural order and in the reversed order, respectively, downstream of the *tac* promoter (Mori, Tobimatsu, Hara, & Toraya, 1997). It should be emphasized that the expression vector pCXV (replication origin of p15A) is compatible with pUSI2E (replication origin of pBR322) that we used for the high-level expression of DD genes (Tobimatsu et al., 1995). Hence, *in situ* functional analysis of reactivating factor genes is possible by the co-expression of enzyme and reactivating factor genes using these expression vectors (Protocol 9.3). For example, when the *ddrA* gene alone was co-expressed with the DD genes, ~7% reactivation was observed, whereas the *ddrB* gene alone almost no reactivation. These results suggested that the DdrA protein is the essential component of DD-R, while DdrB is an auxiliary protein. When *E. coli* JM109 carrying pCXV(5b-6) or pCXV(6/5b) was cultivated at 37 °C and induced by isopropyl-1-thio-β-D-galactopyranoside (IPTG), the production of M_r 64,000 and 14,000 polypeptides was observed by sodium dodecyl sulfate (SDS)-polyacrylamide gel electrophoresis (PAGE), but the band of latter polypeptide was not as intense with pCXV(5b-6), as compared with pCXV(6/5b). For high-level expression of the DD-R genes, therefore, we constructed the expression plasmid pUSI2ENd(6/5b) in which the *ddrAB* genes are placed downstream of the *tac* promoter in the reversed order (Mori, Tobimatsu, Hara, & Toraya, 1997). For the large-scale preparation of DD-R, *E. coli* JM109 carrying this plasmid is cultivated at 30 °C in Luria-Bertani (LB) medium and induced by IPTG (Toraya & Mori, 1999).

3.2 Glycerol dehydratase-reactivating factor (GD-R) genes

A homology search revealed that polypeptides homologous to *K. oxytoca* DdrA and DdrB proteins are encoded by ORF4 (Tobimatsu et al., 1996) or *dhaB4* and *orf2b*, respectively, of *K. pneumoniae* in the vicinity of the GD genes (*gldABC* or *dhaB1B2B3*) (corresponding to *Citrobacter freundii dhaBCE* in Fig. 2B, chapter "Coenzyme B_{12}-dependent eliminases: Diol and glycerol dehydratases and ethanolamine ammonia-lyase" by Toraya et al.) (Tobimatsu, Kajiura, Yunoki, Azuma, & Toraya, 1999). By co-expressing these two ORFs with the GD genes, the *in situ* reactivation of glycerol-inactivated holoGD, as well as the *in situ* activation of inactive GD•CN-Cbl complex, was confirmed (Tobimatsu et al., 1999) (see Protocol 9.2). Co-expression of either gene alone was not effective for GD (re)activation. Thus, we identified these genes of *K. pneumoniae* ATCC 25955 as the genes encoding a reactivating factor for inactivated GD (Tobimatsu et al., 1999) and named them *gdrAB* as they encode the *GD-r*eactivating factor (GD-R). *C. freundii dhaFG* genes, orthologs of *gdrAB*, were later confirmed to encode GD-R by other investigators (Seifert, Bowien, Gottschalk, & Daniel, 2001) (see Fig. 2B in chapter "Coenzyme B_{12}-dependent eliminases: Diol and glycerol dehydratases and ethanolamine ammonia-lyase" by Toraya et al.).

The *gdrAB* genes encode polypeptides consisting of 607 and 117 amino acid residues with predicted molecular weights of 63,594 and 11,994, respectively. For high-level expression of the GD-R genes, we constructed the expression plasmids pCXV(*gdrB*•*gdrA*) and pET(*gdrB*•*gdrA*) in which *gdrAB* genes are placed downstream of a *tac* promoter (Mori, Tobimatsu, Hara, & Toraya, 1997) and a T7 promoter, respectively, in this order. The former plasmid has been routinely used for the co-expression experiments of GD-R genes with the GD genes in pUSI2E(GD) (Tobimatsu et al., 1996). For the large-scale preparation of GD-R, *E. coli* BL21(DE3) carrying the latter plasmid is cultivated at 30 °C and induced by IPTG overnight at 17 °C (Kajiura, Mori, Tobimatsu, & Toraya, 2001).

3.3 Ethanolamine ammonia-lyase-reactivating factor (EAL-R) genes

Among 17 genes in the *S. typhimurium* (see Fig. 2C in chapter "Coenzyme B_{12}-dependent eliminases: Diol and glycerol dehydratases and ethanolamine ammonia-lyase" by Toraya et al.) and *E. coli eut* (*e*thanolamine *ut*ilization) operons (Kofoid, Rappleye, Stojiljkovic, & Roth, 1999), *eutA* is adjacent to the 5′-terminal of the EAL genes (*eutBC*). Genetic evidence shows that *eutA* mutants are sensitive to growth inhibition by CN-Cbl (Roof & Roth, 1989). A homology search indicated that the EutA protein shows no overall similarity but high fragmentary similarities with DdrA and GdrA in the three regions consisting of 15–18 amino acid residues (Mori et al., 2004; Toraya, 2003). Co-expression experiments with permeabilized *E. coli* cells demonstrated that the *eutA* gene is essential for the *in situ* reactivation of ethanolamine-inactivated holoEAL and the *in situ* activation of inactive EAL•CN-Cbl complex (Protocol 9.2). Thus, *eutA* was identified as the gene encoding an EAL-reactivating factor (EAL-R) (Mori et al., 2004).

The *eutA* gene of *E. coli* encodes a polypeptide consisting of 467 amino acid residues with a predicted molecular weight of 49,599. No evidence has been obtained to date that shows the presence of the auxiliary protein(s) potentiating the reactivation or associating with EutA (Mori et al., 2004). For high-level expression of the EAL-R gene, we constructed the expression plasmid pUSI2ENd(eutA) in which the *eutA* gene is placed downstream of a *tac* promoter (Mori et al., 2004). For the large-scale preparation of EAL-R, *E. coli* JM109 carrying this plasmid is cultivated at 37 °C in LB medium and induced by IPTG (Mori et al., 2004).

4. Purification, assays, and protein-chemical properties of recombinant reactivating factors

4.1 Assays of *in vitro* (re)activation

The *in vitro* reactivation of glycerol-inactivated holoenzymes or O_2-inactivated holoenzymes (Protocol 9.4) of DD and GD and the activation of the enzyme•CN-Cbl complexes (Protocols 9.5) by cognate reactivating factors, *i.e.*, (re)activation of DD by DD-R and of GD by GD-R, are assayed using 1,2-propanediol as substrate in the presence of AdoCbl, ATP, and $MgCl_2$ (Kajiura et al., 2001; Mori & Toraya, 1999; Toraya & Mori, 1999). The *in vitro* reactivation of ethanolamine-inactivated holoEAL or holoEAL inactivated in the absence of substrate (Protocol 9.4) and the activation of EAL•CN-Cbl complex (Protocols 9.5) are assayed using ethanolamine as substrate in the presence of AdoCbl, ATP, and $MgCl_2$ (Mori et al., 2004). The amount of propionaldehyde or acetaldehyde formed during incubation for appropriate time periods is determined by the 3-methyl-2-benzothiazolinone hydrazone (MBTH) method (Toraya, Ushio, Fukui, & Hogenkamp, 1977).

4.2 Purification and protein-chemical properties of DD-R and GD-R

Recombinant wild-type DdrA and DdrB proteins of *K. oxytoca* were co-purified to homogeneity from *E. coli* by a conventional method, *i.e.*, $(NH_4)_2SO_4$ fractionation and column chromatography on diethylaminoethyl (DEAE)-cellulose, hydroxyapatite, and Sephadex G-200 (Toraya & Mori, 1999) (Protocol 9.6). Recombinant wild-type GdrA and GdrB proteins of *K. pneumoniae* were also co-purified to homogeneity from overexpressing *E. coli* by a conventional method, *i.e.*, $(NH_4)_2SO_4$ fractionation and Sepharose CL-6B column chromatography (Kajiura et al., 2001) (Protocol 9.7). DD-R consists of equimolar amounts of two subunits with M_rs of 64,000 (α) and 19,000 (β), which correspond to the products of *ddrAB* genes, respectively. GD-R is also composed of equimolar amounts of two subunits with M_rs of 64,000 (α) and 12,000 (β), which correspond to the products of *gdrAB* genes, respectively. The molecular weights of DD-R and GD-R determined by fast protein liquid chromatography (FPLC) with a calibrated Superose 6 column were ~150,000 and ~135,000, respectively (Kajiura et al., 2001; Toraya & Mori, 1999). Thus, the subunit structures of both DD-R and GD-R are $\alpha_2\beta_2$. Purified DD-R and GD-R functioned as reactivating factors for glycerol-inactivated and O_2-inactivated holoenzymes of DD and GD, respectively, in the presence of AdoCbl, ATP, and Mg^{2+} (Kajiura et al., 2001; Toraya & Mori, 1999). They activated respective enzyme•CN-Cbl complexes as well. The capability of reactivating glycerol-inactivated or O_2-inactivated holoenzymes is well correlated with that of activating the enzyme•CN-Cbl complexes. The specificity of DD-R for the base and ribose moieties of nucleoside triphosphates was rather broad (relative effectiveness (%): ATP (100); CTP (83); UTP (75); GTP (57); 3′-deoxyATP (64); 2′-deoxyATP (55)). Its specificity for divalent metal ions was also not strict (relative effectiveness (%): Mg^{2+} (100); Mn^{2+} (185); Co^{2+} (97); Ni^{2+} (48); Ca^{2+}, Cu^{2+}, and Cr^{2+} (<2)) (Mori, Hosokawa, Yoshinaga, & Toraya, 2010).

4.3 Purification and protein-chemical properties of EAL-R

Recombinant wild-type EutA protein of *E. coli* was purified to homogeneity from overexpressing *E. coli* by a conventional method, *i.e.*, $(NH_4)_2SO_4$ fractionation and column chromatography on DEAE-cellulose, hydroxyapatite, and Sephadex G-200 (Mori et al., 2004) (Protocol 9.6). The EutA protein migrated as a single band with a M_r of 50,000 upon SDS-PAGE, whereas it migrated as a very broad band upon nondenaturing PAGE, indicating that EutA has a tendency to self-associate or aggregate. In the presence of AdoCbl, ATP and Mg^{2+}, purified EutA functioned as a reactivating factor for ethanolamine-inactivated holoEAL and holoEAL inactivated in the absence of substrate (Mori et al., 2004). It activated the inactive EAL•CN-Cbl complex as well under the same conditions.

5. Mechanistic studies

5.1 Cobalamin exchange and intermediate formation of apoenzymes

The absolute requirement for free AdoCbl in the (re)activation of inactivated holoenzymes and the inactive enzyme•CN-Cbl complexes of DD and GD strongly suggested that the (re)activation takes place by the exchange of enzyme-bound, damaged cofactor(s) or CN-Cbl for intact AdoCbl. To examine the possibility of DD-R- or GD-R-mediated cobalamin exchange, CN-Cbl and AdePeCbl were used as models of an adenine-lacking, damaged cofactor and intact AdoCbl, respectively (Protocol 9.8). AdePeCbl is an inactive coenzyme analog containing the adenine moiety in the upper axial ligand. When the enzyme•CN-Cbl complexes were

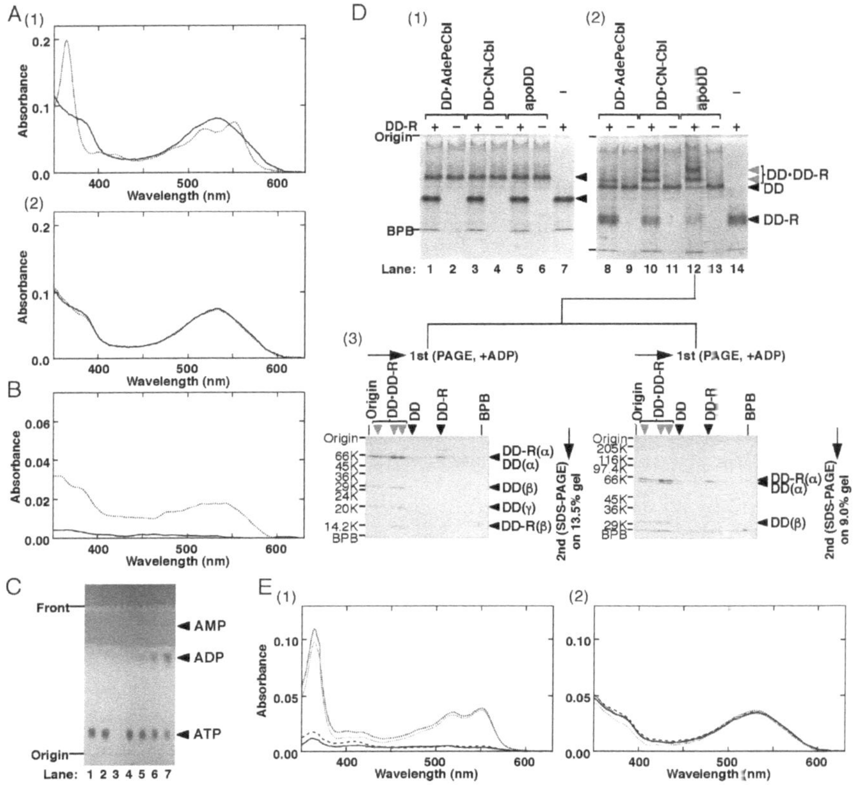

Fig. 2 Selected experimental approaches to investigating the action mechanisms of reactivating factors. (A) Spectral evidence for cobalamin exchange (Protocol 9.8). The exchange of DD-bound CN-Cbl for free AdePeCbl occurs with DD-R (*solid line*), but not without DD-R (*dotted line*) (1), but the reverse is not the case (2). (B) Intermediate formation of apoDD (Protocol 9.10). Spectra of glycerol-inactivated holoDD with (*solid line*) and without (*dotted line*) DD-R after sequential dialysis at 4 °C for 18 h against a buffer containing 20 mM ATP/Mg^{2+} and 50 mM K_2SO_3, and then for additional 2 days against copious buffer without ATP/Mg^{2+} and K_2SO_3. (C) TLC analysis of ATPase activity of DD-R (for TLC conditions, see Text). Lane 1, ATP; lane 2, ATP (60 min at 37 °C); lane 3, DD-R; lanes 4–7, ATP + DD-R (0, 10, 30, and 60 min at 37 °C, respectively). (D) PAGE and SDS-PAGE analyses of tight complex formation between DD and DD-R in the presence of ATP/Mg^{2+} (1) or ADP/Mg^{2+} (2) and of subunit compositions (3) (Protocol 9.11). Lane 12 of (2) containing the bands of DD•DD-R complexes is excised and subjected to SDS-PAGE (second dimension) on 13.5% and 9% gels. (E) Spectral evidence for cobalamin release. DD•CN-Cbl (1) and DD•AdePeCbl (2) complexes are incubated at 37 °C for 1.5 h with DD-R in the absence (*thin dotted line*) and presence of 16 mM ATP/Mg^{2+} (*solid line*) or ADP/Mg^{2+} (*dashed line*) and without DD-R (*dotted line*) and ultrafiltrated.

incubated with DD-R or GD-R in the presence of AdePeCbl, ATP, and Mg^{2+}, followed by dialysis to remove unbound cobalamins, the spectra of dialysates indicated that enzyme-bound CN-Cbl is replaced by AdePeCbl (Fig. 2A(1)) (Kajiura, Mori, Tobimatsu, & Toraya, 2001; Toraya & Mori, 1999). This exchange did not occur without DD-R or GD-R. In contrast, enzyme-bound AdePeCbl was not replaced at all by CN-Cbl even in the presence of a reactivating factors, ATP, and Mg^{2+} (Fig. 2A(2)).

To test the possibility that only the upper axial ligand undergoes exchange, the enzyme•[^{57}Co]CN-Cbl complexes were incubated with and without DD-R or GD-R in the presence of AdePeCbl, ATP, and Mg^{2+} and then subjected to ultrafiltration to remove unbound cobalamins (Protocol 9.9). The ^{57}Co radioactivity bound to DD or GD was almost completely lost in the presence of reactivating factors, whereas the radioactivity corresponding to about 2 mol of CN-Cbl/mol of enzyme remains bound to DD or GD in the absence of reactivating factors (Kajiura et al., 2001; Mori & Toraya, 1999). It was thus concluded that DD-R and GD-R mediate the exchange not of the upper ligand but the whole molecule of enzyme-bound CN-Cbl.

Are apoenzymes formed as intermediates in the (re) activation? The enzyme•CN-Cbl complexes were incubated with DD-R or GD-R in the presence of ATP and Mg^{2+} and then dialyzed to remove unbound cobalamin. However, no significant dissociation of CN-Cbl was observed, possibly because of recombination with apoenzyme. We used the glycerol-inactivated holoenzymes or enzyme•OH-Cbl•5′-deoxyadenosine complexes and added $SO_3{}^{2-}$ to the dialysis buffer to prevent the dissociated OH-Cbl from recombination (Protocol 9.10). $SO_3{}^{2-}$ is known to react with free aqCbl or OH-Cbl forming sulfitocobalamin (Hill, Pratt, & Williams, 1962) which binds to DD only loosely (Toraya, Watanabe, Ushio, Matsumoto, & Fukui, 1983). However, $SO_3{}^{2-}$ does not react with the glycerol-inactivated holoDD or DD•OH-Cbl•5′-deoxyadenosine complex, probably because the *Co*β position is blocked by the enzyme-bound 5′-deoxyadenosine (Toraya et al., 1983). Both the glycerol-inactivated holoenzymes (Fig. 2B) and the enzyme•OH-Cbl•5′-deoxyadenosine complexes (data not shown) underwent dissociation of cobalamins in the presence of DD-R or GD-R, but not in its absence (Kajiura et al., 2001; Mori & Toraya, 1999). The apoenzymes thus formed could be reconstituted into active holoenzymes with added AdoCbl. Thus, it is evident that the cobalamin exchange in the (re)activation takes place through intermediate formation of apoenzyme.

5.2 ATPase activity and tight complex formation of reactivating factors with apoenzymes

Upon incubation of ATP with DD-R or GD-R, a time-dependent decrease of ATP and the formation of ADP were observed by thin-layer chromatography (TLC) on polyethyleneimine-cellulose F with 2 M formic acid containing 0.5 M LiCl as a solvent system (Mangold, 1969) (Fig. 2C) (Kajiura et al., 2001; Mori & Toraya, 1999). The adenine nucleotides were located under ultraviolet (UV) light at 254 nm. For quantitative analysis, the release of [^{32}P]P$_i$ from [γ-^{32}P]ATP upon incubation with DD-R or GD-R was measured by a modification of the reported procedure (Schnebli & Abrams, 1970). The ATP-hydrolyzing activity (k_{cat}) was 1.3–1.5 min^{-1} for DD-R (Mori & Toraya, 1999) and 0.9 (reactivating conditions) or 1.6 (non-reactivating conditions) for GD-R (Kajiura et al., 2001), indicating that both reactivating factors are ATPases with extremely low activity.

Complex formation between DD-R or GD-R and the apoenzyme of DD or GD were analyzed by nondenaturing PAGE (Protocols 9.11). When apoenzymes are incubated with reactivating factors in the presence of ADP, new major band(s) appear (Fig. 2D(2)) (Kajiura et al., 2001; Mori & Toraya, 1999). In the case of DD + DD-R, the development of the lane containing these bands (Fig. 2D *lane* 12) in the second dimension by SDS-PAGE showed that these band(s) contain DD•DD-R complexes consisting of all the three subunits of DD and the two subunits of DD-R (Fig. 2D(3)). Formation of such enzyme•reactivating factor complexes is not detected in the presence of ATP (Fig. 2D(1)). From these findings, it can be concluded that ADP- and ATP-bound DD-R and GD-R are high-affinity and low-affinity forms, respectively, for cognate dehydratases.

The importance of complex formation between enzyme and reactivating factor was investigated by cross-reactivation experiments as well. Cross-reactivation between DD-R and GD-R toward dehydratase substrates was first examined using permeabilized cells of co-expressing *E. coli* cells. DD-R efficiently cross-reactivated glycerol-inactivated GD, whereas GD-R hardly reactivated the inactivated DD (Tobimatsu, Kajiura, & Toraya, 2000). The *in situ* experiments with hybrid reactivating factors showed that GD-R is much more specific for the cognate dehydratase, and that the large subunit (α) principally decides the specificity for a dehydratase substrate. We confirmed *in vitro* with purified proteins that DD-R efficiently cross-reactivates glycerol-inactivated GD, whereas GD-R hardly reactivates the inactivated DD (Kajiura, Mori, Shibata, & Toraya, 2007). DD-R actually formed complexes with both DD and

GD, whereas GD-R formed complexes with GD but not with DD. It was thus shown that the specificities of the reactivating factors for the dehydratase enzymes are determined by their ability to form tight complexes with the apoenzymes.

5.3 Two-step mechanism: Reactivating factors are chaperones

When similar PAGE analyses of complex formation were carried out with the enzyme•cobalamin complexes, instead of apoenzymes in the presence of ADP, most of the enzyme•CN-Cbl complexes were converted to the enzyme•reactivating factor complexes (Fig. 2D(2)), whereas only small amounts of the enzyme•reactivating factor complexes were formed from the enzyme•AdePeCbl complexes (Kajiura et al., 2001; Mori & Toraya, 1999). In contrast, the enzyme•reactivating factor complexes were not formed at all in the presence of ATP from either the enzyme•CN-Cbl complexes or the enzyme•AdePeCbl complexes (Fig. 2D(1)). To examine whether the enzyme•reactivating factor complexes formed from the enzyme•CN-Cbl complexes in the presence of ADP still bind CN-Cbl, the enzyme•CN-Cbl complexes were incubated with DD-R or GD-R in the presence of ATP or ADP, followed by ultrafiltration to remove unbound cobalamin. The spectra of the protein fractions showed that essentially all the enzyme-bound CN-Cbl is released from the complexes by reactivating factors in the presence of either ADP or ATP, but not without the factors (Fig. 2E(1)) (Kajiura et al., 2001; Mori & Toraya, 1999). However, cobalamin release did not occur from the enzyme•AdePeCbl complexes (Fig. 2E(2)). It was thus concluded that reactivating factors release CN-Cbl, an adenine-lacking cobalamin, but not AdePeCbl, an adenine-containing cobalamin, in the presence of ADP or ATP through the complex formation with apoenzymes.

However, it turned out that the resulting "apoenzymes" formed by the dissociation of enzyme-bound CN-Cbl cannot be reconstituted into catalytically active holoenzyme by adding AdoCbl if ADP is present (Kajiura et al., 2001; Mori & Toraya, 1999). This suggested that the apoenzymes in the tight apoenzyme•reactivating factor•ADP complexes cannot bind AdoCbl. Actually, when apoenzymes were preincubated with reactivating factors in the presence of ADP, enzyme activity was strongly inhibited (Kajiura et al., 2001; Mori & Toraya, 1999), and this inhibition was reversed by the addition of ATP. Nondenaturing PAGE showed that ATP dissociates the ternary complexes to apoenzymes and the ATP-bound, low-affinity form of reactivating factors.

It was highly suggested that the (re)activation of inactivated holoenzymes and inactive enzyme•cobalamin complexes takes place in two steps: cobalamin release in the first step and the dissociation of resulting apoenzyme•reactivating factor complexes in the second step, as illustrated schematically in Fig. 3A (Kajiura et al., 2001; Mori & Toraya, 1999). Then, how do reactivating factors distinguish between enzyme-bound damaged cofactor(s) and intact coenzyme? In the case of DD, the presence of the adenosyl group-binding site has been biochemically demonstrated (Toraya, 1985; Toraya et al., 1986). The interaction of this site with the adenine moiety in the upper axial ligand of adenine-containing cobalamin would lower the equilibrium constant K_{eq} in the formation of an intermediate ternary complex between enzyme, cobalamin, and DD-R by a factor

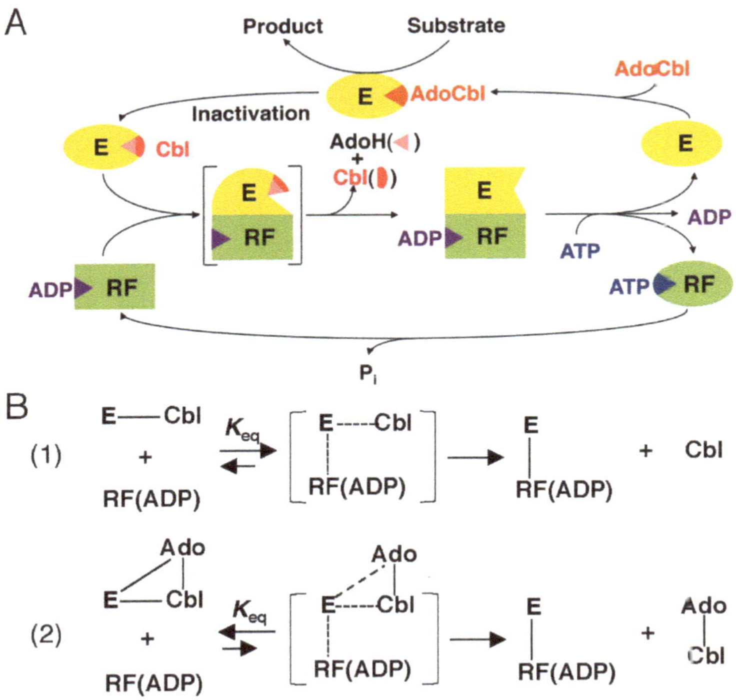

Fig. 3 Two-step mechanism for the reactivation of inactivated holoenzymes by reactivating factors. (A) Schematic representation of cobalamin exchange by chaperone functions of DD-R and GD-R. (B) Plausible mechanism of distinguishment between adenine-lacking damaged cofactor(s) and adenine-containing intact coenzyme. E, apoDD or apoGD; RF, reactivating factor or reactivase (DD-R or GD-R); Cbl, damaged cofactor(s); AdoH, 5′-deoxyadenosine.

of 1.7×10^4 (Mori & Toraya, 1999) when calculated from K_D for adenine (Toraya et al., 1986) (Fig. 3B). This explains why damaged cofactor(s), *i.e.*, an adenine-lacking cobalamin, are preferentially released from enzymes over adenine-containing cobalamins, such as AdoCbl and AdePeCbl.

To confirm the two-step mechanism, the DD•CN-Cbl complex was preincubated with DD-R in the presence of various nucleotides, and the release of enzyme-bound CN-Cbl was estimated from recovered activity with added AdoCbl alone or AdoCbl plus ATP (Table 1) (Mori & Toraya, 1999). The recovered activity with added AdoCbl plus ATP is a measure of the effectiveness of a nucleotide for cobalamin release, while the recovered activity with AdoCbl alone the effectiveness for overall (re) activation, *i.e.*, both cobalamin release and the dissociation of DD•DD-R complex. The ATP-replacing effect of a nucleotide for the dissociation of DD•DD-R complex can be separately estimated by preincubating the DD•CN-Cbl complex with DD-R in the presence of ADP, followed by

Table 1 ATP-replacing effects of nucleotides on overall (re)activation, cobalamin release, and dissociation of DD•DD-R complex.

	ATP-Replacing Effect (%) on		
Nucleotide	**(Expt. *a*) Overall (re)activation**	**(Expt. *b*) Cobalamin release**	**(Expt. *c*) Dissociation of DD•DD-R complex**
ATP	(100)	(100)	(100)
ADP	8	115	4
ATPγS	4	4	65
AMPPNP	2	4	45
AMPPCP	0	17	2
None	0	4	0

The DD•CN-Cbl complex was formed by incubating 6.5 units of apoenzyme at 37 °C for 30 min with 15 μM CN-Cbl in 44 μL of 0.045 M potassium phosphate buffer (pH 8.0). (Expts. *a* and *b*) To 5 μL of the mixture was added 23 μg of DD-R in the presence of a nucleotide and $MgCl_2$ (16 mM each) in 15 μL of 0.035 M potassium phosphate buffer (pH 8.0). After 1.5 h at 37 °C, the mixture was appropriately diluted with Buffer A, and the enzyme activity was measured by adding 15 μM AdoCbl without (*a*) and with (*b*) ATP and $MgCl_2$ (3 mM each). (Expt. *c*) To 9 μL of the above-mentioned mixture was added 42 μg of DD-R in the presence of ADP and $MgCl_2$ (16 mM each) in 27 μL of 0.035 M potassium phosphate buffer (pH 8.0). After 1.5 h at 37 °C, to 4 μL of the mixture was added 0.026 M potassium phosphate buffer (pH 8.0) containing 0.22 M 1,2-propanediol with or without a nucleotide and $MgCl_2$ (20 mM each) in a total volume of 45 μL, followed by incubation at 37 °C for 15 min. AdoCbl (15 μM) was then added to a total volume of 50 μL, and after incubation at 37 °C for 10 min, the reaction was terminated by adding potassium citrate buffer (pH 3.6) to 0.05 M. The amount of propionaldehyde formed was determined after appropriate dilution.

the measurement of enzyme activity with added AdoCbl plus a nucleotide (Table 1). ADP was effective only for cobalamin release. ATP was effective for both, *i.e.*, it plays dual roles, probably because ATP dissociates DD•DD-R complex, and ADP formed by the hydrolysis of ATP serves as an effector for cobalamin release. When ATP-replacing effects of the three non-hydrolysable ATP analogs adenosine 5′-*O*-(3-thiotriphosphate) (ATPγS), 5′-adenylyl imidodiphosphate (AMPPNP), and 5′-adenylyl methylenediphosphonate (AMPPCP) were compared, none of the three were effective in the overall (re)activation. However, AMPPCP was slightly effective for cobalamin release, whereas the other two were essentially not effective. In contrast, for the dissociation of DD•DD-R complex, ATPγS and AMPPNP were effective, *i.e.*, served as an effector for converting DD-R to the low-affinity form, whereas AMPPCP almost not.

Collectively, we proposed a plausible overall mechanism for reactivation of inactivated holoenzymes by DD-R or GD-R (Fig. 3A) (Kajiura et al., 2001; Mori & Toraya, 1999). In this mechanism, reactivating factors bind to the target enzymes and induce their conformational change through tight complex formation with them. This results in the release of tightly-bound, damaged cofactor(s). The reactivating factors themselves do not become a constituent of the final products. Thus, they meet the criteria of molecular chaperones. Their extremely low ATPase activity is not unusual for chaperones. Therefore, reactivating factors for AdoCbl-dependent enzymes can be considered as a new type of chaperones that participate in the reactivation of inactivated holoenzymes. Hence, we also called them "reactivating chaperones" (Toraya, 2014).

5.4 Multiple turnovers: DD-R is a reactivase

To examine whether a reactivating factor mediates multiple turnovers of cobalamin exchange, the replacement of enzyme-bound CN-Cbl (a model of damaged cofactor) for free AdePeCbl (a model of AdoCbl) was investigated in the presence of ATP and Mg^{2+} at a molar ratio of DD-R to DD•CN-Cbl of 1:10. Enzyme-bound CN-Cbl was almost completely replaced by AdePeCbl within 4 h at 37 °C. Thus, DD-R mediates the cobalamin exchange not stoichiometrically but catalytically in the presence of ATP and Mg^{2+} (Mori et al., 2010). In the sense that these reactivating factors are enzymes, they were renamed "reactivases," such as DD-reactivase (DD-R) and GD-reactivase (GD-R), as well.

6. X-ray structures of DD-R

6.1 Structure determination

We obtained the initial structure of DD-R in the ADP bound form at 2.6 Å resolution by the multiwavelength anomalous diffraction (MAD) method (Protocols 9.12–14) using the selenomethionine-substituted derivative purified to homogeneity from overexpressing *E. coli* B834 (Novagen), a methionine-auxotroph, by a conventional method (Protocol 9.6) (Mori, Hieda, Yamanishi, Shibata, & Toraya, 2005; Shibata et al., 2005; Toraya & Mori, 1999). The resolution of the structure was improved later to 2.0 Å using wild-type crystals (Shibata et al., 2005). The structure of nucleotide-free form was solved at 3.0 Å resolution by the molecular replacement (MR) method using that of ADP-bound form as a starting model.

6.2 Overall structure

The crystal structures of DD-R in the ADP-bound and nucleotide-free forms were determined in the $P2_12_12_1$ space group with a dimer of αβ heterodimer (protomer) related by a noncrystallographic 2-fold axis in the asymmetric unit (Fig. 4A). The size exclusion chromatography analyses suggested that DD-R is composed of two α subunits and two β subunits as described in Section 4. The α subunit mainly contributes to the interface between two protomers, whereas the β subunit plays minor role in the interaction between two protomers.

6.3 Structures of the α and β subunits

The α subunit of DD-R has elongated structures consisting of four domains, *i.e.*, ATPase domain, swiveling domain, linker domain, and insert domain (Fig. 4B), named according to the structure of the nucleotide-free form of GD-R (Liao, Reiss, Turner, & Dotson, 2003). DD-R and GD-R contain structural features resembling both GroEL and Hsp70 families of molecular chaperones (Liao, Reiss, Turner, & Dotson, 2003; Shibata et al., 2005). Reactivases for AdoCbl-dependent enzymes and Hsp70 chaperones share three regions consisting of 15–18 amino acid residues in Loops I–III (Fig. 5A and B) with high fragmentary sequence similarity and thus may be considered as high-specificity and low-specificity chaperones, respectively, that are evolved from a common ancestor protein (Mori et al., 2004; Toraya, 2003).

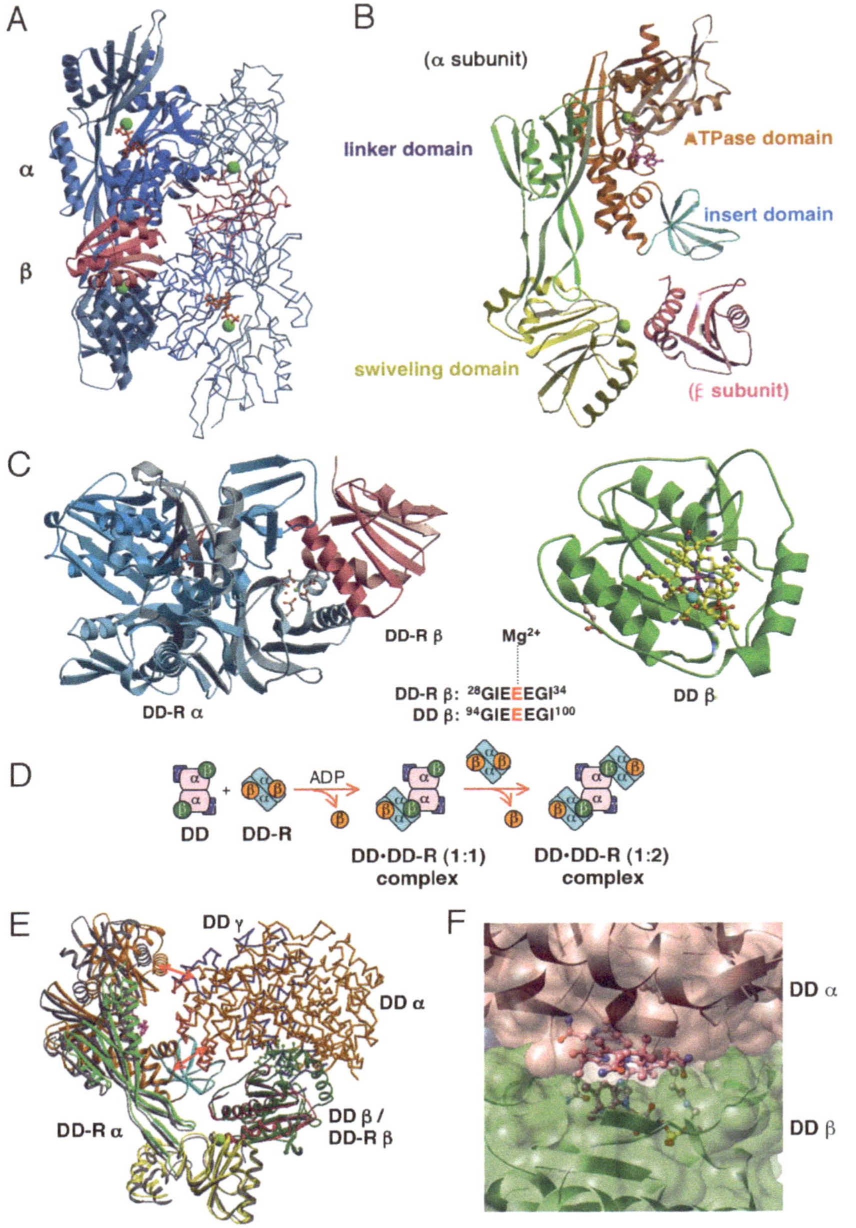

Fig. 4 X-ray structures of DD-R and subunit swapping between DD-R and DD. (A) Overall structure $(\alpha\beta)_2$ of DD-R. (B) Domain structure of heterodimer unit αβ. (C) Similarity of folds between DD-R β and DD β subunits. DD-R β has a Glu cluster site whose central Glu residue (Gluβ31) is coordinated to Mg^{2+}. A similar Glu cluster site is present in DD β as well. (D) Subunit swapping between DD-R and DD. (E) Docking model of DD•DD-R complex. Steric repulsion between amino acid side-chains that come closer than the van der Waals contact is shown by *red arrows*. (F) Preexisting gap between DD α (*pink*) and DD β (*green*). Drawn with Molscript (Kraulis, 1991)/Raster3D (Merritt & Bacon, 1997) (A–C and E) and with Chimera (Pettersen et al., 2004) (F).

The insert domain, which can be regarded as an accessary domain of the ATPase domain, has direct contacts with the β subunits of both protomers. The ATPase domain has no direct interaction with the β subunit, but the insert domain bridges the gap between the ATPase domain and the β subunits in the same and the other protomer.

The swiveling domain has no direct interaction with the ATPase domain of the same α subunit, but it structurally connects the β subunit and the ATPase domain of the other α subunit in the $(\alpha\beta)_2$ heterotetramer. This domain has a metal ion-binding site in the interface between the α and β subunits, and the metal ion bound were assigned to Ca^{2+} in the nucleotide-free forms of DD-R (Shibata et al., 2005) and GD-R (Liao, Reiss, Turner, & Dotson, 2003) and to Mg^{2+} in the ADP-bound form of DD-R (Shibata et al., 2005) (Fig. 4C). Mg^{2+} is thought to be the functional metal ion *in vivo*, but it may be replaced by Ca^{2+} in the purification procedure using a hydroxyapatite column (Toraya & Mori, 1999).

The β subunits are single-domain, globular proteins and have Rossmann-like folds similar to the β subunits of DD and GD, except that they lack some AdoCbl-binding elements (Fig. 4C). Based on the similarities, a "subunit swap" hypothesis was suggested and discussed in Section 7.

7. Molecular mechanism of action of reactivating chaperones

7.1 Cobalamin release by subunit swapping

The binding of ADP to the nucleotide-binding site of DD-R causes a marked conformational change of the ATPase domain of α subunit, which would weaken the interactions between the DD-R α and β subunits and promote the subunit swapping, *i.e.*, the displacement of the DD-R β subunit by DD through the β subunit (Shibata et al., 2005). The subunit swapping hypothesis was biochemically tested with DD-R. When DD is incubated with DD-R in the presence of ADP and Mg^{2+}, DD•DD-R (1:1) and (1:2) complexes are formed (Fig. 4D) (Mori et al., 2010). Upon complex formation, the release of β subunit from the $(\alpha\beta)_2$ heterotetramer of DD-R was observed by nondenaturing PAGE, suggesting that subunit swapping actually takes place upon DD•DD-R complex formation. The dissociation of DD-R $(\alpha\beta)_2$ into α•αβ and β in the presence of ADP and Mg^{2+} occurred even without DD. The space opened by the dissociation of DD-R β subunit would be occupied by the DD β subunit, since these subunits have similar folds (Fig. 4C) (Shibata et al., 1999; Shibata et al., 2005).

The docking model of the DD•DD-R complex indicates that marked steric repulsion is induced between the DD α and DD-R α subunits in the complex (Shibata et al., 2005) (Fig. 4E). The amino acid side-chains that come closer than the van der Waals contact in the modeled structure would push aside each other, resulting in tilting of the DD α subunit with respect to the DD β subunit and widening the preexisting gaps (~5 Å in height and ~15 Å in width) (Fig. 4F) between the α and β subunits of DD or EAL to ~11 Å in height (Mori et al., 2010; Shibata et al., 2010). This size is comparable with that of adenine-lacking cobalamin, and thus damaged cofactors would be allowed to pass through the widened gap from inactivated holoenzymes (Mori et al., 2010). Intact AdoCbl or AdePeCbl is not released by reactivases once it binds to the enzymes. One reason for this difference might be its larger size (Mori et al., 2010); another reason might be the additional interaction at the adenine moiety with the enzyme's adenine-binding pocket (see chapter "Coenzyme B_{12}-dependent eliminases: Diol and glycerol dehydratases and ethanolamine ammonia-lyase" by Toraya et al.) (Fig. 3B). Interaction of AdoCbl or other adenine-containing cobalamins with DD at both the adenine moiety and the cobalamin moiety (Shibata et al., 1999) may stabilize the interaction between the α and β subunits and make the tilting difficult in DD.

7.2 Mutational studies of nucleotide-switch and metal-coordination

The conserved regions among reactivases and Hsp70 chaperones correspond to the three loops at the ADP-binding sites (Fig. 5A,B) and thus may constitute an ADP/ATP switch (Toraya, 2003). Reactivases mediate cofactor exchange by changing its affinity for apoenzymes through a nucleotide-switch (Kajiura et al., 2001; Mori & Toraya, 1999). Functions of selected amino acid residues have been investigated by the technique of site-directed mutagenesis (Protocol 9.15). Neither Dα8N, Dα413N, nor Gα556A/Gα557A mutant of DD-R functioned as a reactivase upon *in situ* reactivation assay with *E. coli* co-expressing mutant DD-R genes with DD genes (Mori et al., 2013). Although the Dα413N mutant showed ATPase activity, it forms a tight complex with apoenzyme even in the presence of ATP. The Dα8N mutant was also an inactive reactivase and showed very low ATPase activity. Therefore, it can be concluded that Aspα413 in the Loop II region is involved in the nucleotide-switch mechanism, whereas Aspα8 in the Loop I region mainly in the ATPase activity (Mori et al., 2013). Since these residues are in the regions conserved among

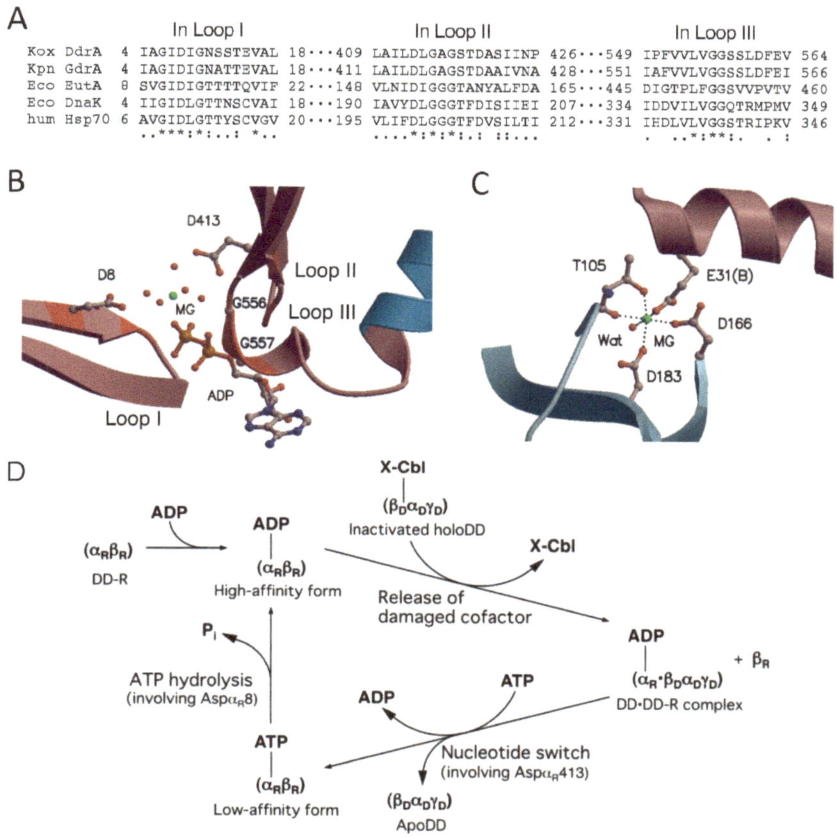

Fig. 5 Essential roles of nucleotide-switch and metal-coordination in the chaperone function of DD-R. (A) Fragmentary sequence homology between reactivating chaperones and Hsp70 family chaperones. Kox, *K. oxytoca*; Kpn, *K. pneumonia*; Eco, *E coli*; hum, human. Identical, highly similar, and similar amino acid residues are indicated by asterisks, colons, and dots, respectively. (B) Structure of nucleotide-binding site. Red and pink indicate conserved and similar residues, respectively. Green ball, Mg^{2+}; red balls, water molecules. (C) Structure of metal ion-binding site between DD-R α (blue) and β (pink) subunits. Residue numbers in DD-R α. (D) Involvement of Aspα8 and Aspα413 residues in the proposed ATP/ADP-switch mechanism for DD-R. X-Cbl, damaged cofactor(s) or adenine-lacking cobalamin; ($\alpha_R\beta_R$) and ($\beta_D\alpha_D\gamma_D$), halves of DD-R and DD, respectively. Suffixes R and D represent DD-R and DD, respectively. Drawn with Molscript (Kraulis, 1991)/Raster3D (Merritt & Bacon, 1997) (B, C).

reactivases and Hsp70 chaperones, such roles of Asp residues suggest the common mechanism of ATP/ADP-switches. Our proposed mechanism of nucleotide-switch in the reactivation of inactivated DD by DD-R is schematically represented in Fig. 5D (Mori et al., 2013). The conformation of

Aspα413-containing loop (Loop II) may be sensitive to a bound nucleotide and thus important for maintaining the radical status of enzyme.

Fig. 5C shows the metal ion-binding site of DD-R between the α and β subunits. Among DD-Rs mutated at the Mg^{2+}-binding site, neither Tα105A, Dα166A, nor Dα183A mutant was an active reactivase when assayed by the *in situ* reactivation experiment described above (Mori et al., 2013). Only Eβ31Q was about 40% as active as wild-type DD-R and forms a tight complex with apoenzyme, indicating that the DD-R β subunit is not absolutely required for reactivation. Gluβ31 of DD-R, a Mg^{2+}-coordinating residue (Fig. 4C), is thus not essential for reactivation. Gluβ97 of DD is a residue occupying the position corresponding to Gluβ31 in the Glu cluster site of DD-R (Fig. 4C). The Eβ97Q mutant of DD was enzymatically as active as wild-type DD, but its complex with CN-Cbl did not undergo *in vitro* activation by wild-type DD-R. No DD•DD-R complex was formed with this mutant DD. It was thus concluded that Gluβ97 of DD is absolutely required for the DD•DD-R complex formation and for undergoing (re)activation (Mori et al., 2013). Its coordination to Mg^{2+} of DD-R would be essential for subunit swapping and thus for (re)activation.

8. Role of reactivating chaperones in coenzyme recycling: Activity-maintaining system

8.1 Coenzyme recycling

Holoenzymes of GD and DD undergo mechanism-based inactivation by glycerol, although either of them is necessary for the anaerobic growth on glycerol. Moreover, holoenzymes of both dehydratases undergo O_2-inactivation in the absence of substrate. Either inactivation is accompanied by the damage of enzyme-bound coenzyme, *i.e.*, the irreversible cleavage of the Co—C bond. The role of reactivating chaperones is to release the tightly bound, damaged cofactor(s) from enzymes and to enable to exchange them for intact AdoCbl (Fig. 6) (Kajiura et al., 2001; Mori & Toraya, 1999; Toraya & Mori, 1999). Then, what happens when free AdoCbl is exhausted in the cell? Even if apoenzyme is formed by a reactivase, it cannot be active without AdoCbl, which leads to the complete cessation of enzyme reaction. For maintaining catalytic activity, the regeneration of AdoCbl from damaged cofactor(s) and coenzyme recycling are absolutely required.

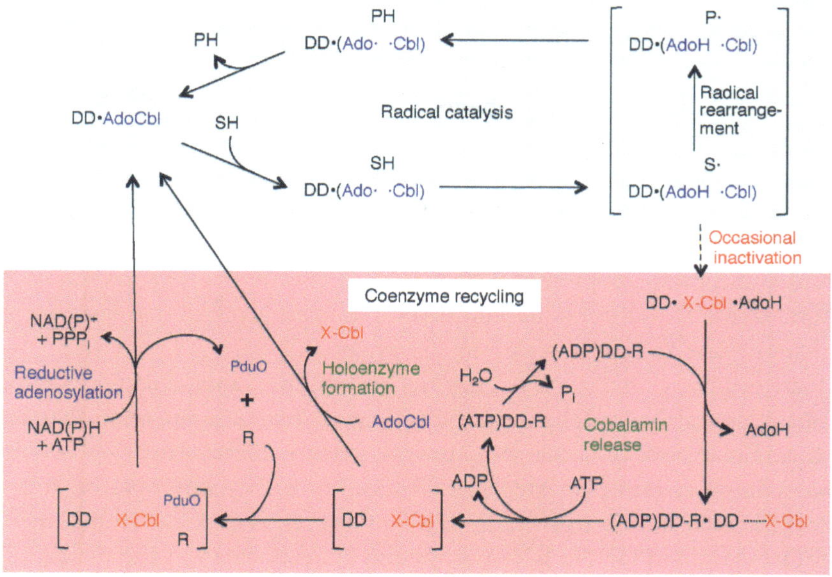

Fig. 6 Essential role of DD-R in coenzyme recycling for DD by an activity-maintaining system. DD, apoenzyme; SH, substrate; S•, substrate radical; PH, product; P•, product radical; Ado•, adenosyl radical; AdoH, 5′-deoxyadenosine; Cbl, cobalamin; •Cbl, cob(II) alamin; X-Cbl, damaged cofactor(s); R, cobalamin reductase(s). DD-R, PduO, and R constitute an activity-maintaining system that enables coenzyme recycling for DD.

8.2 Activity-maintaining system and coenzyme recycling

The requirement for cofactor recycling was supported by our finding that inactivated holoenzymes or the inactive enzyme•CN-Cbl complex of DD were (re)activated by the incubation with NADH, ATP, and Mg^{2+} (or Mn^{2+}) in crude extracts of *K. oxytoca* (Toraya et al., 2016). This finding indicated the presence of an activity-maintaining system for DD in the extract. The requirements for reducing agents and ATP suggested the involvement of *Co*-adenosylation of damaged cofactor(s). The reducing system with NADH was replaceable by $FMNH_2$. When glycerol-inactivated holoenzyme or the enzyme•CN-Cbl complex was incubated with purified recombinant DD-R and an ATP:cob(I)alamin adenosyltransferase in the presence of $FMNH_2$, ATP, and Mg^{2+}, DD activity was restored (Toraya et al., 2016) (Protocol 9.16). Among the three adenosyltransferases (PduO, EutT, and CobA) of this bacterium, PduO and CobA were much more efficient for the (re)activation than EutT, although PduO showed the lowest adenosyltransferase activity toward the free cob(I)alamin (Fig. 1B(6)) substrate. These results indicated that: (1) DD activity is maintained by coenzyme recycling in the bacterium (Fig. 6). (2) Coenzyme recycling of AdoCbl requires the releasing factor DD-R, because AdoCbl is a tightly bound, prosthetic group-type coenzyme. (3) The activity-maintaining system for DD is composed of DD-R, PduO adenosyltransferase, and a reducing system (Toraya et al., 2016). Although $FMNH_2$ was mainly used as a reductant in this study, a natural reducing system may be composed of PduS cobalamin reductase and NADH. All the genes encoding DD, DD-R, PduO, and PduS are present within the *pdu* operon (see Fig. 2A in chapter "Coenzyme B_{12}-dependent eliminases: Diol and glycerol dehydratases and ethanolamine ammonia-lyase" by Toraya et al.).

In summary, radical B_{12} enzymes utilize the super-reactive adenosyl radical for catalyzing chemically challenging reactions in many bacteria. The cellular pool of active B_{12} enzyme status is preserved by an activity-maintaining system that is encoded by genes in the same operon. It is an intriguing question as to whether and how the repair and reactivation mechanisms identified for the bacterial systems are replicated in the mammalian system.

9. Protocol

9.1 Assay of *in situ* reactivation by measuring time-course of reaction with permeabilized cells

1.1 Harvest *K. pneumoniae*, *K. oxytoca*, or *E. coli* cells in the late exponential phase, wash twice with 0.05 M potassium phosphate buffer (Buffer A), and resuspend them in the same buffer to a concentration of 10 mg of dry cells per mL.

1.2 Add 0.2 mL of toluene to 20 mL of cell suspension (final concentration, 1.0% (v/v)), and shake the mixture vigorously at room temperature for 5 min.

1.3 Collect toluene-treated (permeabilized) cells by centrifugation, wash with 40 mL of Buffer A, and finally resuspend them in 20 mL of the same buffer. The cells thus obtained has a faint odor of toluene, but can be used for the *in situ* assays of GD, DD, or EAL.

1.4 Incubate an appropriate amount of permeabilized cells at 37 °C for various time periods in reaction mixtures consisting of substrate (0.1–0.2 M glycerol for GD and DD or 0.01 M ethanolamine for EAL), 0.05 M KCl, 0.03–0.04 M potassium phosphate buffer (pH 8.0), and 15 μM AdoCbl in the presence and absence of ATP and $MgCl_2$ (3 mM each), in a total volume of 1.0 mL.

1.5 Determine the amount of aldehyde formed by the MBTH method (see Protocol 9.7, chapter "Coenzyme B_{12}-dependent eliminases: Diol and glycerol dehydratases and ethanolamine ammonia-lyase" by Toraya et al.).

Note 1. The amounts of permeabilized cells used are: 0.01–0.02 mg of dry cells per tube for *K. pneumoniae* ATCC 25955 and *K. oxytoca* ATCC 8724 for GD and DD assays; 7.2×10^6 cells for nonrecombinant *E. coli* JM109 for EAL assay. In the cases of recombinant *E. coli* cells, the amounts of permeabilized cells should be varied depending upon the activity of cells to be assayed.

2. Since AdoCbl is light-sensitive, all the operations should be done in dim light. This is the case for all the experiments in which organocobalamins are used, even though not indicated so.

9.2 Assay of *in situ* (re)activation of inactivated holoenzymes (A) and inactive enzyme•cobalamin complexes (B)

2.1 Toluene-treated (permeabilized) cells are obtained as described in Protocol 9.1.1–3.

2.2.A (See *Note* 2 to Protocol 9.1) Form glycerol- or ethanolamine-inactivated holoenzymes in permeabilized cells by incubating an appropriate amount of permeabilized cells at 37 °C for 30 min with 15 μM AdoCbl and a substrate (0.2 M glycerol for GD and DD or 0.4 M ethanolamine for EAL) in 5 mL of 0.04 M potassium phosphate buffer (pH 8.0) containing 0.05 M KCl. After washing the cells five times with 25 mL of Buffer A, dialyze the cell suspension overnight against 1000 volumes of the same buffer.

2.2.B Form inactive enzyme•cobalamin complexes in permeabilized cells by incubating an appropriate amount of permeabilized cells at 37 °C for 10–20 min with 19 μM OH-Cbl, CN-Cbl, MeCbl, or PeCbl and a substrate (0.25 M 1,2-propanediol for DD and GD or 0.01 M ethanolamine for EAL) in 0.8 mL of 0.04 M potassium phosphate buffer (pH 8.0) containing 0.063 M KCl.

2.3.A For *in situ* reactivation assays, incubate an appropriate amount of permeabilized cells at 37 °C for various time periods in the mixture consisting of a substrate (0.2 M 1,2-propanediol for GD and DD or 0.01 M ethanolamine for EAL), 0.05 M KCl, 0.04 M potassium phosphate buffer (pH 8.0), and 15 μM AdoCbl in the presence and absence of ATP and $MgCl_2$ (3 mM each), in a total volume of 1.0 mL. Carry out minus coenzyme experiments as controls.

2.3.B For *in situ* activation assays, add 15 μM AdoCbl with and without ATP and $MgCl_2$ (3 mM each) to a final volume of 1.0 mL, and incubate the mixture at 37 °C for various time periods. Carry out minus coenzyme experiments as controls.

2.4 Determine the amount of aldehyde formed by the MBTH method.

Note: For the amounts of permeabilized cells to be used, see *Note* 1 to Protocol 9.1.

9.3 *In situ* functional analysis of reactivating factor genes by co-expression experiments

3.1 Insert the DD-R (*ddrAB*) or GD-R (*gdrAB*) genes or their mutant genes into the *Nde*I-*Eco*RI region of the expression vector pCXV which is compatible with the expression vector pUSI2E in *E. coli* cells.

3.2 Co-transform *E. coli* JM109 cells with pUSI2E(DD) or pUSI2E(GD) and pCXV(6/5b), pCXV(orf2b•ORF4), or their mutant plasmids in any combinations by electroporation.

3.3 Grow *E. coli* carrying these plasmids aerobically at 30 °C or 37 °C in LB medium containing both ampicillin and chloramphenicol (50 μg /mL each), and add 1 mM IPTG for induction at an OD_{600} ~0.9. Harvest cells in the late logarithmic phase, and wash them twice with Buffer A.

3.4 Permeabilize co-expressing *E. coli* cells by toluene treatment, as described in Protocol 9.1.

3.5 (See *Note* 2 to Protocol 9.1) Carry out assays of *in situ* (re)activation of inactivated holoenzymes and inactive enzyme•CN-Cbl complexes with permeabilized cells, as described in Protocols 9.1 and 9.2.

Note: For the amounts of permeabilized cells to be used, see *Note* 1 to Protocol 9.1.

9.4 Assay of *in vitro* reactivation of suicidally-inactivated holoenzymes (A) and holoenzymes inactivated in the absence of substrates (B)

4.1.A (See *Note* 2 to Protocol 9.1) For DD or GD, form glycerol-inactivated holoenzyme by incubating 15 units of apoenzyme with 15 μM AdoCbl at 37 °C for 30 min in 50 μL of 0.03 M potassium phosphate buffer (pH 8.0) containing 0.3 M glycerol and 0.05 M KCl, followed by dialysis at 4 °C for 24 h against 1000 volumes of Buffer A containing 0.3 M 1,2-propanediol with a buffer change. For EAL, form ethanolamine-inactivated holoenzyme by incubating 0.11 unit of apoEAL at 37 °C for 45 min with 20 μM

AdoCbl in 25 μL of 0.04 M potassium phosphate buffer (pH 8.0) containing 0.4 M ethanolamine and 0.4% Brij 35.

4.1.B Form O_2-inactivated holoenzyme of DD or GD by incubating 15 units of substrate-free apoenzyme with 49 μM AdoCbl at 37 °C for 30 min in 0.2 mL of Buffer A.

4.2 For DD or GD, incubate 0.75 unit of glycerol-inactivated holoenzyme or O_2-inactivated holoenzyme at 37 °C for appropriate time periods with and without 24 μg of DD-R or GD-R in 0.03 M potassium phosphate buffer (pH 8.0) containing 21 μM AdoCbl and 1.2 M 1,2-propanediol in the presence and absence of ATP and $MgCl_2$ (24 mM each), in a total volume of 25 μL. For EAL, incubate 0.11 unit of ethanolamine-inactivated holoenzyme at 37 °C for appropriate time periods with and without 33 μg of EAL-R in 0.02 M potassium phosphate buffer (pH 8.0) containing 10 μM AdoCbl and 0.2 M ethanolamine in the presence and absence of ATP and $MgCl_2$ (20 mM each), in a total volume of 50 μL. Carry out similar experiments with non-inactivated enzyme as controls.

4.3 Terminate the reaction by adding potassium citrate buffer (pH 3.6) to a final concentration of 0.05 M.

4.4 Centrifuge the reaction mixture to remove precipitate, and dilute appropriately for determining the amount of propionaldehyde or acetaldehyde formed by the MBTH method.

Note: It should be noted that, unlike *in situ* (re)activation assay, *in vitro* (re) activation assay needs much higher concentrations of reactivating factors and enzymes to ensure the protein-protein interaction between them. The product formed can be determined after appropriate dilution.

9.5 Assay of *in vitro* activation of inactive enzyme•cobalamin complexes

5.1 For DD or GD, form the complex with CN-Cbl or other inactive cobalamin by incubating 15 units of apoenzyme with 11 μM CN-Cbl or other cobalamin at 37 °C for 30 min in 0.2 mL of Buffer A containing 2% 1,2-propanediol. For EAL, form the complex with CN-Cbl by incubating 0.11 unit of apoenzyme at 37 °C for 15 min with 4 μM CN-Cbl in 25 μL of 0.04 M potassium phosphate buffer (pH 8.0) containing 0.4 M ethanolamine and 0.4% Brij 35.

5.2 (See *Note* 2 to Protocol 9.1) For DD or GD, incubate 0.75 unit of the complex with cobalamin at 37 °C for appropriate time periods with and without 24 μg of DD-R or GD-R in 25 μL of 0.03 M potassium phosphate buffer (pH 8.0) containing 21 μM AdoCbl and 1.2 M 1,2-propaediol in the presence and absence of ATP and $MgCl_2$ (24 mM each). For EAL, incubate 0.11 unit of the complex with CN-Cbl at 37 °C for appropriate time periods with and without 33 μg of EAL-R in 50 μL of 0.03 M potassium phosphate buffer (pH 8.0) containing 10 μM AdoCbl and 0.2 M ethanolamine in the presence and absence of ATP and $MgCl_2$ (20 mM each). Carry out similar experiments with non-inactivated enzyme as controls.

5.3 Terminate the reaction by adding potassium citrate buffer (pH 3.6) to a final concentration of 0.05 M.

5.4 Centrifuge the reaction mixture to remove precipitate, and dilute appropriately for determining the amount of propionaldehyde or acetaldehyde formed by the MBTH method.

9.6 Purification of recombinant wild-type DD-R and EAL-R

6.1 Grow *E. coli* JM109 carrying pUSI2ENd(6/5b) (for DD-R) or pUSI2ENd(eutA) (for EAL-R) aerobically at 30 °C and 37 °C, respectively, in LB medium containing ampicillin (50 μg/mL), and add 1 mM IPTG for induction at an OD_{600} ~0.9. Harvest cells in the late logarithmic phase, and wash them twice with Buffer A.

6.2 All the following operations are carried out at 0–4 °C. Suspend 10–16.5 g of wet cells in 40–50 mL of Buffer A containing 2 mM ethylenediaminetetraacetate (EDTA) and 2 mM phenylmethanesulfonyl fluoride (PMSF), and disrupt cells by sonication for 10 min at 240 W with a Kaijo TA-5287 ultrasonic destruction system (Japan), followed by centrifugation at 27,200 × *g* for 30 min.

6.3 For DD-R, add $(NH_4)_2SO_4$ powder to the supernatant combined with a washing to a concentration of 15% saturation. After the removal of a precipitate by centrifugation, $(NH_4)_2SO_4$ is added again to 35% saturation. For EAL-R, add $(NH_4)_2SO_4$ to the supernatant to 35% saturation in one step. In each case, centrifuge the suspension, dissolve the precipitate in 20–60 mL of Buffer A containing 2 mM EDTA and 2 mM PMSF, and dialyze the solution for 1–2 days against 1000 volumes of 5 mM potassium phosphate buffer (pH 8.0) containing 0.5 mM EDTA with a buffer change.

6.4 Load dialysate onto a DEAE-cellulose column (bed volume, 100–200 mL) that has been equilibrated with 0.01 M potassium phosphate buffer (pH 8.0). Wash the column successively with

500–1000 mL of the same buffer and with 600–1000 mL of 5 mM potassium phosphate buffer (pH 8.0) containing 0.1 M (for EAL-R) or 0.13 M (for DD-R) KCl, and then develop the column by increasing the KCl concentration to 0.15 M (for EAL-R) or 0.2 M, and finally to 0.25 M (for DD-R).

6.5 Combine the DdrA/DdrB- or EutA-containing fractions, and lower the buffer concentration by repeated concentration by ultrafiltration through a Diaflo PM-10 membrane (Millipore) and dilution in 1 mM potassium phosphate buffer (pH 8.0) (for DD-R) or by dialysis against 2 L of 7 mM potassium phosphate buffer (pH 8.0) after concentration to 25 mL (for EAL-R).

6.6 For DD-R, load the resulting solution onto a hydroxyapatite column (bed volume, 70–100 mL) which has been equilibrated with 2 mM potassium phosphate buffer (pH 8.0). Wash the column successively with 280 mL of the same buffer and with 200 mL of 6 mM potassium phosphate buffer (pH 8.0), and then develop the column by increasing the buffer concentration to 0.01 M. In the case of EAL-R, load dialysate onto the hydroxyapatite column which has been equilibrated with 7 mM potassium phosphate buffer (pH 8.0). EutA moves slowly and is recovered in 340 mL of the effluent. Combine the DdrA/DdrB- or EutA-containing fractions, and concentrate the solution to about 1.7–3.5 mL by ultrafiltration through a Diaflo PM-10 membrane and Centriplus-10 concentrator (Millipore).

6.7 Load the concentrated solution onto a Sephadex G-200 column (bed volume, 150–180 mL) which has been equilibrated with Buffer A (for DD-R) or 0.01 M potassium phosphate buffer (pH 8.0) (for EAL-R). Develop the column with the same buffer, and combine the DdrA/DdrB- or EutA-containing fractions. Then, concentrate the solution to ~5 mL by ultrafiltration through a Centriplus-10, and store it at −80 °C.

9.7 Purification of recombinant wild-type GD-R

7.1 Grow *E. coli* BL21(DE3) carrying pET(*gdrB/gdrA*) aerobically at 30 °C in Terrific Broth containing ampicillin (50 μg/mL), and add 1 mM IPTG for induction at an OD_{650} ~ 1.0. Harvest cells after cultivation overnight at 17 °C, and wash them with Buffer A.

7.2 All the following operations are carried out at 0–4 °C. Suspend 15 g of wet cells in 150 mL of Buffer A containing 2 mM EDTA and 2 mM

PMSF, and disrupt cells by sonication, followed by centrifugation at 27,200 × *g* for 30 min.

7.3 Add $(NH_4)_2SO_4$ powder to the supernatant to a concentration of 5% saturation. After removal of a precipitate by centrifugation, add $(NH_4)_2SO_4$ again to 25% saturation. Collect the precipitate by centrifugation, and dissolve it in 15 mL of Buffer A containing 2 mM EDTA and 2 mM PMSF. Dialyze the solution against 1000 volumes of 5 mM potassium phosphate buffer (pH 8.0) containing 0.5 mM EDTA.

7.4 Load dialysate onto a Sepharose CL-6B column (bed volume, 600 mL) which has been equilibrated with Buffer A containing 0.3 M 1,2-propanediol, and develop the column with the same buffer. Combine the GdrA/GdrB-containing fractions, concentrate the solution to ~ 5 mL by ultrafiltration through a Centriplus-10, and store it at −80 °C.

9.8 Spectral confirmation of the exchange of enzyme-bound CN-Cbl for AdePeCbl

8.1 (See *Note* 2 to Protocol 9.1) Form enzyme•CN-Cbl and enzyme•AdePeCbl complexes of DD or GD by incubating 100 units of apoenzyme with 33 μM CN-Cbl and AdePeCbl, respectively, at 37 °C for 30 min in 0.6 mL of Buffer A containing 2% 1,2-propanediol.

8.2 To the enzyme•CN-Cbl complex, add 2.5 mg of cognate reactivating factor (DD-R or GD-R) in 0.04 M potassium phosphate buffer (pH 8.0) containing ATP and $MgCl_2$ (20 mM each) and 20 μM AdePeCbl, in a total volume of 1.0 mL. To the enzyme•AdePeCbl complex, add CN-Cbl instead of AdePeCbl. Carry out similar experiments without reactivating factor as controls.

8.3 Incubate the mixtures at 37 °C for 0.5–1 h, and then dialyze the mixtures against 1000 volumes of 0.01 M potassium phosphate buffer (pH 8.0) containing 2% 1,2-propanediol with a buffer change.

8.4 Measure the spectra of dialysates, and correct them for dilution.

9.9 Verification of cobalamin exchange mechanism using radiolabeled CN-Cbl

9.1 Form the labeled enzyme•CN-Cbl complex of DD or GD by incubating 10 units of apoenzyme with 15 μM [^{57}Co]CN-Cbl (5870 cpm) at 37 °C for 30 min in 70 μL of 0.04 M potassium phosphate buffer (pH 8.0).

9.2 (See *Note* 2 to Protocol 9.1) To 30 μL of the enzyme•[^{57}Co]CN-Cbl complex, add 0.14 mg of DD-R or GD-R in 0.035 M potassium phosphate buffer (pH 8.0) containing ATP and $MgCl_2$ (15 mM each) and 5 μM AdePeCbl, in a total volume of 90 μL. Carry out similar experiments without reactivating factors as controls.

9.3 Incubate the mixtures at 37 °C for 1 h, and then remove unbound cobalamins by ultrafiltration on a Microcon-10 microconcentrator (Millipore).

9.4 Wash protein fraction retained on the filter twice with 90 μL of 0.03 M potassium phosphate buffer (pH 8.0) containing ATP and $MgCl_2$ (16 mM each).

9.5 Measure radioactivity in the protein fraction by a γ-counter.

9.10 Intermediate formation of apoenzymes from inactivated holoenzymes (A) and model complexes (B)

10.1.A (See *Note* 2 to Protocol 9.1) Form glycerol-inactivated holoenzymes of DD or GD by incubating 120 units of apoenzyme with 97 μM AdoCbl at 37 °C for 30 min in 0.65 mL of Buffer A containing 2 M glycerol.

10.1.B Form the enzyme•OH-Cbl•5′-deoxyadenosine complex of DD or GD by incubating 120 units of apoenzyme with 0.9 mM 5′-deoxyadenosine at 37 °C for 30 min in 0.65 mL of Buffer A containing 0.5 M glycerol, followed by incubation at 37 °C for additional 30 min with 150 μM aqCbl in a total volume of 0.7 mL.

10.2 Dialyze the mixtures at 4 °C for 2 days against 1500 volumes of the same buffer containing 0.3 M 1,2-propanediol with a buffer change, and then add 1.25 mg of DD-R or GD-R to aliquots of dialysates containing 50 units of inactivated holoenzymes and model complexes in a volume of 0.4 mL. Carry out similar experiments without reactivating factors as controls.

10.3 Dialyze mixtures again at 4 °C for 18 h against 250 volumes of 0.01 M potassium phosphate buffer (pH 8.0) containing 0.3 M 1,2-propanediol, ATP and $MgCl_2$ (20 mM each), and 0.05 M K_2SO_4.

10.4 Dialyze mixtures for additional 2 days against 2500 volumes of 0.01 M potassium phosphate buffer (pH 8.0) containing 0.3 M 1,2-propanediol with a buffer change.

10.5 Measure the spectra of dialysates with a minus cobalamin control as reference, and correct them for dilution.

9.11 Evaluation of complex formation between apoenzymes and reactivating factors

11.1 (See *Note* 2 to Protocol 9.1) Form enzyme•AdePeCbl and enzyme•CN-Cbl complexes of DD or GD by incubating 0.89 unit of apoenzyme at 37 °C for 30 min with 15 μM AdePeCbl and CN-Cbl, respectively, in 6 μL of 0.045 M potassium phosphate buffer (pH 8.0).

11.2 To 2.5 μL of apoenzyme, enzyme•CN-Cbl, or enzyme•AdePeCbl complex, add 12 μg of DD-R or GD-R in the presence of ATP and $MgCl_2$ (16 mM each) (1) or ADP and $MgCl_2$ (16 mM each) (2) in 0.035 M potassium phosphate buffer (pH 8.0) to a total volume of 7.5 μL. Carry out similar experiments without reactivating factor as controls.

11.3 After 1.5 h at 37 °C, analyze the mixtures by nondenaturing PAGE (6% gel) in the presence of ATP and $MgCl_2$ (1 mM each) (1) or ADP and $MgCl_2$ (1 mM each) (2).

11.4 Excise a lane containing new bands of apoenzyme•reactivating factor complexes, and analyze it by SDS-PAGE (second dimension) on 13.5% and 9% gels.

11.5 After staining with Coomassie Brilliant Blue R-250, determine the subunit compositions of enzyme•reactivating factor complexes by densitometric analysis.

9.12 Crystallization

12.1 Incubate purified wild-type and selenomethionine-substituted DD-Rs (20–40 mg/mL in 0.01 M potassium phosphate buffer (pH 8.0)) at 30 °C for 30 min with either 50 mM 2-mercaptoethanol (for the ADP-bound form) or 10 mM dithiothreitol (for the nucleotide-free form). For the ADP-bound form, add ADP and $MgCl_2$ (2 mM each), and incubate the mixture at 20 °C for 20 min.

12.2 Prepare precipitant solution consisting of 15% (w/v) PEG 6000, 90 mM $(NH_4)_2SO_4$, 0.012 M Tris•HCl buffer (pH 8.0), 20% (v/v) PEG 400, and either 50 mM 2-mercaptoethanol (for the ADP-bound form) or 10 mM dithiothreitol (for the nucleotide-free form).

12.3 Put 0.5 mL of the precipitant solution into each of the reservoir wells of a commercially available 24-well crystallization plate for the sandwich-drop vapor-diffusion method.

12.4 Mix 30 μL DD-R solution and 30 μL precipitant solution on the protein well of the crystallization plate, and keep the plate at 4 °C.

9.13 X-ray diffraction data collection

13.1 Picked up a crystal using CryoLoops (Hampton Research Inc.) or LithoLoops (Protein Wave Corp.), and flash-cool it in a 100 K stream of cryogenic nitrogen gas.

13.2 For the MAD data of the selenomethionine-substituted derivative, estimate the absorption edge wavelength of the selenium atoms by measuring X-ray absorption fine structure (XAFS) data.

13.3 Based on the XAFS data, set the wavelengths for the peak, edge, and remote data sets.

13.4 X-ray diffraction data sets are collected from the frozen crystals at 100 K using a synchrotron beamline for protein crystallography. The ADP-bound and nucleotide-free forms of wild-type DD-R crystals diffract up to 2.0 Å and 3.0 Å resolutions, respectively. The best conditions for the diffraction experiment depend on the beamlines, but the followings are typical parameters for the native DD-R crystals: wavelength, 1.0000 Å; crystal to detector distance: 150–250 mm; oscillation range per image, 0.1–1.0°; total oscillation range, 180–360°.

13.5 The collected diffraction data are indexed, integrated and scaled using the HKL2000 (Otwinowski & Minor, 1997) or XDS (Kabsch, 2010).

9.14 Phase determination, model building, and refinement

14.1 The initial phase determination of the ADP-bound form is performed by the MAD method using the data sets of the selenomethionine-substituted derivative. We highly recommend using the latest program packages of CCP4 (Collaborative Computational Project, 1994) or PHENIX (Liebschner et al., 2019) for crystal structure determination. Both programs can perform heavy-atom detection, initial phase determination, phase improvement, model building, and model improvement automatically. The structure of the nucleotide-free form is determined by the MR method using that of the ADP-bound form as a search model. Both CCP4 and PHENIX contain programs for the MR method.

14.2 Refinement of the initial models of the ADP-bound and nucleotide-free forms is performed using a structure refinement program such as phenix.refine in PHENIX and REFMAC (Murshudov et al., 2011).

14.3 After the initial refinement, manual model rebuilding is performed by using a protein modeling program such as COOT (Emsley, Lohkamp, Scott, & Cowtan, 2010).

14.4 Iteration of steps 14.3 and 14.4 is carried out several times to obtain the refined structural model.

9.15 Functional analysis of amino acid residues of DD-R and DD by site-directed mutagenesis

15.1 Mutations are introduced into pCXV(5b-6) (for DD-R) and pUSI2ENd(DD) (for DD) using a QuikChange™ Site-Directed Mutagenesis kit (Stratagene, CA). The mutagenic sense and antisense primers for a mutant are designed according to the manufacturer's instructions (for guidelines, see Protocol 9.21 in chapter "Coenzyme B_{12}-dependent eliminases: Diol and glycerol dehydratases and ethanolamine ammonia-lyase" by Toraya et al.). For example, 5′-ctggcgatcctcaacctcggcgcgg-3′ is designed as the mutagenic sense primer for the Dα413N mutant of DD-R.

15.2 Synthesize nicked mutant strand by temperature cycling using *PfuTurbo* DNA polymerase with mutagenic sense and antisense primers and pCXV(5b-6) or pUSI2ENd(DD) as a template (~16 cycles).

15.3 Digest methylated, non-mutated parental DNA template with *Dpn*I restriction enzyme. Mutated plasmid containing nicked circular strands is not digested.

15.4 The circular, nicked double-stranded DNA is transformed into XL1-Blue supercompetent cells, in which the nicks in the mutated plasmid are repaired.

15.5 Confirm by nucleotide sequencing of the DNA region encompassing the entire *ddrAB* genes (for DD-R) or *pddABC* genes (for DD) that no unintended mutations are incorporated into the insert of the mutant expression plasmid during mutagenesis.

15.6 (See *Note* 2 to Protocol 9.1) Determine the DD-R activity of mutants by *in situ* (re)activation assays using *E. coli* cells co-expressing DD and mutant DD-R genes, as described in Protocol 9.3.

15.7 For *in vitro* assays, purify mutant DD-Rs as described in Protocol 9.6 from *E. coli* JM109 carrying mutant plasmids of pCXV(5b-6), except that chloramphenicol (35 μg/mL) is added to LB medium instead of ampicillin.

15.8 Examine mutant DD-Rs for DD-R activity by *in vitro* (re)activation assays, as described in Protocols 9.4 and 9.5. Measure ATPase activity, and evaluate the ability to form complexes with DD, as described in the text and in Protocol 9.11, respectively.

15.9 Purify the Eβ97Q mutant of DD, and measure its enzyme activity, as described in Protocols 9.2 and 9.7 in chapter "Coenzyme B_{12}-dependent eliminases: Diol and glycerol dehydratases and ethanolamine ammonia-lyase" by Toraya et al., respectively. Examine the Eβ97Q mutant for *in vitro* activation by DD-R, and evaluate the ability to form complexes with DD-R, as described in Protocols 9.5 and 9.11, respectively.

9.16 Assay of (re)activation by coenzyme recycling using purified proteins

16.1 Purify His_6-tagged PduO from the membrane fraction of overexpressing *E. coli* BL21(DE3) carrying pET15b(pduO), as described in the original paper (Toraya et al., 2016).

16.2 (See *Note* 2 to Protocol 9.1) Form glycerol-inactivated holoDD and the DD•CN-Cbl complex by incubating an appropriate amount of apoenzyme at 37 °C for 30 min with 1.3 molar equivalent of AdoCbl in the presence of 4.3 M glycerol and with 30-fold molar excess CN-Cbl in the presence of 0.2 M 1,2-propanediol, respectively. Remove unbound cobalamins by dialysis at 4 °C for 2 days against 100 volumes of 0.01 M potassium phosphate buffer (pH 8.0) containing 1% Brij 35 and 2% 1,2-propanediol with a buffer change.

16.3 Prepare assay mixtures consisting of 2 μM glycerol-inactivated holoDD (A) or DD·CN-Cbl complex (B), 2–6 μM DD-R, 3–14 μM PduO, 1.3 M 1,2-propanediol, 22 mM ATP, 21 mM $MgCl_2$, and 0.15 mM $FMNH_2$ in 50 μL of 2–3 mM potassium phosphate buffer containing 0.05–0.08 M *N*-Cyclohexyl-2-aminoethanesulfonate (CHES) buffer (pH 8.0).

16.4 Incubate the mixtures under a H_2 atmosphere for appropriate time periods.

16.5 Terminate the reaction by adding 50 μL of 0.1 M potassium citrate buffer (pH 3.6), and determine the amount of propionaldehyde formed by the MBTH method after appropriate dilution.

10. Conclusion

AdoCbl-dependent enzymes are inevitably inactivated during catalysis by side reactions, because they use a super-reactive primary-carbon radical for catalysis (see chapter "Coenzyme B_{12}-dependent eliminases: Diol and glycerol dehydratases and ethanolamine ammonia-lyase" by Toraya et al.). How do living organisms deal with inactivation when enzymes are inactivated by natural substrates? Here we have described our combined biochemical and structural approaches to studying reactivating chaperones for AdoCbl-dependent dehydratases and a deaminase. These chaperones serving as releasing factors for damaged cofactor(s) from enzymes and thus mediate their exchange for intact AdoCbl. Together with cobalamin reductase and cob(I)alamin adenosyltransferase, they constitute activity-maintaining systems for their respective AdoCbl-dependent enzymes. Since all of these enzymes and a reactivating chaperone are encoded by genes in the same operon, living organisms foresee the inevitable inactivation of radical B_{12} enzymes and address it by coenzyme recycling. In this chapter, specific methods for investigating reactivating factors are also presented in the Protocol section, which may hopefully be useful for other investigators to start working with these intriguing chaperones.

Acknowledgments

This work was supported by Grants-in-Aid for Scientific Research from Japan Society for the Promotion of Science and the Ministry of Education, Culture, Sports, Science and Technology and other research grants. We are grateful to all the people who worked with us at Okayama University, Kyoto University, and University of Hyogo.

References

Abeles, R. H., Brownstein, A. M., & Randles, C. H. (1960). β-Hydroxypropionaldehyde, an intermediate in the formation of 1,3-propanediol by *Aerobacter aerogenes*. *Biochimica et Biophysica Acta*, *41*, 530–531.

Bachovchin, W. W., Eagar, R. G., Jr., Moore, K. W., & Richards, J. H. (1977). Mechanism of action of adenosylcobalamin: glycerol and other substrate analogues as substrates and inactivators for propanediol dehydratase- -kinetics, stereospecificity, and mechanism. *Biochemistry*, *16*, 1082–1092.

Bachovchin, W. W., Moore, K. W., & Richards, J. H. (1978). Mechanism of action of adenosylcobalamin: Hydrogen transfer in the inactivation of diol dehydratase by glycerol. *Biochemistry*, *17*, 2218–2224.

Bradbeer, C. (1965a). The clostridial fermentations of choline and ethanolamine: I. preparation and properties of cell-free extracts. *Journal of Biological Chemistry*, *240*, 4669–4674.

Bradbeer, C. (1965b). The clostridial fermentations of choline and ethanolamine: II. Requirement for a cobamide coenzyme by an ethanolamine deaminase. *Journal of Biological Chemistry, 240*, 4675–4681.
Carty, T. J., Babior, B. M., & Abeles, R. H. (1974). The mechanism of action of ethanolamine ammonia-lyase, a B_{12}-dependent enzyme. Evidence for two intermediates in the catalytic process. *Journal of Biological Chemistry, 249*, 1683–1688.
Chang, G. W., & Chang, J. T. (1975). Evidence for the B_{12}-dependent enzyme ethanolamine deaminase in *Salmonella*. *Nature, 254*, 150–151.
Collaborative Computational Project, Number 4. (1994). The *CCP4* suite: Programs for protein crystallography. *Acta Crystallographica Section D, 50*, 760–763.
Emsley, P., Lohkamp, B., Scott, W. G., & Cowtan, K. (2010). Features and development of Coot. *Acta Crystallographica Section D, 66*, 486–501.
Forage, R. G., & Foster, M. A. (1982). Glycerol fermentation in *Klebsiella pneumoniae*: Functions of the coenzyme B_{12}-dependent glycerol and diol dehydratases. *Journal of Bacteriology, 149*, 413–419.
Hill, J. A., Pratt, J. M., & Williams, R. J. P. (1962). The corphyrins. *Journal of Theoretical Biology, 3*, 423–445.
Honda, S., Toraya, T., & Fukui, S. (1980). In situ reactivation of glycerol-inactivated coenzyme B_{12}-dependent enzymes, glycerol dehydratase and diol dehydratase. *Journal of Bacteriology, 143*, 1458–1465.
Kabsch, W. (2010). XDS. *Acta Crystallographica Section D, 66*, 125–132.
Kajiura, H., Mori, K., Shibata, N., & Toraya, T. (2007). Molecular basis for specificities of reactivating factors for adenosylcobalamin-dependent diol and glycerol dehydratases. *FEBS Journal, 274*, 5556–5566.
Kajiura, H., Mori, K., Tobimatsu, T., & Toraya, T. (2001). Characterization and mechanism of action of a reactivating factor for adenosylcobalamin-dependent glycerol dehydratase. *Journal of Biological Chemistry, 276*, 36514–36519.
Kaplan, B. H., & Stadtman, E. R. (1968). Ethanolamine deaminase, a cobamide coenzyme-dependent enzyme: I. purification, assay, and properties of the enzyme. *Journal of Biological Chemistry, 243*, 1787–1793.
Kofoid, E., Rappleye, C., Stojiljkovic, I., & Roth, J. (1999). The 17-gene ethanolamine (*eut*) operon of *Salmonella typhimurium* encodes five homologues of carboxysome shell proteins. *Journal of Bacteriology, 181*, 5317–5329.
Kraulis, P. J. (1991). *MOLSCRIPT*: A program to produce both detailed and schematic plots of protein structures. *Journal of Applied Crystallography, 24*, 946–950.
Lee, H. A., Jr., & Abeles, R. H. (1963). Purification and properties of dioldehydrase, an enzyme requiring a cobamide coenzyme. *Journal of Biological Chemistry, 238*, 2367–2373.
Liao, D.-I., Reiss, L., Turner, I., Jr., & Dotson, G. (2003). Structure of glycerol dehydratase reactivase: A new type of molecular chaperone. *Structure, 11*, 109–119.
Liebschner, D., Afonine, P. V., Baker, M. L., Bunkóczi, G., Chen, V. B., Croll, T. I., et al. (2019). Macromolecular structure determination using X-rays, neutrons and electrons: Recent developments in *Phenix*. *Acta Crystallographica Section D, 75*, 861–877.
Mangold, H. K. (1969). Nucleic acids and nucleotides. In E. Stahl (Ed.), *Thin-layer chromatography. A laboratory handbook* (2nd ed., pp. 786–807). Berlin: Springer-Verlag.
Merritt, E. A., & Bacon, D. J. (1997). Raster3D: Photorealistic molecular graphics. *Methods in Enzymology, 277*, 505–524.
Mickelson, M. N., & Werkman, C. H. (1940). The dissimilation of glycerol by coli-aerogenes intermediates. *Journal of Bacteriology, 39*, 709–715.
Mori, K., Bando, R., Hieda, N., & Toraya, T. (2004). Identification of a reactivating factor for adenosylcobalamin-dependent ethanolamine ammonia lyase. *Journal of Bacteriology, 186*, 6845–6854.

Mori, K., Hieda, N., Yamanishi, M., Shibata, N., & Toraya, T. (2005). Crystallization and preliminary X-ray analysis of molecular chaperone-like diol dehydratase-reactivating factor in ADP-bound and nucleotide-free forms. *Acta Crystallographica Section F, 61*, 603–605.
Mori, K., Hosokawa, Y., Yoshinaga, T., & Toraya, T. (2010). Diol dehydratase-reactivating factor is a reactivase- -evidence for multiple turnovers and subunit swapping with diol dehydratase. *FEBS Journal, 277*, 4931–4943.
Mori, K., Obayashi, K., Hosokawa, Y., Yamamoto, A., Yano, M., Yoshinaga, T., et al. (2013). Essential roles of nucleotide-switch and metal-coordinating residues for chaperone function of diol dehydratase-reactivase. *Biochemistry, 52*, 8677–8686.
Mori, K., Tobimatsu, T., Hara, T., & Toraya, T. (1997). Characterization, sequencing, and expression of the genes encoding a reactivating factor for glycerol-inactivated adenosylcobalamin-dependent diol dehydratase. *Journal of Biological Chemistry, 272*, 32034–32041.
Mori, K., Tobimatsu, T., & Toraya, T. (1997). A protein factor is essential for *in situ* reactivation of glycerol-inactivated adenosylcobalamin-dependent diol dehydratase. *Bioscience, Biotechnology, and Biochemistry, 61*, 1729–1733.
Mori, K., & Toraya, T. (1999). Mechanism of reactivation of coenzyme B_{12}-dependent diol dehydratase by a molecular chaperone-like reactivating factor. *Biochemistry, 38*, 13170–13178.
Murshudov, G. N., Skubák, P., Lebedev, A. A., Pannu, N. S., Steiner, R. A., Nicholls, R. A., et al. (2011). *REFMAC5* for the refinement of macromolecular crystal structures. *Acta Crystallographica Section D, 67*, 355–367.
Otwinowski, Z., & Minor, W. (1997). Processing of X-ray diffraction data collected in oscillation mode. *Methods in Enzymology, 276*, 307–326.
Pettersen, E. F., Goddard, T. D., Huang, C. C., Couch, G. S., Greenblatt, D. M., Meng, E. C., et al. (2004). UCSF Chimera–a visualization system for exploratory research and analysis. *Journal of Computational Chemistry, 25*, 1605–1612.
Poznanskaya, A. A., Yakusheva, M. I., & Yakovlev, V. A. (1977). Study of the mechanism of action of adenosylcobalamin-dependent glycerol dehydratase from Aerobacter aerogenes. II. The inactivation kinetics of glycerol dehydratase complexes with adenosylcobalamin and its analogs. *Biochimica et Biophysica Acta, 484*, 236–243.
Roof, D. M., & Roth, J. R. (1989). Functions required for vitamin B_{12}-dependent ethanolamine utilization in *Salmonella typhimurium*. *Journal of Bacteriology, 171*, 3316–3323.
Scarlett, F. A., & Turner, J. M. (1976). Microbial metabolism of amino alcohols. Ethanolamine catabolism mediated by coenzyme B_{12}-dependent ethanolamine ammonia-lyase in *Escherichia coli* and *Klebsiella aerogenes*. *Journal of General Microbiology, 95*, 173–176.
Schnebli, H. P., & Abrams, A. (1970). Membrane adenosine triphosphatase from *Streptococcus faecalis*. Preparation and homogeneity. *Journal of Biological Chemistry, 245*, 1115–1121.
Schneider, Z., & Pawelkiewicz, J. (1966). The properties of glycerol dehydratase isolated from *Aerobacter aerogenes*, and the properties of the apoenzyme subunits. *Acta Biochimica Polonica, 13*, 311–328.
Seifert, C., Bowien, S., Gottschalk, G., & Daniel, R. (2001). Identification and expression of the genes and purification and characterization of the gene products involved in reactivation of coenzyme B_{12}-dependent glycerol dehydratase of *Citrobacter freundii*. *European Journal of Biochemistry, 268*, 2369–2378.
Shibata, N., Masuda, J., Tobimatsu, T., Toraya, T., Suto, K., Morimoto, Y., et al. (1999). A new mode of B_{12} binding and the direct participation of a potassium ion in enzyme catalysis: X-ray structure of diol dehydratase. *Structure, 7*, 997–1008.
Shibata, N., Mori, K., Hieda, N., Higuchi, Y., Yamanishi, M., & Toraya, T. (2005). Release of a damaged cofactor from a coenzyme B_{12}-dependent enzyme: X-ray structures of diol dehydratase-reactivating factor. *Structure, 13*, 1745–1754.

Shibata, N., Tamagaki, H., Hieda, N., Akita, K., Komori, H., Shomura, Y., et al. (2010). Crystal structures of ethanolamine ammonia-lyase complexed with coenzyme B_{12} analogs and substrates. *Journal of Biological Chemistry*, *285*, 26484–26493.

Sobolov, M., & Smiley, K. L. (1960). Metabolism of glycerol by an acrolein-forming Lactobacillus. *Journal of Bacteriology*, *79*, 261–266.

Tobimatsu, T., Azuma, M., Matsubara, H., Takatori, H., Niida, T., Nishimoto, K., et al. (1996). Cloning, sequencing, and high level expression of the genes encoding adenosylcobalamin-dependent glycerol dehydrase of *Klebsiella pneumoniae*. *Journal of Biological Chemistry*, *271*, 22352–22357.

Tobimatsu, T., Hara, T., Sakaguchi, M., Kishimoto, Y., Wada, Y., Isoda, M., et al. (1995). Molecular cloning, sequencing, and expression of the genes encoding adenosylcobalamin-dependent diol dehydrase of *Klebsiella oxytoca*. *Journal of Biological Chemistry*, *270*, 7142–7148.

Tobimatsu, T., Kajiura, H., & Toraya, T. (2000). Specificities of reactivating factors for adenosylcobalamin-dependent diol dehydratase and glycerol dehydratase. *Archives of Microbiology*, *174*, 81–88.

Tobimatsu, T., Kajiura, H., Yunoki, M., Azuma, M., & Toraya, T. (1999). Identification and expression of the genes encoding a reactivating factor for adenosylcobalamin-dependent glycerol dehydratase. *Journal of Bacteriology*, *181*, 4110–4113.

Toraya, T. (1985). The binding site for the adenosyl group of coenzyme B_{12} in diol dehydrase. *Archives of Biochemistry and Biophysics*, *242*, 470–477.

Toraya, T. (2003). Radical catalysis in coenzyme B_{12}-dependent isomerization (eliminating) reactions. *Chemical Reviews*, *103*, 2095–2127.

Toraya, T. (2014). Cobalamin-dependent dehydratases and a deaminase: Radical catalysis and reactivating chaperones. *Archives of Biochemistry and Biophysics*, *544*, 40–57.

Toraya, T., & Fukui, S. (1977). Immunochemical evidence for the difference between coenzyme-B_{12}-dependent diol dehydratase and glycerol dehydratase. *European Journal of Biochemistry*, *76*, 285–289.

Toraya, T., Honda, S., Kuno, S., & Fukui, S. (1978). Coenzyme B_{12}-dependent diol dehydratase: Regulation of apoenzyme synthesis in *Klebsiella pneumoniae* (*Aerobacter aerogenes*) ATCC 8724. *Journal of Bacteriology*, *135*, 726–729.

Toraya, T., Kuno, S., & Fukui, S. (1980). Distribution of coenzyme B_{12}-dependent diol dehydratase and glycerol dehydratase in selected genera of *Enterobacteriaceae* and *Propionibacteriaceae*. *Journal of Bacteriology*, *141*, 1439–1442.

Toraya, T., Matsumoto, T., Ichikawa, M., Itoh, T., Sugawara, T., & Mizuno, Y. (1986). The synthesis of adenine-modified analogs of adenosylcobalamin and their coenzymic function in the reaction catalyzed by diol dehydrase. *Journal of Biological Chemistry*, *261*, 9289–9293.

Toraya, T., & Mori, K. (1999). A reactivating factor for coenzyme B_{12}-dependent diol dehydratase. *Journal of Biological Chemistry*, *274*, 3372–3377.

Toraya, T., Shirakashi, T., Kosuga, T., & Fukui, S. (1976). Substrate specificity of coenzyme B_{12}-dependent diol dehydrase: Glycerol as both a good substrate and a potent inactivator. *Biochemical and Biophysical Research Communications*, *69*, 475–480.

Toraya, T., Tanokuchi, A., Yamasaki, A., Nakamura, T., Ogura, K., & Tobimatsu, T. (2016). Diol dehydratase-reactivase is essential for recycling of coenzyme B_{12} in diol dehydratase. *Biochemistry*, *55*, 69–78.

Toraya, T., Ushio, K., Fukui, S., & Hogenkamp, H. P. C. (1977). Studies on the mechanism of the adenosylcobalamin-dependent diol dehydrase reaction by the use of analogs of the coenzyme. *Journal of Biological Chemistry*, *252*, 963–970.

Toraya, T., Watanabe, N., Ushio, K., Matsumoto, T., & Fukui, S. (1983). Ligand exchange reactions of diol dehydrase-bound cobalamins and the effect of the nucleoside binding. *Journal of Biological Chemistry*, *258*, 9296–9301.

Ushio, K., Honda, S., Toraya, T., & Fukui, S. (1982). The mechanism of *in situ* reactivation of glycerol-inactivated coenzyme B_{12}-dependent enzymes, glycerol dehydratase and diol dehydratase. *Journal of Nutritional Science and Vitaminology*, *28*, 225–236.

Wagner, O. W., Lee, H. A., Jr., Frey, P. A., & Abeles, R. H. (1966). Studies on the mechanism of action of cobamide coenzymes. Chemical properties of the enzyme-coenzyme complex. *Journal of Biological Chemistry*, *241*, 1751–1762.

Yamanishi, M., Kinoshita, K., Fukuoka, M., Saito, T., Tanokuchi, A., Ikeda, Y., et al. (2012). Redesign of coenzyme B_{12} dependent diol dehydratase to be resistant to the mechanism-based inactivation by glycerol and act on longer chain 1,2-diols. *FEBS Journal*, *279*, 793–804.

CHAPTER ELEVEN

Glutamate mutase and 2-methyleneglutarate mutase

Wolfgang Buckel[a,*], Birgitta Beatrix[b], and Oskar Zelder[a,c]
[a]Department of Biology, Philipps University, Marburg, Germany
[b]Gene Center, Department of Biochemistry, Ludwig Maximilian University, München, Germany
[c]Industrial Biotechnology I, BASF SE, Ludwigshafen am Rhein, Germany
*Corresponding author: e-mail address: buckel@staff.uni-marburg.de

Contents

Abstract

Glutamate mutase converts (*S*)-glutamate to (*S*,*S*)-3-methylaspartate, whereas 2-methyleneglutarate mutase isomerizes 2-methyleneglutarate to (*R*)-3-methylitaconate. Both enzymes occur in amino acid fermenting clostridia, require coenzyme B_{12} (adenosylcobalamin) and catalyze a reversible, radical carbon skeleton rearrangement. The enzymes are assayed using the consecutive enzymes in their fermentation pathways: methylaspartase and methylitaconate isomerase, which produce the higher absorbing mesaconate and 2,3-dimethylmaleate, respectively. Although the quaternary structures of glutamate mutase and 2-methyleneglutarate mutase are different, the binding mode of coenzyme B_{12} and their catalytic mechanism appear to be very similar. The crystal structure of glutamate mutase and EPR spectroscopy of both enzymes during catalysis revealed a distance of 6.6 Å between the C-H bond of the substrate to be cleaved and the Co-atom of cob(II)alamin. Methods are described for facile production of both enzymes in *Escherichia coli* and purification.

Methods in Enzymology, Volume 668
ISSN 0076-6879
https://doi.org/10.1016/bs.mie.2021.12.011

Abbreviations

C.	*Clostridium*
DTT	dithiothreitol
E.	*Eubacterium*
EPR	electron paramagnetic resonance
Glm	glutamate mutase from *C. cochlearium*
Mgm	2-methyleneglutarate mutase
Mut	glutamate mutase from *C. tetanomorphum*

1. Introduction

Glutamate and 2-methyleneglutarate mutases either contain coenzyme B_{12} (adenosylcobalamin) or require added coenzyme for activity. The enzymes play central roles in the anaerobic degradation of glutamate and nicotinate, respectively. Both enzymes catalyze the reversible radical rearrangement of a 1,5-dicarboxylic acid to a branched 1,4-dicarboxylic acid, which allows the subsequent two-electron aldol cleavage to pyruvate and acetate or propionate. Glutamate mutase catalyzes the carbon skeleton rearrangement of (*S*)-glutamate to (2*S*,3*S*)-3-methylaspartate. $K_{eq} = [(S)\text{-glutamate}] \times [(2S,3S)\text{-3-methylaspartate}]^{-1} = 10.5$ (Barker, Suzuki, Iodice, & Rooze, 1964) (Fig. 1). 2-Methyleneglutarate mutase mediates the reversible equilibration between 2-methyleneglutarate (2-methylideneglutarate) and (*R*)-3-methylitaconate (2-methylidene-3-methylsuccinate), $K_{eq} = [\text{2-methyleneglutarate}] \times [(R)\text{-3-methylitaconate}]^{-1} = 4.3$ (Kung, Cederbaum, Tsai, & Stadtman, 1970; Kung & Stadtman, 1971) (Fig. 2). In both rearrangements, a residue migrates from carbon 3 to carbon 2 of propionate with inversion of configuration at C2. With glutamate mutase, the residue is glycine and with 2-methyleneglutarate mutase the residue is acrylate (Hartrampf & Buckel, 1984; Sprecher, Switzer, & Sprinson, 1966). The formation of both methyl groups of propionate occurs with complete racemization (Ciceri et al., 2000; Hartrampf & Buckel, 1984). Binding of the substrate to its corresponding enzyme induces homolytic cleavage of the cobalt-carbon bond of coenzyme B_{12} yielding cob(II)alamin and the 5′-deoxyadenosyl radical. This radical abstracts the 4-*Si* hydrogen either from glutamate (Fig. 1) or the 4-*Re* hydrogen from 2-methyleneglutarate (Fig. 2) and is converted to 5′-deoxyadenosine. The resulting methine radical at C4 rearranges to the methylene radical at C3, which re-abstracts a hydrogen atom from

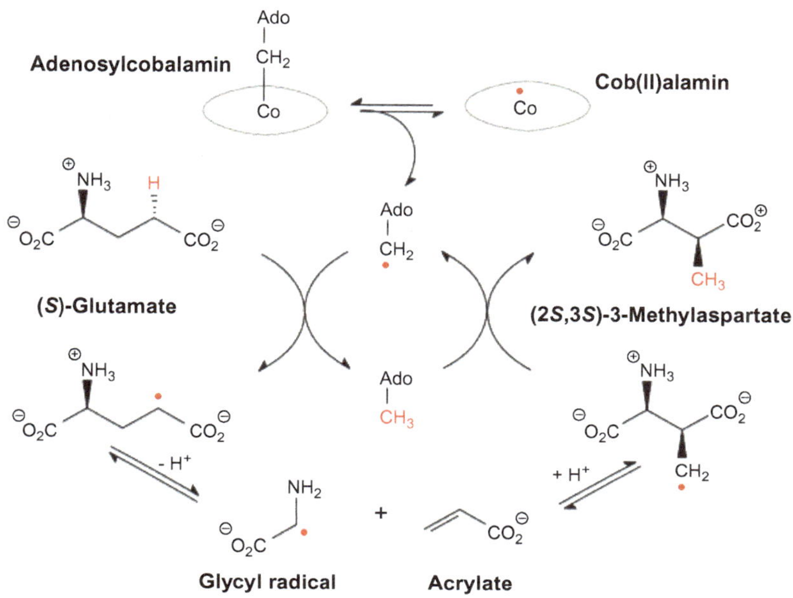

Fig. 1 Proposed mechanism of glutamate mutase. *From Buckel, W., & Golding, B. T. (2006). Radical enzymes in anaerobes.* Annual Review of Microbiology, 60, *27-49. doi:10.1146/annurev.micro.60.080805.142216.*

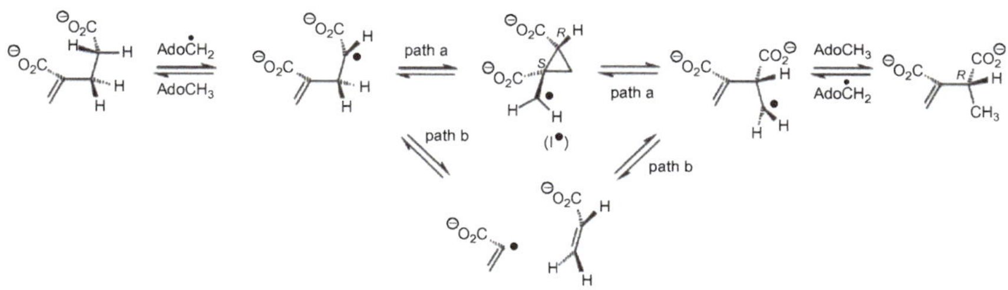

Fig. 2 Possible mechanisms of 2-methyleneglutarate mutase. 2-Methyleneglutarate (left), (*R*)-3-methylitaconate (right); path a: addition–elimination mechanism; path b: fragmentation–recombination mechanism. AdoCH$_3$, 5′deoxyadenosine; I$^•$, proposed intermediate radical. *From Pierik, A. J., Ciceri, D., Lopez, R. F., Kroll, F., Bröker, G., Beatrix, B., … Golding, B. T. (2005). Searching for intermediates in the carbon skeleton rearrangement of 2-methyleneglutarate to (*R*)-3-methylitaconate catalyzed by coenzyme B$_{12}$-dependent 2-methyleneglutarate mutase from* Eubacterium barkeri. Biochemistry, *44(31), 10541–10551. doi:10.1021/bi050049n. Copyright American Chemical Society.*

5′-deoxyadenosine to give the methyl group of 3-methylaspartate or 3-methylitaconate, whereby the 5′-deoxyadenosyl radical is regenerated. Cob(II)alamin and the 5′-deoxyadenosyl radical reform the cobalt-carbon bond of the coenzyme and complete the catalytic cycle, for reviews see (Buckel & Golding, 1996; Buckel & Golding, 2006) (Figs. 1 and 2).

2. Roles of glutamate and 2-methyleneglutarate mutases in metabolism

Both mutases belong to anaerobic degradation pathways for glutamate and nicotinic acid in clostridia. Glutamate mutase is the first enzyme of the fermentation pathway of glutamate to ammonia, CO_2, acetate and pyruvate (Eq. 1) in *Clostridium tetanomorphum*, *Clostridium cochlearium*, *Clostridium tetani* and other related clostridia (Buckel, 2021; Buckel & Barker, 1974).

$$\text{Glutamate}^- + H_2O \rightarrow NH_4^+ + \text{acetate}^- + \text{pyruvate}^- \quad (1)$$

The coenzyme B_{12}-dependent rearrangement of glutamate to 3-methylaspartate enables facile β-elimination of ammonia to mesaconate (methylfumarate). The whole fermentation of glutamate to acetate and pyruvate is shown in Fig. 3. For the further fate of pyruvate, see (Buckel, 2021).

2-Methyleneglutarate mutase is involved in the fermentation of nicotinate by *Eubacterium barkeri*, formerly called *Clostridium barkeri*, to ammonia, CO_2, acetate and propionate (Eq. 2; Fig. 4) (Kung et al., 1970; Kung & Stadtman, 1971).

$$\text{Nicotinate}^- + 4\,H_2O \rightarrow NH_4^+ + CO_2 + \text{Acetate}^- + \text{Propionate}^- \quad (2)$$

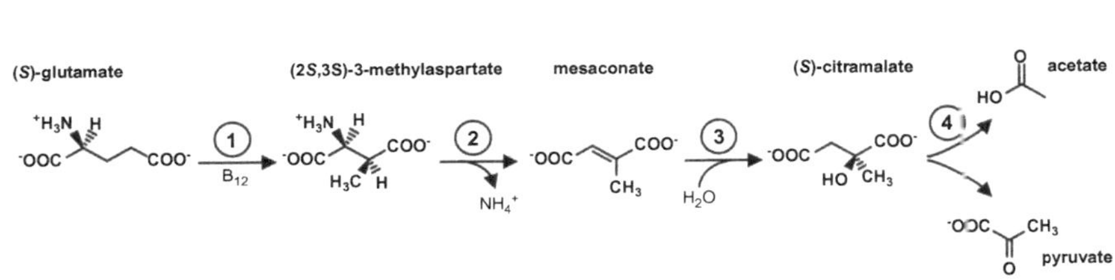

Fig. 3 Pathway of glutamate fermentation in *C. tetanomorphum*. Enzymes: 1, Glutamate mutase; 2, methylaspartate ammonia lyase; 3, mesaconase (fumarase); 4, citramalate lyase.

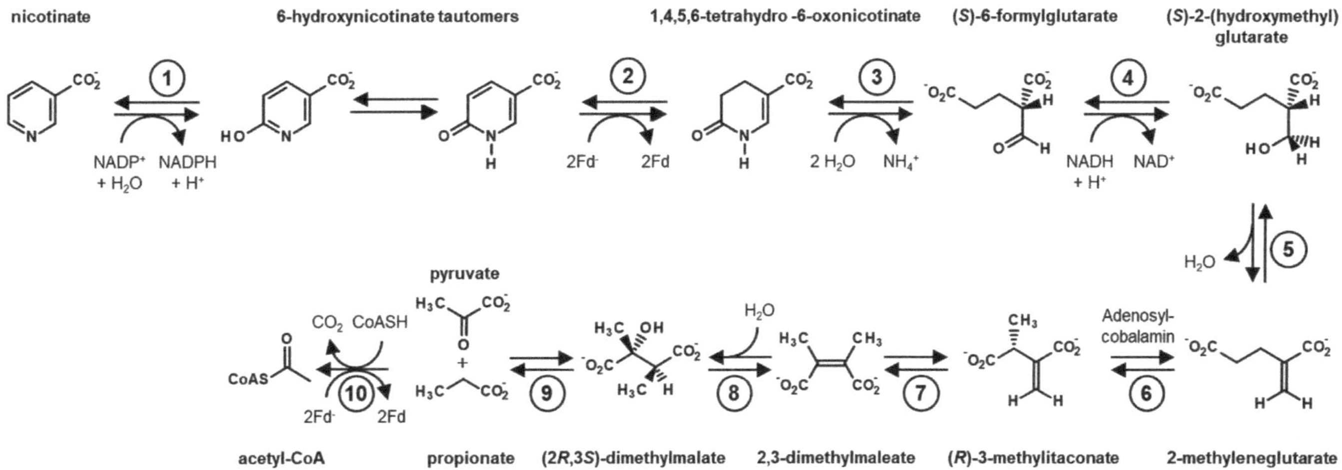

Fig. 4 Pathway of nicotinate fermentation in *E. barkeri*. Enzymes: 1, nicotinate dehydrogenase; 2, 6-hydroxynicotinate reductase; 3, enamidase; 4, 2-(hydroxymethyl)-glutarate dehydrogenase; 5, 2-(hydroxymethyl)glutarate dehydratase; 6, 2-methyleneglutarate mutase; 7, (*R*)-3-methylitaconate isomerase; 8, (2*R*,3*S*)-dimethylmalate dehydratase; 9, (2*R*,3*S*)-dimethylmalate lyase; 10, pyruvate ferredoxin oxidoreductase (PFOR); Fd, ferredoxin; Fd^- reduced ferredoxin. *Modified from Alhapel, A., Darley, D. J., Wagener, N., Eckel, E., Elsner, N., & Pierik, A. J. (2006). Molecular and functional analysis of nicotinate catabolism in* Eubacterium barkeri. Proceedings of the National Academy of Sciences of the United States of America, 103*(33), 12341–12346. doi:10.1073/pnas.0601635103. Copyright Proceedings National Academy of Science USA.*

3. Properties of glutamate mutase

Glutamate mutase (EC 5.4.99.1) was purified from cell extracts of *C. tetanomorphum* (Barker, Rooze, Suzuki, & Iodice, 1964) and later of *C. cochlearium* (Leutbecher, Böcher, Linder, & Buckel, 1992). It consists of two components, E and S (Table 1). The enzyme is assayed with help of the consecutive enzyme in the pathway of glutamate fermentation, 3-methylasparate ammonia lyase, short methylaspartase (EC 4.3.1.2). The product mesaconate has a higher absorption at 240 nm than 3-methylaspartate, $\Delta\varepsilon = 3.85\,mM^{-1}\,cm^{-1}$ (Barker, Rooze, et al., 1964). An assay in the direction from 3-methylaspartate to glutamate uses the auxiliary enzymes glutamate oxaloacetate aminotransferase (EC 2 6.1.11) and NADH-dependent (*R*)-2-hydroxyglutarate dehydrogenase (EC 1.1.99.39) (Yu et al., 2012). Kinetic constants for the *C. cochlearium* mutase are: $K_m = 1.5\,mM$ (*S*)-glutamate, 0.5 mM (2*S*,3*S*)-3-methylaspartate; for competitive inhibitors: $K_i = 0.07\,mM$ (2*S*,4*R*)-4-fluoroglutamate, 0.4 mM 2-methyleneglutarate, 0.1 mM (*S*)-3-methylitaconate, 1.2 mM itaconate. Other stereoisomers of these compounds were either not substrates or not inhibitory (Leutbecher et al., 1992). The initial rate of the reaction from 3-methylaspartate to glutamate decreased to 60%, 42%, and 20% in the presence of equimolar mixtures of acrylate and glycine (5, 10, and 15 mM, each). No significant inhibition was observed with glycine or acrylate alone (Beatrix et al., 1995).

The genes encoding glutamate mutase (Table 1) are located in operons; *mutS, mutL, mutE* and *bma* in *C. tetanomorphum* (Holloway & Marsh, 1993) and *glmS, glmL, glmE* and *bma* in *C. cochlearium* (Zelder, Beatrix, et al., 1994). The genes *mutL* and *glmL* probably encode chaperons, which remove inactive coenzyme species from ES complexes. The gene *bma* encodes methylaspartase. MutE and MutS share 94% and 80% sequence identities with GlmE and GlmS, respectively (Zelder, Beatrix, et al., 1994). Comparison of the sequences of MutS and GlmS with sequences of coenzyme B_{12}-dependent 2-methyleneglutarate mutase (Beatrix et al., 1994), methylmalonyl-CoA mutases from *Propionibacterium shermanii*, mouse and human, as well as *E. coli* methionine synthase (MetH) revealed a cobalamin binding motif: **D**-x-**H**-x-x-**G** in MutS (Banerjee, Johnston, Sobeski, Datta, & Matthews, 1989; Marsh & Holloway, 1992). Mutation of these aspartate and histidine residues in MutS and in 2-methyleneglutarate mutase resulted in enzymes with 0.01–0.2 % activities as compared of those of the

Table 1 Subunit composition of glutamate mutase and 2-methyleneglutarate mutase enzymes

Enzyme and Organism	Subunits	Molecular Mass (Da)	Amio acid residues	Gene	Tertiary Structure	References
Glutamate mutase *C. tetanomorphum*	Component E (ε_2)	$2 \times 53{,}708$	2×485	*mutE*	Heterotetra-mer, $\varepsilon_2\sigma_2$ 136,912 Da	Holloway and Marsh (1994)
	Component S (σ)	14,748	137	*mutS*		
Glutamate mutase *C. cochlearium*	Component E (ε_2)	$2 \times 53{,}504$	2×483	*glmE*	Heterotetra-mer, $\varepsilon_2\sigma_2$ 136,634 Da	Zelder, Beatrix, Leutbecher, and Buckel (1994)
	Component S (σ)	14,813	138	*glmS*		
2-Methylene-glutarate mutase *E. barkeri*	Mgm	$4 \times 66{,}650$	4×614	*mgm*	Tetramer 267,200 Da	Beatrix, Zelder, Linder, and Buckel (1994)

wild-types (Chen & Marsh, 1997). Later it turned out that upon binding of cobalamin to the enzymes, the conserved histidine replaced dimethylbenzimidazole, the lower ligand of the cobalt atom (Drennan, Huang, Drummond, Matthews, & Ludwig, 1994; Zelder, Beatrix, Kroll, & Buckel, 1995).

Together with coenzyme B_{12}, MutE (ε) and MutS (σ), produced in *Escherichia coli* (Table 1), gave an active enzyme with a loose heterotetrametric quaternary structure and complex kinetic properties that were protein concentration-dependent. To simplify the kinetics, MutS was fused to the C-terminus of MutE, which resulted in a dimeric enzyme, GlmES $(\varepsilon\sigma)_2$. This modification to the quaternary structure of glutamate mutase did not appreciably alter k_{cat} or the affinity of the enzyme for substrates and coenzyme, but removed the concentration dependence from them, thereby facilitating subsequent mechanistic studies (Chen & Marsh, 1997b; Holloway, Harding, & Marsh, 1996). (*S*)-2-Hydroxyglutarate was found to be a substrate, but reacted at a 100-times lower rate than (*S*)-glutamate; the product was (*S*,*S*)-3-methylmalate (Roymoulik, Moon, Dunham, Ballou, & Marsh, 2000). 2-Oxoglutarate caused homolysis of the cobalt-carbon bond, but the resulting 4-oxoglutaryl-radical did not fragment (Roymoulik, Chen, & Marsh, 1999). 2-Thiolglutarate was fragmented by glutamate mutase but the resulting thioglycolyl radical was too stable to react further, resulting in inactivation of the enzyme (Yoon et al., 2006).

GlmS (σ) and $(GlmE)_2$ (ε_2) from *C. cochlearium*, produced in *E. coli*, together with coenzyme B_{12} gave a pink tight complex composed of 2 σ, ε_2 and only 1 coenzyme B_{12} (Table 1). This recombinant glutamate mutase exhibited normal Michaelis-Menten enzyme kinetics, a molecular mass of 137 kDa and a specific activity of 15 U/mg protein (Zelder, Beatrix, & Buckel, 1994; Zelder, Beatrix, et al., 1994); the specific activity of the glutamate mutase purified from cell-extracts of *C. cochlearium* was 13 U/mg (Leutbecher et al., 1992).

4. Properties of 2-methyleneglutarate mutase

2-Methyleneglutarate mutase was purified to homogeneity from cell extracts of *E. barkeri*. The enzyme was assayed with 2-methyleneglutarate by conversion of the product (*R*)-3-methylitaconate to 2,3-dimethylmaleate with an increased absorbance at 256 nm ($\Delta\varepsilon = 0.66\,mM^{-1}\,cm^{-1}$) catalyzed by 3-methylitaconate Δ-isomerase (Fig. 4) (Michel, Hartrampf, & Buckel, 1989; Velarde et al., 2009). 2-Methyleneglutarate mutase was found to consist of only one subunit (70 kDa) forming a homotetramer, which contained adenosylcobalamin and cob(II)alamin in an approximately 1:1 ratio (Michel, Albracht, & Buckel, 1992; Zelder & Buckel, 1993). Thus pure 2-methylenglutarate mutase samples prepared in the dark or under dim red light needed no addition of the coenzyme to develop an activity of 5.4 U/mg protein (Michel, Buckel, & Linder, 1991). The gene encoding the 2-methyleneglutarate mutase subunit was identified as *mgm* in the genome of *E. barkeri* (Table 1). It is followed by the genes *mii* encoding 3-methylitaconate Δ-isomerase and *mgmL* encoding a putative repair enzyme for 2-methyleneglutarate mutase related to *mutL* and *glmL* of glutamate mutase. The recombinant tetrameric 2-methyleneglutarate mutase produced in *E. coli* was inactive but matured to a full active enzyme after addition of coenzyme B_{12} with the much higher specific activity of 28 U/mg protein (Beatrix et al., 1994; Pierik et al., 2005). The active enzyme contained only 2 mol coenzyme B_{12} per tetramer; the same ratio was found for glutamate mutase. K_m for 2-methyleneglutarate varied between preparations from 1.0 to 3.0 mM; $K_m = 0.4$ mM for (*R*)-3-methylitaconate was estimated discontinuously by HPLC (Michel et al., 1989; Pierik et al., 2005). 2-Methyleneglutarate mutase was inhibited by acrylate. A plot of 1/v (v = initial velocity) versus the square of the acrylate concentration gave a straight line (Dixon & Webb, 1964). This linear correlation indicates that two acrylate molecules are involved in the inhibition of the enzyme (Beatrix et al., 1995). (*Z*)-Glutaconate (0.2 mM) inactivated 2-methyleneglutarate mutase; 50% of the activity was lost within 1 min, whereas in the presence of 4 mM (*E*)-glutaconate, (*E*)-2-methylglutaconate, (*Z*)-2-methylglutaconate, or butadiene-2, 3-dicarboxylate the activity of the enzyme remained almost unchanged within 2 min (Michel et al., 1989; Pierik et al., 2005; Plett, 2005).

5. Crystal structure of glutamate mutase

Recombinant inactive enzyme preparations from *C. cochlearium* reconstituted either with cyanocobalamin or methylcobalamin were crystallized in 50 mM DL-tartrate buffer pH 4.6. The obtained crystals were soaked briefly in a solution containing 6% PEG 4000, 31% glycerol and 100 mM DL-tartaric acid (pH 4.6) and flash cooled in liquid nitrogen. The monoclinic crystals contained one heterotetrameric molecule of the composition $\epsilon_2\sigma_2$ $(B_{12})_2$ per asymmetric unit (Reitzer et al., 1999). Contrary to biochemical data indicating an adenosylcobalamin content of about 1 mol/$\epsilon_2\sigma_2$ (137 kDa) (Zelder, Beatrix, et al., 1994), the crystal structure showed each heterodimer to contain a full methyl- or cyanocobalamin molecule plus a tartrate ion, the latter in a location presumed to be the active site (Reitzer et al., 1999). The structures confirmed the replacement of the dimethylimidazole moiety of the coenzyme by the conserved histidine residue. In the overall structure, the ε-subunit forms a TIM barrel, which is tightly connected to one σ-subunit, whereas the loose connection between the two heterodimers $(\varepsilon\sigma)_2$ is only based on two helices from each ε-subunit. The (*S,S*)-tartrate molecule has been selected by the enzyme from the racemic buffer during crystallization and is fixed by 14 hydrogen bonds from seven amino acid side chains and one water molecule. Remarkably, the carboxylate distal to the coenzyme is two-fold double hydrogen bonded to two arginine side-chains in a Y-like arrangement. The proximal carboxylate is connected to a third arginine, which led to the name "arginine claw." The (*S,S*)-tartrate probably mimics (*S,S*)-3-methylaspartate, which needs to be tightly bound to the enzyme to avoid errors in the radical rearrangement. Furthermore, the (*S,S*)-tartrate molecule is 6.6 Å apart from the Co-atom; the same distance was found with (*S*)-glutamate by EPR-spectroscopy. PDB codes for the structures are 1CCW with cyanocobalamin and 1CB7 with methylcobalamin.

The structures of the isolated components S of glutamate mutase from *C. tetanomorphum* (Hoffmann et al., 2001; Tollinger et al., 2001; Tollinger, Konrat, Hilbert, Marsh, & Kräutler, 1998) and *C. cochlearium* (Hoffmann, Konrat, Bothe, Buckel, & Kräutler, 1999) were determined by NMR spectroscopy. By comparing the crystal structure of glutamate mutase from *C. cochlearium* with the solution structures of B_{12}-free GlmS and MutS, a consistent picture on the mechanism of B_{12} binding has been obtained. Important elements of the binding site only become structured upon binding B_{12}; these include the cobalt-coordinating histidine residue, and an alpha helix that forms one side of the cleft accommodating the nucleotide "tail" of the coenzyme.

6. EPR spectroscopy

EPR spectra of glutamate mutase were recorded after a short incubation of the recombinant enzyme $(\varepsilon_2\sigma_2)$ with coenzyme B_{12} and (*S*)-glutamate (Fig. 5C) (Zelder, Beatrix, et al., 1994). The EPR signals ($g_{xy} = 2.1$, $g_z = 1.985$) show a typical hyperfine splitting derived from Co(II), but are different from cob(II)alamin ($g_{xy} = 2.25$) (Fig. 5A). The spin concentrations were 34-50 % of the concentration of the enzyme $(\varepsilon_2\sigma_2)$-coenzyme complex. The inhibitors (2*S*,4*S*)-4-fluoroglutamate and

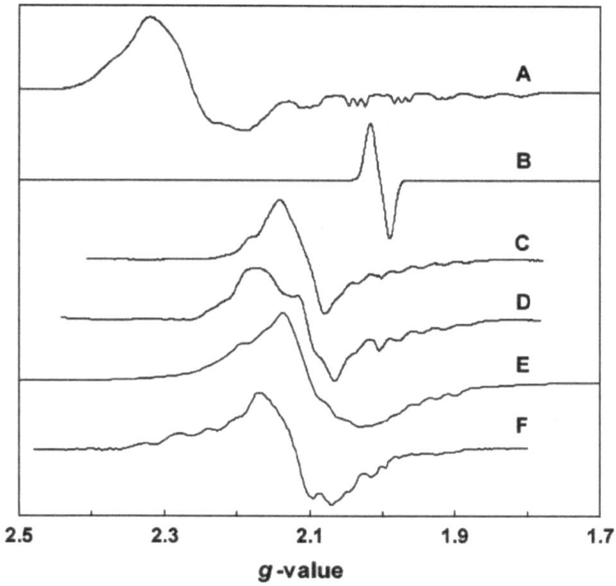

Fig. 5 Compilation of EPR spectra of enzyme-bound cob(II)alamins and radicals (Pierik et al., 2005). (A) Uncoupled cob(II)alamin in methanol:coenzyme M methyltransferase MtaC subunit. (B) A simulated malonyl radical ($g_{xyz} = 2.0$). (C) Glutamate mutase and (*S*)-glutamate. (D) 2-Methyleneglutarate mutase and 2-methyleneglutarate (Pierik et al., 2005). (E) Methylmalonyl-CoA mutase and succinyl-CoA. (F) Ribonucleotide triphosphate reductase, reductant and dGTP. The spectra have been converted to an x-axis with *g*-values to allow better comparison. Copyright American Chemical Society. *Panel (A) Adapted from Sauer, K., & Thauer, R. K. (1998). Methanol:coenzyme M methyltransferase from* Methanosarcina barkeri*—Identification of the active-site histidine in the corrinoid-harboring subunit MtaC by site-directed mutagenesis.* European Journal of Biochemistry, 253*(3), 698-705. doi:10.1046/j.1432-1327.1998.2530698.x. Panel (C) Adapted from Bothe, H., Darley, D. J., Albracht, S. P., Gerfen, G. J., Golding, B. T., & Buckel, W. (1998). Identification of the 4-glutamyl radical as an intermediate in the carbon skeleton rearrangement catalyzed by coenzyme B_{12}-dependent glutamate mutase from* Clostridium cochlearium. Biochemistry, 37*(12), 4105-4113. doi:10.1021/bi971393q. Panel (E) adapted from Padmakumar, R., & Banerjee, R. (1995). Evidence from electron paramagnetic resonance spectroscopy of the participation of radical intermediates in the reaction catalyzed by methylmalonyl-coenzyme A mutase.* The Journal of Biological Chemistry, 270*(16), 9295-9300. doi:10.1074/jbc.270.16.9295. (F) Adapted from Licht, S., Gerfen, G. J., & Stubbe, J. (1996). Thiyl radicals in ribonucleotide reductases.* Science, 271*(5248), 477-481. doi:10.1126/science.271.5248.477.*

2-methyleneglutarate induced similar but not identical signals with spin concentrations of 134–148% of the enzyme concentration. (*S*)-[2,3,3,4,4-$^2H_5$]Glutamate induced a signal significantly different to that of unlabeled (*S*)-glutamate. The paramagnetic species probably represents a radical pair composed of Co(II) and an organic radical. "There are two radicals talking to each other." (Brian Hoffmann, Northwestern University, USA). Overnight incubation of glutamate mutase with substrate or (2*S*,4*S*)-4-fluoroglutamate resulted in an inactive enzyme and cob(II)alamin, which was identified by EPR spectroscopy (g_{xy} =2.25). The spectrum of cob(II)alamin (Fig. 5A) appeared to be very similar to that of the cob(II)amide present in inactive glutamate mutase preparations from *C. cochlearium*. Later it turned out that the apparent competitive inhibitor 2-methyleneglutarate acted as inactivator, because it caused homolysis of the cobalt-carbon bond and the 5′-deoxyadenosyl radical added slowly at the double bond of 2-methyleneglutarate (Huhta, Ciceri, Golding, & Marsh, 2002).

The g_z signal of cob(II)alamin is split into 8 lines, each of which is split further into three lines due to interaction with ^{14}N of the axial ligand. If this ligand is ^{15}N, the three lines merge into two (Fig. 6). This phenomenon was used to identify the axial ligand when the coenzyme was bound to the enzyme.

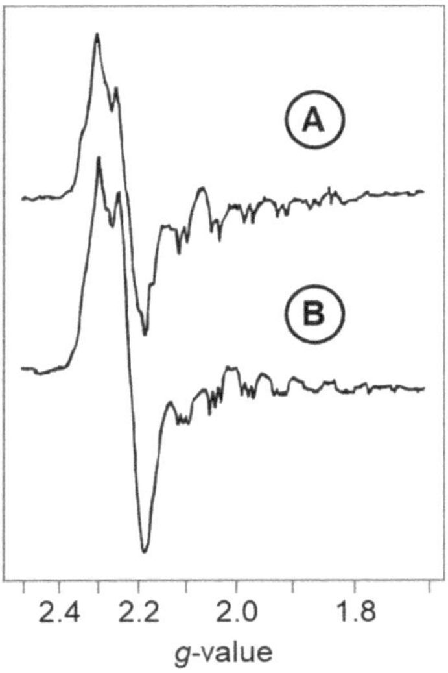

Fig. 6 EPR spectra of glutamate mutase in presence of the inhibitor 2-methyleneglutarate (for details see text and Zelder et al., 1995). (A) All nitrogens of the enzyme were ^{15}N-labeled revealing two lines in the superhyperfine structure. (B) All amino acids were ^{15}N-labeled, except the histidines which contained ^{14}N.

To identify the presumed organic radical, which was thought to interact with cob(II)alamin in the reaction catalyzed by glutamate mutase, a series of ^{2}H- and ^{13}C-labeled glutamates were used as substrates. These compounds contained the isotopes at C-2, C-3, or C-4 of the carbon chain: [2-^{2}H], [3,3-$^2H_2$], [4,4-$^2H_2$], [2,3,3,4,4-$^2H_5$], [2-^{13}C], [3-^{13}C], and [4-^{13}C]glutamate. Each reaction was monitored by EPR spectroscopy and revealed a similar signal characterized by $g'_{xy} = 2.1$, $g'_z = 1.985$, and $A' = 5.0$ mT. The interpretation of the spectral data was aided by simulations, which gave close agreement with experiment. This approach underpinned the idea of formation of a radical pair, consisting of cob(II)alamin interacting with an organic radical at a distance of 6.6 ± 0.9 Å. Comparison of the hyperfine couplings observed with unlabeled glutamate with those from the labeled glutamates enabled one partner of the radical pair to be identified as 4-glutamyl radical (Bothe et al., 1998).

Short incubation (5 min) of 2-methyleneglutarate mutase with 40 mM 2-methyleneglutarate under anaerobic conditions generated an EPR signal ($g_{xy} \approx 2.1$; $g_z \approx 2.0$), which by analogy with the findings on glutamate mutase from *C. cochlearium* (Bothe et al., 1998) was assigned to cob(II)alamin coupled to a carbon-centered radical (Fig. 5D) (Pierik et al., 2005). At longer incubation times (>1 h), inactivation of the mutase occurred concomitant with the formation of oxygen-insensitive cob(II)alamin ($g_{xy} \approx 2.25$; $g_z \approx 2.0$). Broadening (0.5 mT) of the EPR signal around $g = 2.1$ was observed only when C2 and/or C4 of 2-methyleneglutarate was labeled with ^{13}C or ^{2}H. A spectrum very similar to that obtained with 2-methyleneglutarate was induced with 40 mM acrylate; [2,3,3-$^2H_3$]acrylate intensified the hyperfine structure in the $g_z = 2.0$ region (Plett, 2005). No effect on the EPR signals was seen when [5′-^{13}C]adenosylcobalamin was used as coenzyme (Pierik et al., 2005).

7. On the mechanism of glutamate and 2-methyleneglutarate mutases

A long standing idea of the mechanism of glutamate mutase was the fragmentation of the 4-glutamyl radical into the glycyl radical and acrylate followed by recombination to the (2*S*,3*S*)-3-methyleneaspartate radical (Buckel & Golding, 1996). The latter would abstract a hydrogen atom from the transiently formed 5′-deoxyadenosine to generate the product (2*S*,3*S*)-3-methylaspartate and the 5′-deoxyadenosyl radical which together with cob(II)alamin reforms the Co-C bond. The combined inhibition of glutamate mutase by acrylate and glycine seems to confirm the fragmentation hypothesis (Fig. 1). Hence acrylate and glycine act synergistically and mimic the intermediate state of the glutamate mutase reaction, which enables them to induce homolytic cleavage of the Co-C bond of the coenzyme (Beatrix et al., 1995). A direct proof of the fragmentation mechanism was obtained by E.N.G. Marsh and H-W. Chih, who quenched the reaction with trichloroacetic acid and detected acrylate and glycine by HPLC (Chih & Marsh, 2000).

In analogy, 2-methyleneglutarate mutase should catalyze the fragmentation of the 2-methlene-4-glutaryl radical into acrylate and 2-acrylyl radical (Fig. 2). The observed inhibition with 1/v proportional to $[\text{acrylate}]^2$ supports this view. For 2-methyleneglutarate mutase also an alternate mechanism is possible; instead of fragmentation and recombination, the mechanism involves addition and elimination (Fig. 2). Intramolecular addition of the 4-methyleneglutarate radical at the double bond of the methylene side chain would lead to the trans or cis-1-methylene-1,2-cyclopropanedicarboxylate radical, which could open to the 3-methyleneitocanate radical. All attempts failed, however, to demonstrate such an intermediate (Pierik et al., 2005).

During the investigation of the mechanism of adenosylcobalamin-dependent enzymes, often the question arose, whether the coenzyme acted as radical generator and cob(II)alamin as spectator of the rearrangement rather than participating in the reaction. For the carbon skeleton mutases, the answer appears to be spectator, because almost the same rearrangement with the B_{12}-independent radical generator S-adenosyl-methionine (SAM) is known. It is involved in the biosynthesis of the 22nd amino acid pyrrolysine, the isomerization of (*S*)-lysine to (2*R*,3*R*)-3-methylornithine in *Methanosarcina barkeri* (Gaston, Zhang, Green-Church, & Krzycki, 2011). This rearrangement appears to be identical to that of glutamate mutase, except the inversion of configuration at C2 (Quitterer, List, Eisenreich, Bacher, & Groll, 2012), whereas glutamate mutase works with retention at C2. Finally, calculations with methylmalonyl-CoA mutase confirm that "Hydrogen bonds (from the enzyme and not from cob(II)alamin) guide the short-lived 5′-deoxyadenosyl radical to the place of action." (Bucher, Sandala, Durbeej, Radom, & Smith, 2012; Buckel, Friedrich, & Golding, 2012).

8. Preparation of active glutamate mutase and 2-methyleneglutarate mutase

Production, purification and reconstitution of glutamate mutase from *C. cochlearium* (Zelder, Beatrix, & Buckel, 1994; Zelder, Beatrix, et al., 1994) and 2-methyleneglutarate mutase from *E. barkeri* (Beatrix et al., 1994; Pierik et al., 2005) can readily be performed by the described procedures. DNA and primers are available for expression of the *glmS* and *glmE* genes as well as for expression of the *mgm* gene (ERC protocol). As indicated in Sections 3 and 4, the catalytic assays of the mutases require auxiliary enzymes, 3-methylaspartase for glutamate mutase and 3-methylitaconate isomerase for 2-methyleneglutarate mutase. The DNA of *C. cochlearium* and *E. barkeri* can also be used for the production of these enzymes (Goda, Minton, Botting, & Gani, 1992; Velarde et al., 2009), because the conventional methods described in earlier publications require growth of the bacteria and laborious purifications from cell-extracts.

9. Key resources table (ERC)

Reagent or resource	Source
Biological Samples	
DNA from *Clostridium cochlearium* DSMZ 1285	Deutsche Sammlung von Mikroorganismen und Zellkulturen (DSMZ); German collection of microorganisms and cell cultures
Optional: *C. cochlearium* DSMZ 1285	
DNA from *Eubacterium barkeri* DSMZ 1223	
Optional: *E. barkeri* DSMZ 1223	
Escherichia coli DH5α	
pJFll8HE	
pBluescript sk +	Stratagene
Chemicals	
Nutrient broth	Merck, Darmstadt, Germany
Optional: Growth medium for *C. cochlearium*	(Buckel & Barker, 1974)
Optional: Growth medium for *E. barkeri*	(Pirzer, Lill, & Eggerer, 1979)

Continued

—cont'd

Reagent or resource	Source
isopropyl-1-thio-β-D-galactoside (IPTG)	
DL-*threo*-β-Methylaspartic acid or (2*RS*,3*S*)-3-methylaspartate	M6126 Sigma Aldrich
Custom syntheses on request: 2-Methyleneglutaric acid and (2*RS*,3*R*)-3-methylitaconic acid	Professor Bernard T. Golding BiBerChem Research Limited, Floor 3, The Biosphere, Drayman's Way, Newcastle Helix, Corporation Street, Newcastle upon Tyne, NE4 5BX, UK.
All 4 dNTPs, ATP,	
Taq DNA polymerase	
*Xba*I, *Hin*dIII, *Bam*HI	
oligonucleotide primers for component S of glutamate mutase fw: 5′-AAG ACT ATT GTT CTT GGA GTT ATT GG-3′ rv: 5′-GTT TCT GGA GAT GTT CCT GG-3′	
oligonucleotide primers for component E of glutamate mutase fw: 5′-CC AGT GGA TCC TGG TGT ACA AAG TAC G-3′ rv: 5′-GAA AAA GGA TCC AGT TAG AGT TTG ATA GG-3′	The underlined part represents a *Bam*HI restriction site
oligonucleotide primers for 2-methyleneglutarate mutase in *E. barkeri* fw: 5′-GAT AGA G*AA GCT T*AT GTT GAA GAG AGT GG-3′ rv: 5′-GGT CCC TGC *AAG CTT* GAT GAC ACA GG-3′	The underlined part represents a *Hin*dIII restriction site
pOZ1: Plasmid for expression of *glmS*	Zelder, Beatrix, & Buckel (1994) and Zelder, Beatrix, et al. (1994)
pOZ5: Plasmid for expression of *glmE*	Zelder, Beatrix, & Buckel (1994) and Zelder, Beatrix, et al. (1994)

—cont'd

Reagent or resource	Source
pBB2: Plasmid for expression of *mgm*	Beatrix et al. (1994)
Buffer 1: 20 mM potassium phosphate buffer (PPB) pH 7.4; 1 mM EDTA, 3 mM DTT	
Buffer 2: 20 mM PPB pH 7.4; 1 mM EDTA, 1 M $(NH_4)_2SO_4$	
Buffer 3: 20 mM PPB pH 7.4; 1 mM EDTA	
Buffer 4: 20 mM PPB pH 7.4	
Buffer 5: 20 mM PPB pH 7.4, 1 mM EDTA, 1 M NaCl	
Buffer 6: 20 mM PPB pH 7.4, 1 mM PMSF, 1 mM EDTA, DNase I (10 mg/L)	
Equipment	
Branson ultrasonics-sfx250-sonifier	
600 mL Superdex 200 Biopilot column	Merck
Phenyl Sepharose (HiLoad 26/10)	Merck
HiLoad 26/10 Q-Sepharose	Merck
G25-fine, gel filtration column (15 × 250 mm)	Merck

10. Expression of *glmS, glmE* and *mgm* without tag in *E. coli* DH5α

Timing: 12 h

1. Competent cells of *E. coli* DH5α were transformed with the respective plasmid. Cultures of up to 4 L were induced with 1 mM IPTG after growth to OD_{578} of 1. The cultures were stirred vigorously at 37 °C over night and harvested by centrifugation.

10.1 Purification of GlmS and GlmE from cultures of recombinant *E. coli*

Timing: 1 day

All purification steps occur at 4°C. Starting weight wet cells: 5–30 g

2. Preparation of crude cell extracts

(a) Dissolution of wet cells in 3 volumes of buffer 1. Cell disruption by sonication on ice (5 min, 70% power, pulse, 50)

(b) Ultracentrifugation at 80,000 × *g* for 30 min.

(c) $(NH_4)_2SO_4$ precipitation

Component S: Addition of solid $(NH_4)_2SO_4$ to supernatant, 50% saturation. Centrifugation 20 min at 20,000 × *g*, suspend pellet in buffer 1.

Component E: Addition of solid $(NH_4)_2SO_4$ to supernatant, 45% saturation. Centrifugation 20 min at 20,000 × *g*, suspend pellet in buffer 1.

3. Chromatographic steps

- **(a)** Component S: gel filtration chromatography with Superdex 200. Apply complete sample and elute with buffer 1. Flow rate: 4 mL/min. Identify fractions containing component S by enzyme assay. Dialyze active fractions against 5 L buffer 4.
- **(b)** Component E: Hydrophobic interaction chromatography with Phenyl-Sepharose. Equilibrate column with buffer 2. Apply complete sample. To remove *E. coli* proteins, elute with decreasing $(NH_4)_2SO_4$ gradient from buffer 2 to buffer 3 in 160 mL. Flow rate: 2 mL/min. Elute component E with 12 mL water. Adjust pH to 7.4 by addition of PPB to a final concentration of 20 mM. Concentrate via Centricon (MWCO: 30 kDa) to the required protein concentration.

4. GlmE and GlmS can be stored at 4 °C over night for further experiments.

10.2 Complementation of GlmS + GlmE with adenosylcobalamin: Glutamate mutase

All steps in the dark or dim red light. GlmE and GlmS (each 1 mg/mL) are incubated 10 min at 37 °C, with 0.5 mM adenosylcobalamin in buffer 4, 1 mM DTT, 200 μL final volume. Unbound coenzyme is removed on a G25 gel filtration column (15 x 250 mm) equilibrated with buffer A at 4 °C. The enzyme can be stored at 4 °C overnight.

10.3 Purification of Mgm from cultures of recombinant *E. coli*

Timing: 2 days

All purification steps occur at 4°C. Start with 25 g wet cellpaste.

1. Preparation of crude cell extracts

(a) Dissolve wet cells in 60 mL buffer 6. Cell disruption by sonication on ice (3 × 5 min, 70% power)

(b) Ultracentrifugation at 80,000 × *g* for 30 min.

2. Chromatographic steps

(a) Clear supernatant through 45 μm sterile filter and load onto HiLoad 26/10 Q-Sepharose equilibrated with buffer 4.

(b) Elute with linear gradient (800 mL) of buffers 4 and 5, flow rate 3 mL/min

(c) Identify fractions with Mgm activity by enzymatic assay (elute between 340 and 450 mM NaCl) and add solid $(NH_4)_2SO_4$ to a final concentration of 1 M, stir for 10 min at 4°C.

(d) Clear again through 45 μm sterile filter and load onto a HiLoad 26/10 phenyl-Sepharose HP column equilibrated with buffer 2. Wash with a 200 mL linear gradient of buffer 2 to buffer 4 and continue to elute with 100% buffer 4 for another 200 mL, flow rate 3 mL/min. Mgm elutes at 100 % buffer 4. Concentrate via Centricon (MWCO: 30 kDa) to 8 mL.

(e) Size exclusion chromatography on Superdex 200 (prep grade 35/600) equilibrated with buffer 4, flow rate 4 mL/min. Identify fractions with Mgm activity by enzymatic assay and concentrate again via Centricon to a final concentration of 6–7 mg/mL.

10.4 Complementation of Mgm with adenosylcobalamin: 2-Methyleneglutarate mutase

(a) 2.5 mg Mgm in buffer 4 (400 μL) are supplemented with 1 mM DTT and 1 mM adenosylcobalamin (final volume 500 μL) and incubated for 10 min at 37 °C.

(b) Remove unbound coenzyme on a G25-fine gel filtration column (15 × 250 mm) equilibrated with buffer 4 at 4 °C. Concentrate 2-methyleneglutarate mutase containing fractions via a Centricon (MWCO: 30 kDa) to the original volume. The enzyme can be stored at 4 °C overnight.

The standard continuous ASSAY OF THE MUTASES measures activities (0.05–0.1 units/mL) with 10 mM (*S*)-glutamate or 10 mM 2-methyleneglutarate as substrate in 100 mM potassium phosphate (pH 7.4, 25 °C) and 1 mM DTT using 3-methylaspartase or 3-methylitaconate isomerase (10–30 units/mL) as auxiliary enzymes, respectively (total volume 0.5 mL). The formed mesaconate causes an UV absorbance increase ($\Delta\varepsilon_{256\ nm} = 3.85\ mM^{-1}\ cm^{-1}$) as does 2,3-dimethylmaleate ($\Delta\varepsilon_{256\ nm} = 0.66\ mM^{-1}\ cm^{-1}$). With low activities, the assay can be performed in an anaerobic chamber under red light with an upper limit of 30 min because of the instability of the coenzyme in the mutases. The reaction with 2-methyleneglutarate mutase is stopped with 0.25 mL 1 M HCl, which converts dimethylmaleate to its cyclic anhydride ($\Delta\varepsilon_{256\ nm} = 4.3\ mM^{-1}\ cm^{-1}$).

References

Banerjee, R. V., Johnston, N. L., Sobeski, J. K., Datta, P., & Matthews, R. G. (1989). Cloning and sequence analysis of the *Escherichia coli metH* gene encoding cobalamin-dependent methionine synthase and isolation of a tryptic fragment containing the cobalamin-binding domain. *The Journal of Biological Chemistry*, *264*(23), 13888–13895.

Barker, H. A., Rooze, V., Suzuki, F., & Iodice, A. A. (1964). The glutamate mutase system assays and properties. *The Journal of Biological Chemistry*, *239*, 3260–3266.

Barker, H. A., Suzuki, F., Iodice, A., & Rooze, V. (1964). Glutamate mutase reaction. *Annals of the New York Academy of Sciences*, *112*, 644–654. https://doi.org/10.1111/j.1749-6632.1964.tb45041.x.

Beatrix, B., Zelder, O., Kroll, F., Örlygsson, G., Golding, B. T., & Buckel, W. (1995). Evidence for a mechanism involving transient fragmentation in carbon skeleton rearrangements dependent on coenzyme B_{12}. *Angewandte Chemie, International Edition*, *34*, 2398–2401.

Beatrix, B., Zelder, O., Linder, D., & Buckel, W. (1994). Cloning, sequencing and expression of the gene encoding the coenzyme B_{12}-dependent 2-methyleneglutarate mutase from *Clostridium barkeri* in *Escherichia coli*. *European Journal of Biochemistry*, *221*(1), 101–109. https://doi.org/10.1111/j.1432-1033.1994.tb18718.x.

Bothe, H., Darley, D. J., Albracht, S. P., Gerfen, G. J., Golding, B. T., & Buckel, W. (1998). Identification of the 4-glutamyl radical as an intermediate in the carbon skeleton rearrangement catalyzed by coenzyme B_{12}-dependent glutamate mutase from *Clostridium cochlearium*. *Biochemistry*, *37*(12), 4105–4113. https://doi.org/10.1021/bi971393q.

Bucher, D., Sandala, G. M., Durbeej, B., Radom, L., & Smith, D. M. (2012). The elusive 5'-deoxyadenosyl radical in coenzyme-B_{12}-mediated reactions. *Journal of the American Chemical Society*, *134*(3), 1591–1599. https://doi.org/10.1021/ja207809b.

Buckel, W. (2021). Energy conservation in fermentations of anaerobic bacteria. *Frontiers in Microbiology*, *12*. https://doi.org/10.3389/fmicb.2021.703525, 703525

Buckel, W., & Barker, H. A. (1974). Two pathways of glutamate fermentation by anaerobic bacteria. *Journal of Bacteriology*, *117*(3), 1248–1260. https://doi.org/10.1128/JB.117.3.1248-1260.1974.

Buckel, W., Friedrich, P., & Golding, B. T. (2012). Hydrogen bonds guide the short-lived 5'-deoxyadenosyl radical to the place of action. *Angewandte Chemie, International Edition*, *51*(40), 9974–9976. https://doi.org/10.1002/anie.201205299.

Buckel, W., & Golding, B. T. (1996). Glutamate and 2-methyleneglutarate mutase: From microbial curiosities to paradigms for coenzyme B_{12}-dependent enzymes. *Chemical Society Reviews*, *25*, 329–337.

Buckel, W., & Golding, B. T. (2006). Radical enzymes in anaerobes. *Annual Review of Microbiology*, *60*, 27–49. https://doi.org/10.1146/annurev.micro.60.080805.142216.

Chen, H. P., & Marsh, E. N. G. (1997). How enzymes control the reactivity of adenosylcobalamin: Effect on coenzyme binding and catalysis of mutations in the conserved histidine-aspartate pair of glutamate mutase. *Biochemistry*, *36*(25), 7884–7889. https://doi.org/10.1021/bi970169y.

Chen, H. P., & Marsh, E. N. G. (1997b). Adenosylcobalamin-dependent glutamate mutase: Characterization of an engineered enzyme with simplified quaternary structure and kinetics. *Biochemistry*, *36*, 14939–14945.

Chih, H. W., & Marsh, E. N. G. (2000). Mechanism of glutamate mutase: Identification and kinetic competence of acrylate and glycyl radical as intermediates in the rearrangement of glutamate to methylaspartate. *Journal of American Chemical Society*, *122*, 10732–10733.

Ciceri, D., Pierik, A. J., Hartrampf, G., Bröker, G., Speranza, G., Buckel, W., et al. (2000). Stereochemistry of the methyl group in (*R*)-3-methylitaconate derived by rearrangement of 2-methylideneglutarate catalysed by a coenzyme B_{12}-dependent mutase. *Helvetica Chimica Acta*, *83*, 2550–2561.

Dixon, M., & Webb, E. C. (1964). *Enzymes* (2nd ed.). London: Longmans.

Drennan, C. L., Huang, S., Drummond, J. T., Matthews, R. G., & Ludwig, M. L. (1994). How a protein binds B_{12}: A 3.0 A X-ray structure of B_{12}-binding domains of methionine synthase. *Science*, *266*(5191), 1669–1674. https://doi.org/10.1126/science.7992050.

Gaston, M. A., Zhang, L., Green-Church, K. B., & Krzycki, J. A. (2011). The complete biosynthesis of the genetically encoded amino acid pyrrolysine from lysine. *Nature*, *471*(7340), 647–650. https://doi.org/10.1038/nature09918.

Goda, S. K., Minton, N. P., Botting, N. P., & Gani, D. (1992). Cloning, sequencing, and expression in *Escherichia coli* of the *Clostridium tetanomorphum* gene encoding beta-methylaspartase and characterization of the recombinant protein. *Biochemistry*, *31*(44), 10747–10756. https://doi.org/10.1021/bi00159a015.

Hartrampf, G., & Buckel, W. (1984). The stereochemistry of the formation of the methyl group in the glutamate mutase-catalysed reaction in *Clostridium tetanomorphum*. *FEBS Letters*, *171*(1), 73–78. https://doi.org/10.1016/0014-5793(84)80462-7.

Hoffmann, B., Konrat, R., Bothe, H., Buckel, W., & Kräutler, B. (1999). Structure and dynamics of the B_{12}-binding subunit of glutamate mutase from *Clostridium cochlearium*. *European Journal of Biochemistry*, *263*(1), 178–188. https://doi.org/10.1046/j.1432-1327.1999.00482.x.

Hoffmann, B., Tollinger, M., Konrat, R., Huhta, M., Marsh, E. N., & Kräutler, B. (2001). A protein pre-organized to trap the nucleotide moiety of coenzyme B_{12}: Refined solution structure of the B_{12}-binding subunit of glutamate mutase from *Clostridium tetanomorphum*. *Chembiochem*, *2*(9), 643–655. https://doi.org/10.1002/1439-7633(20010903)2:9<643::AID-CBIC643>3.0.CO;2-J.

Holloway, D. E., Harding, S. E., & Marsh, E. N. (1996). Adenosylcobalamin-dependent glutamate mutase: Properties of a fusion protein in which the cobalamin-binding subunit is linked to the catalytic subunit. *The Biochemical Journal*, *320*(Pt. 3), 825–830. https://doi.org/10.1042/bj3200825.

Holloway, D. E., & Marsh, E. N. G. (1993). Cloning and sequencing of glutamate mutase component E from *Clostridium tetanomorphum*. Organization of the *mut* genes. *FEBS Letters*, *317*(1-2), 44–48. https://doi.org/10.1016/0014-5793(93)81488-l.

Holloway, D. E., & Marsh, E. N. G. (1994). Adenosylcobalamin-dependent glutamate mutase from *Clostridium tetanomorphum*. Overexpression in *Escherichia coli*, purification, and characterization of the recombinant enzyme. *The Journal of Biological Chemistry*, *269*(32), 20425–20430.

Huhta, M. S., Ciceri, D., Golding, B. T., & Marsh, E. N. G. (2002). A novel reaction between adenosylcobalamin and 2-methyleneglutarate catalyzed by glutamate mutase. *Biochemistry*, *41*(9), 3200–3206. https://doi.org/10.1021/bi011965d.

Kung, H. F., Cederbaum, S., Tsai, L., & Stadtman, T. C. (1970). Nicotinic acid metabolism. V. A cobamide coenzyme-dependent conversion of alpha-methyleneglutaric acid to dimethylmaleic acid. *Proceedings of the National Academy of Sciences of the United States of America*, *65*(4), 978–984. https://doi.org/10.1073/pnas.65.4.978.

Kung, H. F., & Stadtman, T. C. (1971). Nicotinic acid metabolism. VI. Purification and properties of alpha-methyleneglutarate mutase (B_{12}-dependent) and methylitaconate isomerase. *The Journal of Biological Chemistry*, *246*(10), 3378–3388.

Leutbecher, U., Böcher, R., Linder, D., & Buckel, W. (1992). Glutamate mutase from *Clostridium cochlearium*. Purification, cobamide content and stereospecific inhibitors. *European Journal of Biochemistry*, *205*(2), 759–765. https://doi.org/10.1111/j.1432-1033.1992.tb16840.x.

Marsh, E. N. G., & Holloway, D. E. (1992). Cloning and sequencing of glutamate mutase component S from *Clostridium tetanomorphum*. Homologies with other cobalamin-dependent enzymes. *FEBS Letters*, *310*(2), 167–170. https://doi.org/10.1016/0014-5793(92)81321-c.

Michel, C., Albracht, S. P., & Buckel, W. (1992). Adenosylcobalamin and cob(II)alamin as prosthetic groups of 2-methyleneglutarate mutase from *Clostridium barkeri*. *European Journal of Biochemistry*, *205*(2), 767–773. https://doi.org/10.1111/j.1432-1033.1992.tb16841.x.

Michel, C., Buckel, W., & Linder, D. (1991). Purification of the coenzyme B_{12}-containing 2-methyleneglutarate mutase from *Clostridium barkeri* by high-performance liquid chromatography. *Journal of Chromatography*, *587*(1), 93–99. https://doi.org/10.1016/0021-9673(91)85202-q.

Michel, C., Hartrampf, G., & Buckel, W. (1989). Assay and purification of the adenosylcobalamin-dependent 2-methyleneglutarate mutase from *Clostridium barkeri*. *European Journal of Biochemistry*, *184*(1), 103–107. https://doi.org/10.1111/j.1432-1033.1989.tb14995.x.

Pierik, A. J., Ciceri, D., Lopez, R. F., Kroll, F., Bröker, G., Beatrix, B., et al. (2005). Searching for intermediates in the carbon skeleton rearrangement of 2-methyleneglutarate to (*R*)-3-methylitaconate catalyzed by coenzyme B_{12}-dependent 2-methyleneglutarate mutase from *Eubacterium barkeri*. *Biochemistry*, *44*(31), 10541–10551. https://doi.org/10.1021/bi050049n.

Pirzer, P., Lill, U., & Eggerer, H. (1979). Nicotinic acid metabolism. 2,3-Dimethylmalate lyase. *Hoppe-Seyler's Zeitschrift für Physiologische Chemie*, *360*(12), 1693–1702. https://doi.org/10.1515/bchm2.1979.360.2.1693.

Plett, S. (2005). *Neue Inaktivatoren der Coenzym B_{12}-abhängigen 2-Methylenglutarate-mutase aus Eubacterium barkeri* (Master of Science). Marburg, Germany: Philipps-Universität.

Quitterer, F., List, A., Eisenreich, W., Bacher, A., & Groll, M. (2012). Crystal structure of methylornithine synthase (PylB): Insights into the pyrrolysine biosynthesis. *Angewandte Chemie, International Edition*, *51*(6), 1339–1342. https://doi.org/10.1002/anie.201106765.

Reitzer, R., Gruber, K., Jogl, G., Wagner, U. G., Bothe, H., Buckel, W., et al. (1999). Glutamate mutase from *Clostridium cochlearium*: The structure of a coenzyme B_{12}-dependent enzyme provides new mechanistic insights. *Structure*, 7(8), 891–902. https://doi.org/10.1016/s0969-2126(99)80116-6.

Roymoulik, I., Chen, H. P., & Marsh, E. N. (1999). The reaction of the substrate analog 2-ketoglutarate with adenosylcobalamin-dependent glutamate mutase. *The Journal of Biological Chemistry*, *274*(17), 11619–11622. https://doi.org/10.1074/jbc.274.17.11619.

Roymoulik, I., Moon, N., Dunham, W. R., Ballou, D. P., & Marsh, E. N. (2000). Rearrangement of L-2-hydroxyglutarate to L-*threo*-3-methylmalate catalyzed by adenosylcobalamin-dependent glutamate mutase. *Biochemistry*, *39*(33) 10340–10346. https://doi.org/10.1021/bi000121b.

Sprecher, M., Switzer, R. L., & Sprinson, D. B. (1966). Stereochemistry of the glutamate mutase reaction. *The Journal of Biological Chemistry*, *241*(4), 864–867.

Tollinger, M., Eichmüller, C., Konrat, R., Huhta, M. S., Marsh, E. N., & Kräutler, B. (2001). The B_{12}-binding subunit of glutamate mutase from *Clostridium tetanomorphum* traps the nucleotide moiety of coenzyme B_{12}. *Journal of Molecular Biology*, *309*(3), 777–791. https://doi.org/10.1006/jmbi.2001.4696.

Tollinger, M., Konrat, R., Hilbert, B. H., Marsh, E. N., & Kräutler, B. (1998). How a protein prepares for B_{12} binding: Structure and dynamics of the B_{12}-binding subunit of glutamate mutase from *Clostridium tetanomorphum*. *Structure*, *6*(8), 1021–1033. https://doi.org/10.1016/s0969-2126(98)00103-8.

Velarde, M., Macieira, S., Hilberg, M., Bröker, G., Tu, S. M., Golding, B. T., et al. (2009). Crystal structure and putative mechanism of 3-methylitaconate-delta-isomerase from *Eubacterium barkeri*. *Journal of Molecular Biology*, *391*(3), 609–620. https://doi.org/10.1016/j.jmb.2009.06.052.

Yoon, M., Patwardhan, A., Qiao, C., Mansoorabadi, S. O., Menefee, A. L., Reed, G. H., et al. (2006). Reaction of adenosylcobalamin-dependent glutamate mutase with 2-thiolglutarate. *Biochemistry*, *45*(38), 11650–11657. https://doi.org/10.1021/bi061067n.

Yu, X., Bresser, J., Schall, I., Djurdjevic, I., Buckel, W., Wang, X., et al. (2012). Development of a satisfactory and general continuous assay for aminotransferases by coupling with (*R*)-2-hydroxyglutarate dehydrogenase. *Analytical Biochemistry*, *431*(2), 127–131. https://doi.org/10.1016/j.ab.2012.09.009.

Zelder, O., Beatrix, B., & Buckel, W. (1994). Cloning, sequencing and expression in Escherichia coli of the gene encoding component S of the coenzyme B_{12}-dependent glutamate mutase from *Clostridium cochlearium*. *FEMS Microbiology Letters*, *118*(1–2), 15–21. https://doi.org/10.1111/j.1574-6968.1994.tb06797.x.

Zelder, O., Beatrix, B., Kroll, F., & Buckel, W. (1995). Coordination of a histidine residue of the protein-component S to the cobalt atom in coenzyme B_{12}-dependent glutamate mutase from *Clostridium cochlearium*. *FEBS Letters*, *369*(2-3), 252–254. https://doi.org/10.1016/0014-5793(95)00762-x.

Zelder, O., Beatrix, B., Leutbecher, U., & Buckel, W. (1994). Characterization of the coenzyme-B_{12}-dependent glutamate mutase from *Clostridium cochlearium* produced in *Escherichia coli*. *European Journal of Biochemistry*, *226*(2), 577–585. https://doi.org/10.1111/j.1432-1033.1994.tb20083.x.

Zelder, O., & Buckel, W. (1993). On the role of two different cobalt(II) species in coenzyme B_{12}-dependent 2-methyleneglutarate mutase from *Clostridium barkeri*. *Biological Chemistry Hoppe-Seyler*, *374*(1), 85–90. https://doi.org/10.1515/bchm3.1993.374.1-6.85.

CHAPTER TWELVE

Human B_{12}-dependent enzymes: Methionine synthase and Methylmalonyl-CoA mutase

Romila Mascarenhas[†], Harsha Gouda[†], Markus Ruetz, and Ruma Banerjee*

Department of Biological Chemistry, University of Michigan Medical School, Ann Arbor, MI, United States
*Corresponding author: e-mail address: rbanerje@umich.edu

Contents

Abstract

Humans have only two known cobalamin or B_{12}-dependent enzymes: cytoplasmic methionine synthase and mitochondrial methylmalonyl-CoA mutase. A complex intracellular B_{12} trafficking pathway, comprising a multitude of chaperones, process and deliver cobalamin to the two target enzymes. Methionine synthase catalyzes the transfer of a methyl group from N^5-methytetrahydrofolate to homocysteine, generating tetrahydrofolate and methionine. Cobalamin serves as an intermediate methyl group carrier and cycles between methylcobalamin and cob(*I*)alamin. Methylmalonyl-CoA mutase uses the 5′-deoxy-adenosylcobalamin form of the cofactor and catalyzes the 1,2 rearrangement of methylmalonyl-CoA to succinyl-CoA. Two chaperones, CblA (or MMAA) and CblB (or MMAB, also known as adenosyltransferase), serve the mutase and ensure that the fidelity of the cofactor loading and unloading processes is maintained. This chapter focuses on assays for purifying and measuring the activities of methionine synthase and methylmalonyl-CoA mutase.

[†] These authors contributed equally.

Methods in Enzymology, Volume 668
ISSN 0076-6879
https://doi.org/10.1016/bs.mie.2021.12.012

1. Introduction

Humans have two B_{12} or cobalamin-dependent enzymes: methionine synthase (MS) and methylmalonyl-CoA mutase (MCM) (Banerjee & Ragsdale, 2003). MS and MCM represent the termination of the cytoplasmic and mitochondrial branches of the B_{12} trafficking pathway, respectively, that processes and delivers the cofactor to these enzymes (Banerjee, 2006; Banerjee, Gherasim, & Padovani, 2009; Gherasim, Lofgren, & Banerjee, 2013). Redox-linked coordination chemistry plays a key role in the transfer of cobalamin through its intracellular trafficking pathway (Banerjee, Gouda, & Pillay, 2021). MS converts the circulating form of folic acid, N^5-methytetrahydrofolate (CH_3–H_4 folate), to tetrahydrofolate (H_4 folate), making it available as a one-carbon carrier for essential purine and amino acid biosynthesis reactions (Banerjee & Matthews, 1990). In the process, MS recycles homocysteine to methionine. MS uses the methylcobalamin (MeCbl) form of the cofactor and catalyzes two successive methyl transfer reactions from MeCbl to homocysteine (Eq. 1) and from CH_3–H_4 folate to cob(*I*)alamin (Eq. 2), generating methionine and H_4folate (Fig. 1).

$$E \bullet CH_3 - cob(III)alamin + homocysteine \rightarrow E \bullet cob(I)alamin + methionine \quad (1)$$

$$CH_3 - H_4\ folate + E \bullet cob(I)alamin \rightarrow H_4folate + E \bullet CH_3 - cob(III)alamin \quad (2)$$

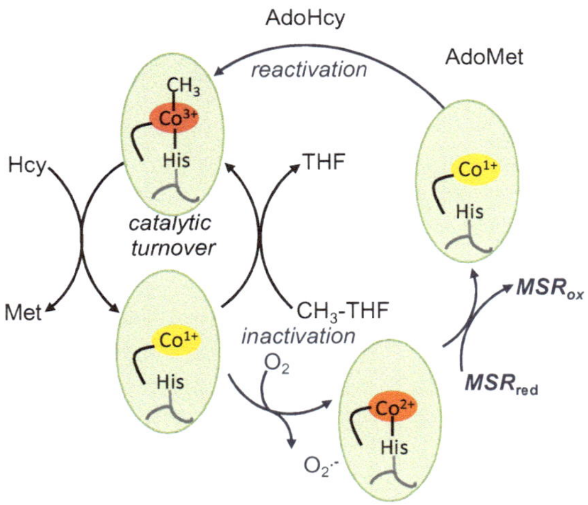

Fig. 1 Reaction catalyzed by methionine synthase. THF, Hcy and Met denote tetrahydrofolate, homocysteine, and methionine, respectively.

Demethylation of MeCbl generates a supernucleophilic cob(*I*)alamin intermediate (Banerjee, Frasca, Ballou, & Matthews, 1990), which is susceptible to oxidative inactivation approximately once every 2000 turnovers (Drummond, Huang, Blumenthal, & Matthews, 1993). Inactive MS is repaired in a reductive methylation reaction, requiring NADPH-dependent methionine synthase reductase (MSR) as an electron donor and S-adenosylmethionine (AdoMet) as a methyl donor (Banerjee, Harder, Ragsdale, & Matthews, 1990; Olteanu, Wolthers, Munro, Scrutton, & Banerjee, 2004). The reactivation cycle results in reformation of MeCbl, enabling reentry of MS to the catalytic cycle (Fig. 1). MSR is a diflavin oxidoreductase that is a dedicated redox partner for MS (Olteanu & Banerjee, 2001). Functional deficiency of either MS (Gulati et al., 1996) or MSR (Leclerc et al., 1998) leads to homocystinuria, an inborn error of metabolism that is inherited as an autosomal recessive disorder (Watkins & Rosenblatt, 2011).

Human MS is a 140 kDa monomeric protein that is predicted to be modular in its organization like its *E. coli* homolog (Goulding, Postigo, & Matthews, 1997). The N-terminal homocysteine and CH_3–H_4 folate binding domains engage with the central B_{12} domain during the catalytic cycle, alternately demethylating MeCbl, and remethylating cob(*I*)alamin, respectively. The C-terminal AdoMet domain engages with the B_{12} domain during the reactivation cycle.

Human MCM belongs to the class of 5′-deoxyadenosylcobalamin (AdoCbl)-dependent isomerases and catalyzes the 1,2 rearrangement of methylmalonyl-CoA to succinyl-CoA, which can enter the TCA cycle (Fig. 2). The MCM reaction is important in the catabolism of odd-chain fatty acids, branched chain amino acids and cholesterol (Banerjee, 2001, 2003). AdoCbl serves as a radical reservoir and substrate binding accelerates the rate of cobalt-carbon bond cleavage, initiating the radical-based carbon skeleton rearrangement reaction (Chowdhury & Banerjee, 2000) (Fig. 2). The 5′-deoxyadenosine is occasionally lost from the catalytic cycle, leading to cob(*II*)alamin accumulation in inactive enzyme, which triggers cofactor off-loading (Padovani & Banerjee, 2009a). Both AdoCbl loading onto MCM and cob(*II*)alamin off-loading from MCM rely on the chaperones CblA and ATR in the mitochondrial B_{12} trafficking pathway (Fig. 3A and B).

AdoCbl is synthesized by ATP:cob(*I*)alamin adenosyltransferase (ATR, also known as MMAB) (Yamanishi, Labunska, & Banerjee, 2005), which also functions as a chaperone, loading AdoCbl onto MCM (Padovani & Banerjee, 2009b; Padovani, Labunska, Palfey, Ballou, & Banerjee, 2008)

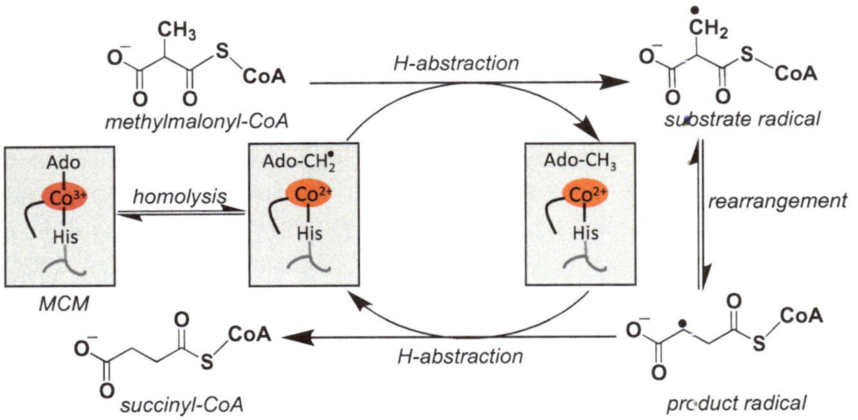

Fig. 2 Reaction catalyzed by MCM. A histidine ligand donated by MCM serves as the lower axial ligand while the DMB nucleotide tail of the corrin is displaced to the side.

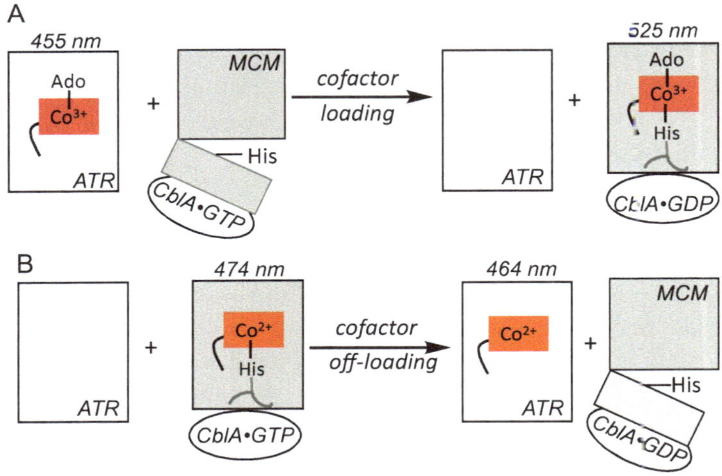

Fig. 3 AdoCbl loading onto MCM (A) and cob(*II*)alamin off-loading from MCM (B) require the chaperones ATR and CblA. Redox-linked coordination changes lead to differences in the absorption spectra of cobalamin bound to ATR and MCM, as indicated. The DMB tail is shown attached to the corrin ring.

and off-loading cob(*II*)alamin from MCM (Padovani & Banerjee, 2009a; Ruetz et al., 2019). AdoCbl bound to MCM is 6-coordinate and has a histidine ligand donated by the protein at the lower axial position. The 5,6 dimethylbenzimidizole (DMB) tail displaced in a side pocket (Froese et al., 2010). In ATR on the other hand, AdoCbl is bound as a 5-coordinate species

(Campanello et al., 2018). The vacancy at the lower axial position of AdoCbl together with the DMB tail of the cofactor are important for its translocation from ATR to MCM (Mascarenhas, Ruetz, McDevitt, Koutmos, & Banerjee, 2020). Cofactor movement between MCM and ATR in either direction requires a second chaperone CblA (also known as MMAA). CblA uses both GTP binding and hydrolysis energy to gate the movement of the cofactor (Padovani & Banerjee, 2009a; Ruetz et al., 2019) (Fig. 3). Functional deficiency of CblA (Dobson et al., 2002), ATR (Dobson et al., 2002) or MCM (Ledley, Lumetta, Nguyen, Kolhouse, & Allen, 1988) results in methylmalonic aciduria, an inborn error of metabolism that is inherited as an autosomal recessive disease (Watkins & Rosenblatt, 2011). In this chapter, the purification and assays for MS and MCM are described. Purification and assays for the B_{12} chaperones ATR and CblA are described in a companion chapter.

1.1 Expression and purification of human MS from insect cells

1.1.1 Expression of MS

Materials and Equipment

- pFastBac Dual LIC cloning vector (Addgene plasmid #30121)
- Bac-to-Bac expression system (ThermoFisher #10359016)
- *Spodoptera frugiperda* (Sf21) insect cells (Invitrogen)
- High Five™ Insect Cells (ThermoFisher)
- Insect-Xpress Medium (Lonza)
- Antibiotic-Antimycotic-100 × (Gibco)
- Fisherbrand™ Shaker Flasks 250 mL (PBV250)
- 2.8 L Nalgene Fernbach Flask
- Hemocytometer
- Biosafety cabinet
- Temperature controlled shaker-incubator that holds 250 mL and 2.8 L flasks

Procedure

1. Human MS cDNA inserted into the pFastBac Dual LIC cloning vector with an N-terminal $HisX_6$-tag followed by a maltose binding protein (MBP) tag and a TEV cleavage site between MBP and MS, is used for recombinant protein expression
2. Generate recombinant baculovirus for MS expression using the Bac-to-Bac expression system following the manufacturer's protocol. Generate low titer virus (P1) by transfecting recombinant bacmids into

Sf21 cells. Amplify the low titer virus for two more rounds to get a high virus titer (P3) following the manufacturer's protocol

3. For growth and maintenance of insect cell lines (Sf21 or High Five™) follow the vendor's protocols. Briefly, warm Insect-Xpress medium (1 L) to 27 °C in a temperature-controlled water bath and move into biosafety cabinet. Add 10 mL of Antibiotic-Antimycotic to each 1 L bottle and shake well. The cells are maintained at a density ~30–400 × 10^4 cells/mL in 250 mL flasks at 27 °C. The cell density is estimated everyday by checking cell count using a 10 μL aliquot of the culture in a hemocytometer. If the cell density is >400 × 10^4 cells/mL, change medium by mixing 18 mL of cell stock with 18 mL fresh medium in a new 250 mL flask
4. For protein expression, prepare a 1 L High Five™ cell stock with ~300 × 10^4 cells/mL in Insect-Xpress Medium with 10 mL Antibiotic-Antimycotic -100 ×. From this 1 L cell stock, prepare 6 × 1 L Hi5 cell cultures containing ~70 × 10^4 cells/mL by diluting with medium (233 mL of 300 × 10^4 cells/mL with 767 mL medium). Incubate for ~24 h in a 2.8 L sterile flask at 27 °C with shaking at 110 rpm. The cell count should be between 100 and 200 × 10^4 cells/mL before infecting with 10–50 mL of a high titer P3 virus. Incubate cell cultures for an additional 44 h at 27 °C, 110 rpm
5. Harvest cells by centrifugation at 5500 × *g* for 30 min in 1 L centrifuge bottles. Transfer the cell pellet into a 50 mL falcon tube and store at −80 °C until use
6. *Note*: Maintain sterile practices when handling insect cells. All equipment must be autoclaved and sprayed down with 70% ethanol before moving into the bio-safety cabinet. Cover windows on the shaker incubator with aluminum foil to ensure that no light enters during protein expression

1.1.2 Purification of human MS

Materials and Equipment

- Buffer A: 50 mM Tris-HCl pH 8, containing 300 mM NaCl
- Buffer B: 50 mM Tris-HCl pH 8, containing 300 mM NaCl and 10% glycerol
- Lysozyme from chicken egg white (Sigma)
- EDTA-free complete Protease Inhibitor Cocktail (Roche)
- β-mercaptoethanol

- Purified His_6-tagged tobacco etch virus (TEV) protease (1 mg/mL)
- Bio-Rad Protein Assay Kit
- Amylose Resin (New England Biolabs)
- Maltose
- 1 M Dithiothreitol (DTT)
- Superdex 200 (S200, 120 mL, GE Healthcare)
- Amicon Ultra-15 50 kDa centrifugal filters (Millipore-Sigma)

Procedure

1. Thaw and suspend the Hi5 cell pellet infected with recombinant baculovirus in 400 mL Buffer A containing 1 tablet of EDTA-free complete Protease Inhibitor Cocktail, 20 mg of lysozyme and 3 mM β-mercaptoethanol. Divide the cell suspension into 50 mL falcon tubes and lyse cells by 4 cycles of freezing in liquid N_2 and thawing in a water bath at room temperature
2. Centrifuge the cell lysates at 38,000 × *g* at 4 °C for 2 h
3. Pool the supernatant in a beaker and load onto an amylose resin column (2.5 × 5 cm) equilibrated with Buffer B at 4 °C and then wash with 200 mL Buffer B
4. Elute His-MBP-tagged MS with Buffer B containing 10 mM maltose and collect 10 mL fractions. Measure the 595 nm absorbance using 5 μL of each fraction and 200 μL of Bio-Rad Protein Assay Reagent in a 96-well plate. Use 0.3 absorbance units as a cut-off to pool fractions containing His-MBP-tagged MS. The purity of His-MBP-tagged MS is shown in Fig. 4A

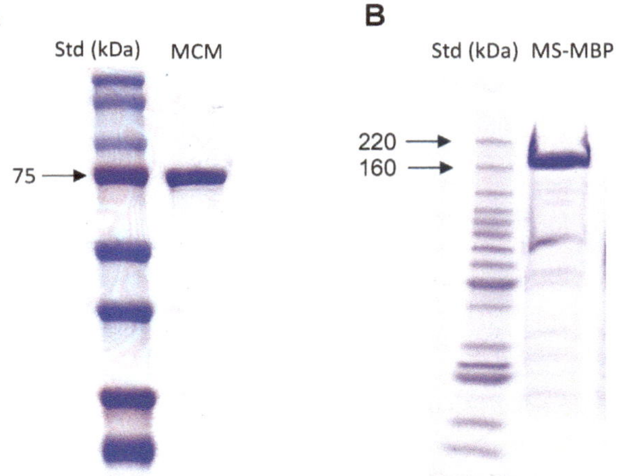

Fig. 4 SDS-PAGE gel analysis of purified methylmalonyl-CoA mutase(A) and maltose binding protein-tagged methionine synthase (B).

5. Concentrate His-MBP-tagged MS to 20 mL using Amicon Ultra-15 (MWCO = 50 kDa) filters. Transfer the concentrated protein to an appropriate length of SnakeSkin Dialysis Tubing (MWCO = 10 kDa) and add 1 mL of TEV-protease. Dialyze (overnight at 4 °C) with stirring in a 500 mL beaker containing 400 mL Buffer B supplemented with 0.4 mL DTT
6. Load the supernatant onto an amylose column (2.5 × 5 cm) and wash with Buffer B. Pool the flow-through and wash and concentrate tag-less MS to 5 mL using Amicon Ultra-15 (MWCO = 50 kDa) filters
7. Load MS onto a Superdex 200 column (S200, 100 mL, GE Healthcare) equilibrated with Buffer B. Monitor MS-containing fractions by their absorbance at 280 nm, pool and concentrate to 0.5 mL using an Amicon Ultra-15 50 kDa centrifugal filter
8. Estimate the concentration of MS concentration using $\varepsilon_{280} = 116{,}550\ M^{-1}\ cm^{-1}$. Separate 25 μL aliquots in 750 μL microcentrifuge tubes, flash-freeze in liquid nitrogen and store at −80 °C till further use. A typical yield is ~3–5 mg MS from 6 L culture

1.2 Assays for methionine synthase activity

Several methods have been developed for measuring the activity of MS (Chen, Chakraborty, & Banerjee, 1995; Drummond, Jarrett, Gonzalez, Huang, & Matthews, 1995; Frasca, Banerjee, Dunham, Sands, & Matthews, 1988; Yamada, Strahler, Andrews, & Matthews, 2005). Here, we describe two assays. In the first, transfer of the methyl group from 5-$^{14}CH_3$-H_4 folate to the product methionine, is monitored. In the second, the product H_4 folate, is derivatized with a formylating agent under acidic conditions to generate methenyl-H_4 folate ($CH^+ = H_4$folate), which is detected at 350 nm (Eq. 3). Unused 5-CH_3-H_4 folate does not contribute to the absorbance at this wavelength.

$$H_4\ \text{folate} + HCOOH \rightarrow CH^+ = H_4\text{folate} + 2\ H_2O \tag{3}$$

1.2.1 *Radioactive MS assay*

Materials

- 5 mM $^{14}CH_3$-H_4 folate (1200 dpm/nmol)
- 1 mM AdoMet
- 25 mM homocysteine
- 1 M dithiotreitol (DTT)

- 0.5 mM aquocobalamin (OH_2Cbl)
- 0.2 M potassium phosphate buffer, pH 7.2
- AG 1-X8 anion exchange resin (Bio Rad)

Procedure

Prepare a 250 μL reaction by mixing 10 μL homocysteine, 25 μL AdoMet, 6.25 μL DTT, 25 μL OH_2Cbl and MS in 0.2 M potassium phosphate buffer pH 7.2. Incubate at 37 °C for 5 min in a temperature-controlled water bath. Start the reaction by adding 12.5 μL $^{14}CH_3$-H_4 folate and incubate at 37 °C for 10 min. Quench at 90 °C for 3 min to terminate the reaction followed by cooling on ice for 2 min. Load 200 μL onto a 0.5 × 6 cm column of Bio-Rad AG 1-X8 (chloride form). Wash the column with 1.5 mL water and measure ^{14}C radioactivity associated with methionine in the combined flow-through and wash using a scintillation counter. One unit of MS is defined as the amount of enzyme required to catalyze the formation of 1 μmol of methionine/min at 37 °C.

1.2.2 Non-radioactive MS assay

Materials

- 1 M potassium phosphate buffer, pH 7.2
- 500 mM DTT
- 3.8 mM AdoMet
- 100 mM L-homocysteine
- 500 μM OH_2Cbl
- 250 μM (*6R,S*)-CH_3–H_4 folate
- 5 N HCl in aqueous 60% formic acid

Procedure

Prepare reagents as described by Drummond et al. (Drummond et al., 1995). Combine distilled water (544 μL, less the volume of enzyme to be added) with the following reagents in the given order: 80 μL potassium phosphate buffer, pH 7.2, 40 μL DTT, 4 μL AdoMet, 4 μL L-homocysteine, and MS. Prepare a blank lacking MS. Next, add 80 μL of OH_2Cbl and incubate at 37 °C. DTT and OH_2Cbl catalyze the reduction of O_2 to H_2O_2, creating a semi-anaerobic environment. Timing is critical from this point onwards, as the time required for activation of MS must be balanced by its slow inactivation as H_2O_2 builds up. After exactly 5 min, add 48 μL of (*6R,S*)-CH_3–H_4 folate to initiate turnover and incubate at 37 °C for 10 min. The concentration of H_4 folate must fall within the linear range of the assay, optimally between 0.1 and 1 absorbance units at 350 nm. Add 200 μL of the acidic derivatization solution (5 N HCl in aqueous 60% formic acid) to quench the reaction and heat immediately at 80 °C for 10 min to convert H_4 folate to $CH^+ = H_4$ folate. Cool the reaction mixture to room temperature and record the absorbance in a spectrophotometer, with a temperature-controlled cuvette holder maintained at 4 °C. The concentration of $CH^+ = H_4$ folate is determined using $\Delta\varepsilon_{350} = 26{,}500\,M^{-1}\,cm^{-1}$.

2. Expression and purification of MCM

2.1 Expression of MCM

Materials

- *E. coli* expression strain BL21(DE3)
- Terrific broth (TB) and Luria Bertani (LB) media
- LB agar plates containing 50 μg/mL of kanamycin
- Dimethyl sulfoxide (DMSO)
- Kanamycin, 50 mg/mL
- 1 M IPTG
- Erlenmeyer flasks (6 × 2.8 L), temperature-controlled shaker

Procedure

1. The human MCM gene was codon optimized for expression in *E. coli* and cloned into a pET-28b(+) vector with a TEV cleavable C-terminal His_6 tag between restriction sites Nco1 and Xho1 (Campanello et al., 2018)
2. Transform BL21(DE3) cells with the pET28b(+)-MCM vector and plate on LB agar containing 50 μg/mL kanamycin. Incubate at 37 °C overnight
3. Inoculate 10 mL sterilized LB medium containing 50 μg/mL kanamycin with a single colony and incubate at 37 °C overnight with shaking at 220 rpm
4. Inoculate culture flasks (6 × 2.8 L) containing 1 L sterilized TB medium and 50 μg/mL of kanamycin with 1 mL of the overnight culture. Incubate at 37 °C with shaking at 220 rpm until $OD_{600} = 1.5$–1.8 (in ~6–8 h)
5. Add 30 mL of DMSO to 1 L culture (3% final concentration) and cool the culture to 16 °C by shaking for 40 min
6. Induce expression of MCM by adding 0.25 mL IPTG (250 μM final concentration) to each flask and incubate at 16 °C overnight (22–24 h) with shaking at 220 rpm
7. Pellet cells by centrifugation at 5500 × *g* for 20 min in 1 L centrifuge bottles. Transfer cell pellet into 50 mL falcon tubes and store at −80 °C until use. A typical yield is 80–100 g of wet cell pellet from a 6 L culture

2.2 Purification of MCM

Materials

- Misonix Sonicator (Model XL, 1.2 cm diameter probe)
- Temperature-controlled centrifuge
- Ni-NTA resin (Qiagen)
- Amicon 50 kDa centrifugal tube (Millipore)
- Ion exchange column: Source Q (Omnifit)
- FPLC system (GE Healthcare)
- His-tagged TEV protease (Sigma)
- Complete EDTA-free protease inhibitor cocktail tablet (Roche)
- 100 mM Phenylmethylsulfonyl fluoride in DMSO
- Lysis buffer: 50 mM Tris pH 8.0, 500 mM NaCl, 20 mM imidazole and 5% glycerol
- TEV buffer: 50 mM Tris pH 8.0, 300 mM NaCl, 0.5 mM EDTA, 1 mM DTT and 5% glycerol
- Buffer A: 50 mM HEPES pH 7.3, 25 mM NaCl and 5% glycerol
- Buffer B: 50 mM HEPES pH 7.3, 500 mM NaCl and 5% glycerol
- Storage Buffer: 50 mM HEPES pH 7.5, 150 mM KCl, 2 mM TCEP, 2 mM $MgCl_2$, 5% glycerol

Procedure

1. Thaw the cell pellet and resuspend in Lysis Buffer (6–8 mL per g of cell pellet) in a 500 mL beaker. Add 0.1 mL PMSF per 100 mL of cell suspension (100 μM final concentration) and one protease inhibitor cocktail tablet and stir at 4 °C for 30 min
2. Lyse cells by sonication using a 1.2 cm diameter probe at 30% power with cycles (15 s on, 1 min off) for a total of 20 min on ice
3. Transfer the cell lysate to 50 mL tubes and centrifuge (38,500 × g for 45 min, 4 °C)
4. Load the supernatant onto a Ni-NTA column and wash with 150 mL Lysis Buffer. Elute His_6-MCM using a gradient mixer containing 150 mL Lysis Buffer and 150 mL Lysis Buffer with 200 mM imidazole
5. Check protein concentration by mixing 10 μL of each fraction with 990 μL of the Bradford reagent in a plastic cuvette and measure the absorbance at 595 nm. Pool and concentrate fractions using an Amicon centrifugal device (50 kDa cut-off)
6. Add His-tagged TEV protease (0.02 mg/mg) to His_6-MCM in a dialysis bag (10 kDa cut-off) and dialyze overnight in 1 L TEV Buffer at 4 °C to cleave the N-terminal His_6-tag on MCM
7. The following day, add additional TEV protease (0.01 mg/mg) to the protein in the dialysis bag and dialyze against TEV buffer for an additional 3–4 h

8. Load the TEV treated MCM solution onto a Ni-NTA column (equilibrated with TEV Buffer) to remove His-tagged TEV. Collect the flow through and wash the column with Lysis Buffer (100–150 mL) until no protein is present in the flow through as detected by the Bradford assay
9. Combine flow through and wash fractions containing MCM and exchange with Buffer A using an Amicon centrifugal device (MWCO= 50 kDa). Alternatively, concentrate MCM to 20–30 mL and dialyze against 1 L Buffer A overnight at 4 °C
10. Equilibrate the SourceQ column (Omnifit) with Buffer A on an FPLC system. Load the protein onto the column and wash with 50 mL Buffer A. MCM is then eluted with the following gradient: 0–70% Buffer B in 35 min, followed by 100% Buffer B in 1 min and maintained at 100% Buffer B for 5 min. Monitor absorbance at 280 nm and collect 5 mL fractions. Concentrate and dialyze MCM-containing fractions against 1 L Storage Buffer overnight at 4 °C
11. Concentrate MCM to 2–3 mL and determine the MCM (dimer) concentration using $\varepsilon_{280nm} = 133{,}620\,M^{-1}\,cm^{-1}$. Aliquot, flash freeze in liquid N_2 and store at −80 °C. A typical yield is ~90 mg MCM from a 6 L culture. The purity of MCM is shown in Fig. 4B

Note: Alternatively, a Mono Q column (GE Healthcare) can be used with the same gradient.

2.3 Determination of MCM activity

MCM activity is determined in a coupled enzyme assay with thiokinase (also known as succinyl-CoA synthase). Thiokinase catalyzes the conversion of succinyl-CoA to CoA and succinic acid in the presence of GDP and phosphate (Fig. 5). The free thiol in CoA is detected using Ellman's reagent, 5,5-dithio-bis-(2-nitrobenzoic acid or DTNB) and the concentration of succinyl-CoA formed in the MCM reaction is estimated using $\varepsilon_{412nm} = 14{,}150\,M^{-1}\,cm^{-1}$ (Ellman, 1959).

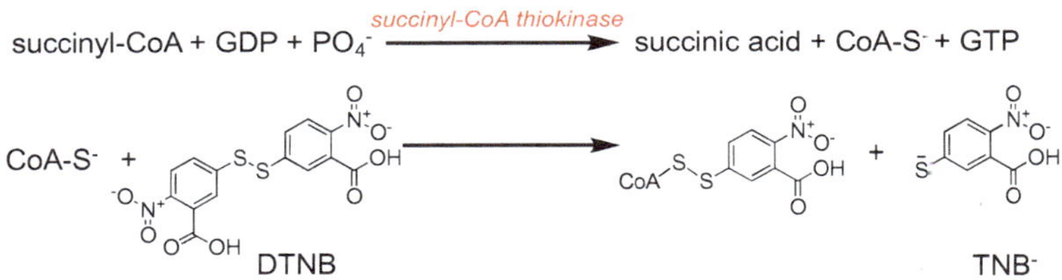

Fig. 5 MCM activity as detected in the coupled thiokinase assay.

Materials

- Buffer A: 100 mM potassium phosphate, 3 mM $MgCl_2$, pH 7.5
- 1 M potassium phosphate pH 7.5
- 100 mM magnesium chloride ($MgCl_2$)
- 100 mM guanosine diphosphate (GDP)
- 5 mg/mL succinyl-CoA thiokinase (Sigma)
- 10 mM DTNB in ethanol
- 50 mM methylmalonyl-CoA in water
- 50 μM MCM dimer
- 500 μM AdoCbl

Procedure

The catalytic activity of MCM is monitored using a UV–visible spectrophotometer with a temperature-controlled cuvette holder. In a 1.5 mL Eppendorf tube, add 38 μL Buffer A, 2 μL AdoCbl and 10 μL MCM to prepare a 10 μM stock of holo-MCM. Incubate at 30 °C using a temperature-controlled Eppendorf Thermostat. After 15 min, transfer the sample to ice. Dilute the sample 1:100 by adding 2 μL of MCM to 198 μL Buffer A to obtain a 100 nM stock of holo-MCM.

In a quartz cuvette, add 178.6 μL Buffer A, 6 μL GDP, 1.4 μL DTNB, 2 μL AdoCbl and 4 μL of a 1:1 (*v*/v) dilution of the methylmalonyl-CoA stock with 1 M potassium phosphate pH 7.5 (see note below). Place the cuvette in a spectrophotometer for 10 min at 30 °C. Add 4 μL thiokinase and monitor the change in absorbance at 412 nm for 5 min to determine background rate. Initiate the reaction by adding 4 μL of 100 nM holo-MCM (2 nM final concentration) and record the increase in absorbance at 412 nm absorbance. The initial rate is used to determine MCM activity using $\varepsilon_{412nm} = 14{,}150\, M^{-1}\, cm^{-1}$ after subtracting the background thiokinase rate. The specific activity of MCM is typically 67 ± 2 μM succinyl-CoA formed $min^{-1}\, mg^{-1}$ at 30 °C.

Note: The methylmalonyl-CoA stock solution is acidic and the 1:1 mixture with 1 M potassium phosphate pH 7.5, is made immediately before use in the assay.

2.4 Assay for AdoCbl loading onto MCM

ATR is tasked with both the synthesis and delivery of AdoCbl to the mitochondrial target, MCM. The transfer can be readily monitored by the large absorbance change that accompanies the movement of 5-coordinate AdoCbl from ATR to 6-coordinate AdoCbl bound to MCM (Fig. 3).

AdoCbl transfer is gated by the GTPase CblA. The non-hydrolysable GTP analog β,γ-methyleneguanosine 5′-triphosphate (GMPPCP) can be used as a GTP mimic (Ruetz et al., 2019) (Fig. 6A).

Materials

- Buffer A: 50 mM HEPES pH 7.4, 150 mM KCl, 2 mM TCEP, 2 mM $MgCl_2$, 5% glycerol
- 250 μM MCM
- 500 μM ATR
- 1 mM AdoCbl
- 750 μM CblA
- 100 mM GTP
- 100 mM GMPPCP

Note: Commercial GTP is sometimes contaminated with PPPi, which leads to formation of the 440 nm species under anaerobic conditions (corresponding to a weakened cobalt-carbon bond in AdoCbl) or to the homolytic cleavage of this bond under aerobic conditions (Campanello et al., 2018).

Equipment

UV/visible spectrophotometer with a temperature-controlled cuvette holder

Ultra-micro quartz glass cuvette

Note: GMPPCP or β,γ-imidoguanosine 5′-triphosphate (GMPPNP).

Procedure

In a quartz cuvette, add 165 μL Buffer A, 3 μL AdoCbl and 6 μL ATR and record the spectrum between 300 and 700 nm. Estimate the

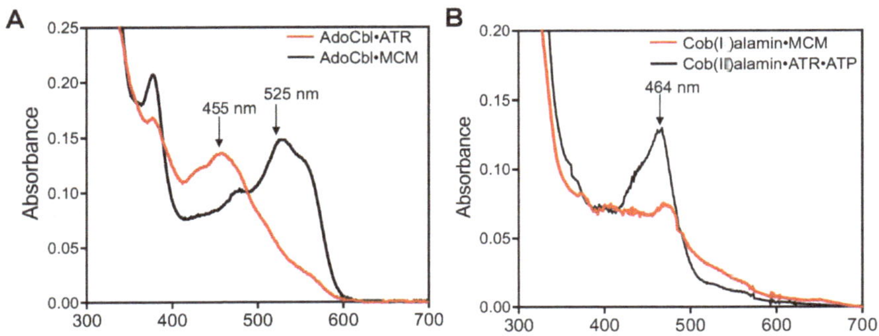

Fig. 6 Spectral analysis of cofactor translocation. (A) Transfer of AdoCbl from ATR (5-coordinate base-off) to MCM (6-coordinate base-off, His-on). (B) Transfer of cob(*II*) alamin from MCM (5-coordinate) to ATR (4-coordinate) in the presence of ATP. Buffer A was used in these experiments.

concentration of ATR bound AdoCbl using $\varepsilon_{455nm} = 8000\,M^{-1}\,cm^{-1}$ (Fig. 6A). Place the cuvette in a spectrophotometer for 10 min at 25 °C. Meanwhile, mix 12 µL MCM, 12 µL CblA and 2 µL GTP or 2 µL GMPPCP in a 0.5 ml Eppendorf tube. Initiate the transfer by adding 26 µL of the MCM-CblA-GTP or MCM-CblA-GMPPCP mixture to the cuvette and record the spectrum every minute for 40 min. In the presence of GTP, an increase at 525 nm corresponding to AdoCbl transfer to MCM is observed, whereas GMPPCP inhibits transfer, resulting in a slight increase at 525 nm. Plot the change in absorbance at 525 nm and fit the data to a single exponential to determine the rate of AdoCbl transfer from ATR to MCM. The concentration of AdoCbl transferred from ATR to MCM is determined using $\Delta\varepsilon_{525nm} = 7500\,M^{-1}\,cm^{-1}$.

2.5 Assay for cob(*II*)alamin off-loading from MCM to ATR

In the catalytic cycle of MCM, the 5′deoxyadeonsyl radical is occasionally lost, resulting in inactive cob(*II*)alamin on MCM (Padovani & Banerjee, 2009a). Cob(*II*)alamin is transferred back to ATR for repair with the help of the CblA chaperone in the presence of GTP (Fig. 3). The UV/visible spectrum of cob(*II*)alamin bound to MCM is indistinguishable from that of 5-coordinate cob(*II*)alamin in solution ($\lambda_{max} = 474$ nm), whereas ATR bound cob(*II*)alamin is 4-coordinate in the presence of ATP ($\lambda max = 464$ nm) (Fig. 6B).

Materials

- Buffer A: 50 mM HEPES pH 7.4, 150 mM KCl, 2 mM TCEP, 2 mM $MgCl_2$, 5% glycerol
- 1 mM cob(*II*)alamin
- 200 µM MCM
- 600 µM CblA
- 500 µM ATR
- 100 mM GTP
- 250 mM ATP

Note: All solutions are made anaerobic by purging with N_2 for 15 min prior before transferring to an anaerobic chamber.

Method

Prepare 2 mL of a 1 mM AdoCbl stock solution in a glass cuvette sealed with a rubber stopper and parafilm and purge with N_2 gas for 15 min. Then, place the AdoCbl sample on ice 10 cm away from a halogen lamp (38 W, G.E.) After 2–3 h, move the sample into an anaerobic chamber.

Complete photolysis of AdoCbl is confirmed by the absorption spectrum of cob(*II*)alamin ($\lambda_{max} = 473$ nm, $\varepsilon_{473nm} = 9.2$ mM^{-1} cm^{-1}).

Note: The pH of the ATP stock, which is acidic, is adjusted to 7.5 with 1 M HEPES pH 7.5.

Equipment

- Anaerobic chamber (<0.4 O_2 ppm)
- UV/visible spectrophotometer with a temperature-controlled cuvette holder
- Ultra-micro UV/visible quartz glass cuvette

Procedure

Cob(*II*)alamin off-loading from MCM (Fig. 6B) is monitored using a UV–visible spectrophotometer in an anaerobic chamber with a temperature-controlled cuvette holder set to 25 °C. In a quartz cuvette, add 166 μL Buffer A and 3 μL of cob(*II*)alamin. Add 9 μL MCM and incubate the cuvette in the spectrophotometer for 15 min. In a 0.5 mL Eppendorf tube, prepare the "repair mixture" by adding 4 μL ATP, 6 μL of ATR, 10 μL CblA and 2 μL GTP. Initiate the cob(*II*)alamin transfer reaction by adding the repair mixture to the cuvette. Record the spectrum between 300 and 750 nm every 15 s for 20 min. Transfer of cob(*II*)alamin from MCM to ATR results in an increase in absorbance at 464 nm ($\varepsilon_{464nm} = 14{,}100$ M^{-1} cm^{-1}) (Campanello et al., 2018), which corresponds to 4-coordinate cob(*II*)alamin bound to ATR. Fit the change in absorbance at 464 nm to a single exponential equation to determine the rate of cob(*II*)alamin transfer from MCM to ATR. Estimate the concentration cob(*II*)alamin transferred using $\Delta\varepsilon_{464nm} = 5900$ M^{-1} cm^{-1}.

Note: If a spectrophotometer inside an anaerobic chamber is not available, the anaerobic reaction mixture is prepared in a cuvette sealed with a rubber septum and the anaerobic repair mixture prepared separately, is added with a gas-tight Hamilton syringe.

Acknowledgment

This work was supported in part by a grant from the National Institutes of Health RO1-DK045776.

References

Banerjee, R. (2001). Radical peregrinations catalyzed by coenzyme B_{12}-dependent enzymes. *Biochemistry*, *40*, 6191–6198.

Banerjee, R. (2003). Radical carbon skeleton rearrangements: Catalysis by coenzyme B_{12}-dependent mutases. *Chemical Reviews*, *103*, 2083–2094.

Banerjee, R. (2006). B_{12} trafficking in mammals: A case for coenzyme escort service. *ACS Chemical Biology*, *1*, 149–159.
Banerjee, R. V., Frasca, V., Ballou, D. P., & Matthews, R. G. (1990). Participation of cob(I) alamin in the reaction catalyzed by methionine synthase from Escherichia coli: A steady-state and rapid reaction kinetic analysis. *Biochemistry*, *29*, 11101–11109.
Banerjee, R., Gherasim, C., & Padovani, D. (2009). The tinker, tailor, soldier in intracellular B_{12} trafficking. *Current Opinion in Chemical Biology*, *13*, 484–491.
Banerjee, R., Gouda, H., & Pillay, S. (2021). Redox-linked coordination chemistry directs vitamin B_{12} trafficking. *Accounts of Chemical Research*, *54*, 2003–2013.
Banerjee, R. V., Harder, S. R., Ragsdale, S. W., & Matthews, R. G. (1990). Mechanism of reductive activation of cobalamin-dependent methionine synthase: An electron paramagnetic resonance spectroelectrochemical study. *Biochemistry*, *29*, 1129–1137.
Banerjee, R. V., & Matthews, R. G. (1990). Cobalamin-dependent methionine synthase. *The FASEB Journal*, *4*, 1450–1459.
Banerjee, R., & Ragsdale, S. W. (2003). The many faces of vitamin B_{12}: Catalysis by cobalamin-dependent enzymes. *Annual Review of Biochemistry*, *72*, 209–247.
Campanello, G. C., Ruetz, M., Dodge, G. J., Gouda, H., Gupta, A., Twahir, U. T., et al. (2018). Sacrificial cobalt-carbon bond homolysis in coenzyme B_{12} as a cofactor conservation strategy. *Journal of the American Chemical Society*, *140*, 13205–13208.
Chen, Z., Chakraborty, S., & Banerjee, R. (1995). Demonstration that the mammalian methionine synthases are predominantly cobalamin-loaded. *The Journal of Biological Chemistry*, *270*, 19246–19249.
Chowdhury, S., & Banerjee, R. (2000). Thermodynamic and kinetic characterization of co-C bond homolysis catalyzed by coenzyme B_{12}-dependent methylmalonyl-CoA mutase. *Biochemistry*, *39*, 7998–8006.
Dobson, C. M., Wai, T., Leclerc, D., Kadir, H., Narang, M., Lerner-Ellis, J. P., et al. (2002). Identification of the gene responsible for the *cbl*B complementation group of vitamin B_{12}-dependent methylmalonic aciduria. *Human Molecular Genetics*, *11*, 3361–3369.
Dobson, C. M., Wai, T., Leclerc, D., Wilson, A., Wu, X., Dore, C., et al. (2002). Identification of the gene responsible for the *cbl*A complementation group of vitamin B_{12}-responsive methylmalonic acidemia based on analysis of prokaryotic gene arrangements. *Proceedings of the National Academy of Sciences of the United States of America*, *99*, 15554–15559.
Drummond, J. T., Huang, S., Blumenthal, R. M., & Matthews, R. G. (1993). Assignment of enzymatic function to specific protein regions of cobalamin-dependent methionine synthase from Escherichia coli. *Biochemistry*, *32*, 9290–9295.
Drummond, J. T., Jarrett, J., Gonzalez, J. C., Huang, S., & Matthews, R. G. (1995). Characterization of nonradioactive assays for cobalamin-dependent and cobalamin-independent methionine synthase enzymes. *Analytical Biochemistry*, *228*, 323–329.
Ellman, G. L. (1959). Tissue sulfhydryl groups. *Archives of Biochemistry and Biophysics*, *82*, 70–77.
Frasca, V., Banerjee, R. V., Dunham, W. R., Sands, R. H., & Matthews, R. G. (1988). Cobalamin-dependent methionine synthase from *Escherichia coli* B: Electron paramagnetic resonance spectra of the inactive form and the active methylated form of the enzyme. *Biochemistry*, *27*, 8458–8465.
Froese, D. S., Kochan, G., Muniz, J. R., Wu, X., Gileadi, C., Ugochukwu, E., et al. (2010). Structures of the human GTPase MMAA and vitamin B_{12}-dependent methylmalonyl-CoA mutase and insight into their complex formation. *The Journal of Biological Chemistry*, *285*, 38204–38213.
Gherasim, C., Lofgren, M., & Banerjee, R. (2013). Navigating the B_{12} road: Assimilation, delivery and disorders of cobalamin. *The Journal of Biological Chemistry*, *288*, 13186–13193.

Goulding, C. W., Postigo, D., & Matthews, R. G. (1997). Cobalamin-dependent methionine synthase is a modular protein with distinct regions for binding homocysteine, methyltetrahydrofolate, cobalamin and adenosylmethionine. *Biochemistry*, *36*, 8082–8091.

Gulati, S., Baker, P., Li, Y. N., Fowler, B., Kruger, W., Brody, L. C., et al. (1996). Defects in human methionine synthase in cblG patients. *Human Molecular Genetics*, *5*, 1859–1865.

Leclerc, D., Wilson, A., Dumas, R., Gafuik, C., Song, D., Watkins, D., et al. (1998). Cloning and mapping of a cDNA for methionine synthase reductase, a flavoprotein defective in patients with homocystinuria. *Proceedings of the National Academy of Sciences of the United States of America*, *95*, 3059–3064.

Ledley, F. D., Lumetta, M., Nguyen, P. N., Kolhouse, J. F., & Allen, R. H. (1988). Molecular cloning of L-Methylmalonyl CoA mutase: Gene transfer and analysis of Mut cell lines. *Proceedings. National Academy of Sciences. United States of America*, *85*, 3518–3521.

Mascarenhas, R., Ruetz, M., McDevitt, L., Koutmos, M., & Banerjee, R. (2020). Mobile loop dynamics in adenosyltransferase control binding and reactivity of coenzyme B12. *Proceedings of the National Academy of Sciences of the United States of America*, *117*, 30412–30422.

Olteanu, H., & Banerjee, R. (2001). Human methionine synthase reductase, a soluble P-450 reductase-like dual flavoprotein, is sufficient for NADPH-dependent methionine synthase activation. *The Journal of Biological Chemistry*, *276*, 35558–35563.

Olteanu, H., Wolthers, K. R., Munro, A. W., Scrutton, N. S., & Banerjee, R. (2004). Kinetic and thermodynamic characterization of the common polymorphic variants of human methionine synthase reductase. *Biochemistry*, *43*, 1988–1997.

Padovani, D., & Banerjee, R. (2009a). A G-protein editor gates coenzyme B_{12} loading and is corrupted in methylmalonic aciduria. *Proceedings of the National Academy of Sciences of the United States of America*, *106*, 21567–21572.

Padovani, D., & Banerjee, R. (2009b). A rotary mechanism for coenzyme B_{12} synthesis by Adenosyltransferase. *Biochemistry*, *48*, 5350–5357.

Padovani, D., Labunska, T., Palfey, B. A., Ballou, D. P., & Banerjee, R. (2008). Adenosyltransferase tailors and delivers coenzyme B_{12}. *Nature Chemical Biology*, *4*, 194–196.

Ruetz, M., Campanello, G. C., McDevitt, L., Yokom, A. L., Yadav, P. K., Watkins, D., et al. (2019). Allosteric regulation of oligomerization by a B_{12} trafficking G-protein is corrupted in methylmalonic aciduria. *Cell Chemical Biology*, *26*(960–969), e964.

Watkins, D., & Rosenblatt, D. S. (2011). Inborn errors of cobalamin absorption and metabolism. *American Journal of Medical Genetics. Part C, Seminars in Medical Genetics*, *157C*, 33–44.

Yamada, K., Strahler, J. R., Andrews, P. C., & Matthews, R. G. (2005). Regulation of human methylenetetrahydrofolate reductase by phosphorylation. *Proceedings of the National Academy of Sciences of the United States of America*, *102*, 10454–10459.

Yamanishi, M., Labunska, T., & Banerjee, R. (2005). Mirror "base-off" conformation of coenzyme B_{12} in human adenosyltransferase and its downstream target, methylmalonyl-CoA mutase. *Journal of the American Chemical Society*, *127*, 526–527.

CHAPTER THIRTEEN

Heterologous production and biophysical characterization of catabolic *Nitratireductor pacificus* pht-3B reductive dehalogenase

Tom Halliwell, Karl Fisher, Stephen E.J. Rigby, and David Leys*
School of Chemistry, University of Manchester, Manchester, United Kingdom
*Corresponding author: e-mail address: david.leys@manchester.ac.uk

Contents

Methods in Enzymology, Volume 668
ISSN 0076-6879
https://doi.org/10.1016/bs.mie.2022.01.004

Abstract

Reductive dehalogenases provide a possible route to the biotechnological remediation of widespread anthropogenic environmental organohalide contamination. These bacterial enzymes employ cobalamin and an internal electron transfer chain of two [4Fe-4S] clusters to remove halide ions from organohalides, leaving an organic molecule more amenable to further transformations. Detailed protocols for the cloning, heterologous expression, purification, crystallization and characterization of the catabolic dehalogenase from *Nitratireductor pacificus* pht-3B (NpRdhA) are presented, together with insight into enzyme turnover, substrate selectivity and the use of electron paramagnetic resonance (EPR) spectroscopy as an active site probe.

1. Introduction

The presence of organohalides in the environment, arising from both anthropogenic and natural processes (Gribble, 2015; Kodavanti & Loganathan, 2017) has driven the evolution of bacterial species that utilize such compounds as terminal electron acceptors (Atashgahi, Häggblom, & Smidt, 2018; Leys, Adrian, & Smidt, 2013). The key enzymes in this organohalide respiration process are the reductive dehalogenases (RdhAs) that cleave the carbon-halide bond, replacing the halide with a hydrogen atom and releasing the halide ion (Jugder, Ertan, Lee, Manefield, & Marquis, 2015). In addition to the respiratory dehalogenases, evolutionarily related catabolic reductive dehalogenases have evolved in certain bacterial catabolic pathways. Exemplars of the two distinct RdhA types show structural similarity despite exhibiting different substrate specificities (Dobbek & Leys, 2016). Both enzymes employ a cobalamin molecule at the bottom of a deep active site cleft and two [4Fe-4S] clusters that form an electron transfer chain from the protein surface to the buried cobalamin. However, the presence of a TAT (twin-arginine translocation) signal that supports translocation into the periplasm, and subsequent association with a complementary membrane anchor protein, distinguishes the membrane-associated respiratory dehalogenases from the soluble, more oxygen-tolerant, cytosolic catabolic enzymes. Catabolic RdhAs thus present a possible biotechnological solution to anthropogenic organohalide pollution of the biosphere (Jugder et al., 2016). To that end we have cloned and heterologously expressed the catabolic dehalogenase from *N. pacificus* pht-3B (NpRdhA) (Halliwell, Fisher, Payne, Rigby, & Leys, 2021; Payne et al., 2015). The sections below present this methodology and subsequent NpRdhA characterization, structure determination, mechanistic and spectroscopic studies and kinetics in a NADPH-driven electron transfer chain, summarized in Scheme 1 (Collins et al., 2018; Halliwell, Fisher, Payne, Rigby, & Leys, 2020).

2. *N. pacificus* pht-3B (NpRdhA) cloning and heterologous expression hosts

The heterologous expression of NpRdhA has been discussed at length previously (Halliwell et al., 2021). The expression systems described each exhibit different advantages and plasmid requirements as shown in Table 1.

2.1 Preparation of vectors for heterologous expression

Numerous bacterial strains have been studied for heterologous expression of NpRdhA with success depending crucially on expression vector design, particularly the promoter system (Halliwell et al., 2021). Vector requirements for these strains are summarized in Table 1 and methods for production of these plasmids are standard. *E. coli* (HMS174(DE3)) and *Vibrio natrigenes*, (Vmax^{TM}), use simple, well characterized plasmids such as pET vectors. *Shimwellia blattae* expression involves the use of a single vector, pASK-IBA63c-plus, with the gene under the control of a tetracycline promoter. In contrast, *Bacillus megaterium* plasmid design is more complex requiring either single or multiple plasmids for heterologous expression. *B. megaterium* harbours an endogenous xylose operon, containing a strong xylose promoter, *xylA*. Genes under the control of this promoter are derepressed in the presence of xylose through the regulatory element XylR (Vary et al., 2007) and therefore *B. megaterium* heterologous expression vectors capitalize on this xylose-inducible promoter. Both *B. megaterium* MS941 and DSM319 have been used for expression of NpRdhA and related enzymes; however, *xylA* is used differently in the two strains. DSM319 uses a single vector, such as pN-HIS-TEV1622, containing both XylR and *xylA* allowing for direct transcription of the NpRdhA gene upon xylose addition. In contrast, MS941 uses two vectors. The first, pT7-RNAP, contains the T7 RNA polymerase (T7-RNAP), not endogenous to *B. megaterium*, under control of *xylA* and pre-transformed into this strain. The second MS941 vector, such as pPT7, contains NpRdhA under the control of the T7-RNAP promoter. This allows tighter control of expression, due to the requirement of T7-RNAP translation before transcription of NpRdhA can occur.

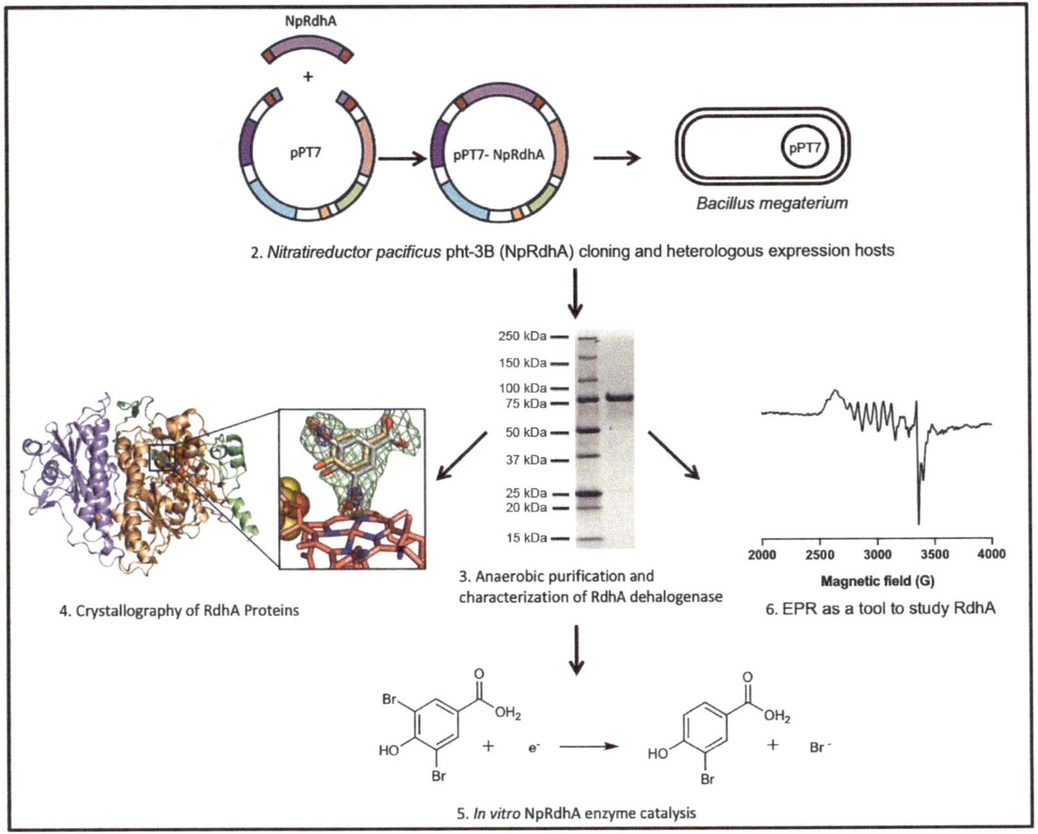

Scheme 1 Schematic summary of the methodologies used in NpRdhA characterization, structure determination, spectroscopic and kinetic studies, discussed in Sections 2–6 of this chapter.

Table 1 Heterologous expression strains for reductive dehalogenases.

Expression strain	Dehalogenase vector backbone	Antibiotic resistance	Promoter	Inducer final concentration
Bacillus megaterium MS491[a]	pPT7	Chloramphenicol/ Tetracycline	*xylA*/T7	0.1% Xylose
E. coli HMS174(DE3)[b]	pET-30a (+)/pPT7[c]	Kanamycin/Ampicillin	T7	0.2–1 mM IPTG
Shimwellia blattae ATCC 33430[d]	pASK-IBA63c (+)	Ampicillin	tetR/tetA	20 ng/mL anhydrotetracycline (hydrochloride)
Vmax™ express[b]	pET-30a (+)/pPT7[c]	Kanamycin/Ampicillin	T7	0.2–1 mM IPTG

[a] *B. megaterium* MS491 contains the pT7-RNAP plasmid (chloramphenicol resistance), enabling expression of T7 RNA-polymerase, under control of the *xylA* promoter, then allowing expression of the gene of interest under control of the T7 promoter.
[b] *E. coli* HMS174(DE3) can be co-transformed with pET-3a-BtuB (ampicillin resistance), containing the *E. coli* cobalamin transporter BtuB increasing cobalamin uptake.
[c] The *Bacillus megaterium* expression vector pPT7 can be used for heterologous expression due to the presence of a T7 promoter and *E. coli* origin of replication.
[d] Protein induction performed anaerobically under a continuous nitrogen gas flow of 30 L/min− 1.
Heterologous expression in all strains has been successful in terrific broth with growth at 37 °C and induction at 18 °C. Inducer is added alongside a final concentration of 50 μM ammonium iron (II) sulfate and 1 μM vitamin B_{12}.

2.1.1 Equipment

- Thermal cycler
- 37 °C static and shaking incubator
- NanoDrop spectrophotometer (ThermoFisher)
- Gel-electrophoresis tank

2.1.2 Reagents and chemicals

- DNA polymerase
- Template insert and destination DNA
- PCR primers
- Autoclaved water
- Agarose
- DpnI (methylation-sensitive restriction enzyme) (NEB)
- PCR clean-up kit (Takara)
- NEBuilder® HiFi DNA assembly
- NEB5α competent cells
- Autoclaved Luria-Bertani (LB) broth and agar
- Sterile antibiotics

2.1.3 Procedure

- PCR reactions will vary depending on polymerases used; perform according to manufacturer's specifications using a suitable T_m for designed primers and extension time for fragment size
- Identify PCR products by agarose gel electrophoreses, visualize and then DpnI digest positive reactions
- Perform clean-up of PCR reaction/DpnI digest and quantify DNA concentration with a NanoDrop
- Assemble vector and insert fragments, e.g., using NEBuilder® HiFi DNA assembly and transform directly into competent cells using the manufacturers protocol
- Culture transformed cells and extract plasmid DNA and confirm presence of insert by DNA sequencing

2.2 Competent cell production and transformation of plasmids

Commercial strains of *E. coli* (HMS174(DE3)) and *Vibrio natrigenes*, (Vmax™), were used for heterologous expression of NpRdhA, using transformation protocols described by the manufacturers, i.e., heat-shock and electroporation, respectively (Genomics, 2017; Hausjell, Weissensteiner, Molitor, Halbwirth, & Spadiut, 2018). More soluble NpRdhA with a higher cobalamin content can be obtained in *E. coli* through co-expression of the outer membrane cobalamin transporter, *BtuB* (Halliwell et al., 2020; Nou & Kadner, 2000). *S. blattae* and *B. megaterium* have no commercial suppliers and the user must make their own competent cells (Inoue, Nojima, & Okayama, 1990; Moore et al., 2013).

2.3 Heterologous expression of reductive dehalogenases

Heterologous expression of NpRdhA has been previously discussed at length (Halliwell et al., 2021). It is affected by cell line, medium (terrific broth (TB), 2xYT, minimal medium) and induction temperature, with optimisation advised for individual RdhAs. Initial expression can be attempted in TB as standard, at 37 °C with agitation. Induction is performed at an OD_{600} of 0.8–1 adding appropriate inducer for the strain (Table 1) and leaving overnight at 18 °C. Media are supplemented with ammonium iron (II) sulfate and vitamin B_{12} upon induction, at a final concentration of 50 and 1 μM, respectively (Halliwell et al., 2021; Payne, Quezada, et al., 2015). *Shimwellia blattae* protein expression also benefits from the addition of 10 μM final concentration 5,6-dimethylbenzimidazole as increased reductive dehalogenase activity and higher cell cobalamin concentration has been reported (Mac Nelly, Kai, Svatoš, Diekert, & Schubert, 2014). Harvested cells can be used immediately or stored at −80 °C.

3. Anaerobic purification and characterization of RdhA reductive dehalogenase

Optimum recombinant protein purification depends on the cell host, the cloning strategy, the concentration of the inducer and induction temperature used for protein expression, and the RdhA being purified. Achieving good protein yield and purity depends on highly selective and efficient capture of the protein from crude cell lysates under conditions that retain associated cofactors required for activity. Various N- or C-terminal affinity purification tags can be used to increase protein stability and solubility, and using an expression vector that includes a fusion tag further facilitates recombinant protein purification.

3.1 Affinity chromatography

This requires selective binding of RdhA to a resin followed by washing to remove non-specific proteins, and finally elution by weakening the protein–resin interaction. The resin chosen will depend on the engineered affinity tag: e.g., hexahistidine (6xHis), Strep-tag II, Maltose-binding protein, etc. By far the most common tag used is 6xHis, which can be used with nitrilotriacetic acid (NTA) coupled agarose resin charged with Ni^{2+}, Co^{2+} or Zn^{2+} ions. Although NpRdhA is relatively tolerant to oxygen, purification is usually performed under anaerobic conditions. A summary of a typical RdhA-6xHis purification with Ni-NTA is given below.

3.1.1 Equipment

- Ultracentrifuge and rotor
- Cell disruptor (e.g., French press)
- Ni-NTA agarose resin
- Econopac 10-DG column
- Anaerobic glove box (Belle Technology)

3.1.2 Buffers and reagents

- 1 L binding buffer: 200 mM NaCl, 50 mM Tris-HCl, pH 7.9
- 0.5 L elution buffer: 200 mM NaCl, 50 mM Tris-HCl, 200 mM imidazole pH 7.9
- 0.5 L 10-DG buffer: 100 mM NaCl, 50 mM Tris-HCl, pH 7.7
- All buffers rendered anaerobic by sparging with nitrogen gas

3.1.3 Procedure

- Resuspend 100 g harvested cells in 200 mL anaerobic binding buffer under a continuous stream of nitrogen gas. Supplement with SigmaFast protease inhibitor cocktail, DNase and RNase and stir until homogenous
- Pass the cell suspension through a French press cell at 1500 psi and collect broken cells in an anaerobic vessel on ice
- Remove cell debris by centrifugation at 125,000*g* at 4 °C for 60 min. During the centrifugation, equilibrate a 5 mL Ni^{2+}-NTA agarose column with binding buffer inside the anaerobic chamber
- Decant the centrifugation supernatant onto the Ni^{2+}-NTA agarose column in the anaerobic chamber (<1 ppm oxygen)
- Successively wash the column with 5 column volumes (CV) each of binding buffer containing 10 and 30 mM imidazole, made by diluting the elution buffer with binding buffer, and collect the flow through. Elute the protein with 4 CV elution buffer by adding 1 mL at a time and collect fractions. Subject all brown colored fractions to SDS-PAGE analysis to locate RdhA containing fractions and pool those containing pure protein

- Remove imidazole from the protein using a 10-DG column equilibrated with 10-DG buffer and flash freeze in liquid nitrogen in bead form. Beads can be stored in liquid nitrogen long-term with no detrimental effects
- Purified protein concentration can be determined from its absorbance at 280 nm using its extinction coefficient, $\varepsilon_{280} = 77{,}810\,M^{-1}\,cm^{-1}$ in case of NpRdhA, calculated from the primary amino acid sequence using the ProtParam program on the ExPASy proteomics server

3.1.4 Note

Higher purity protein can be obtained if the affinity resin is saturated with protein. If expression levels are low, use less resin for the purification or increase the amount of cells taken for crude extract preparation.

3.2 UV–visible spectroscopy of purified NpRdhA

The UV–visible spectrum of NpRdhA (concentration ~50 μM) can be measured between 250 and 750 nm, correcting the baseline using the buffer the protein is stored in before scanning. The aerobic UV–visible spectrum of purified NpRdhA has very broad features with two clearly distinguishable shoulders at 310 and 420 nm (Fig. 1), arising from cob(*II*)alamin and [4Fe-4S] clusters, respectively. The inset shows the unique spectrum obtained when cobalamin is extracted from NpRdhA with cyanide (see Section 3.3.1).

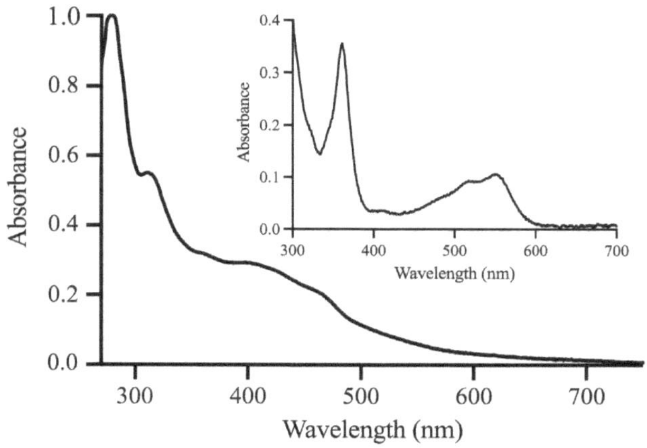

Fig. 1 UV–visible spectra of NpRdhA purified protein. Spectra were recorded on a Cary 50 UV–vis spectrophotometer between 200 and 800 nm. The inset shows a typical cyanocobalamin spectrum observed from cobalamin extraction using cyanide, from NpRdhA.

3.3 Metal analysis

The ability to accurately determine the stoichiometry of cobalamin and iron bound to reductive dehalogenases is an essential prerequisite for any mechanistic study. These enzymes bind one cobalamin cofactor and eight Fe atoms (as 2 [4Fe-4S] clusters per protein) and their stoichiometry influences enzymatic activity.

The following protocols describe our strategy for determining the stoichiometry of bound cobalamin and iron per RdhA polypeptide.

3.3.1 Cobalamin estimation after extraction by dicyano-complex formation

- Mix a known concentration of protein (0.2–0.4 mM) with potassium cyanide to give a final concentration of 10 mM in an Eppendorf tube
- Heat to 80 °C in fume cupboard for 20 min to release the dicyanocobalamin from the enzyme
- Centrifuge at 13,000 rpm to remove protein precipitate
- Record the UV–visible spectrum of the cyanocobalamin sample after blanking with protein buffer. Determine the B_{12} concentration from the absorbance at 550 nm using $\varepsilon_{550} = 8.7\,mM^{-1}\,cm^{-1}$ and the total sample dilution

3.3.2 Iron determination with bathophenanthroline

- Dilute protein to 0.5 mL and add 0.5 mL 2 M HCl. Boil at 80 °C for 10 min in an Eppendorf tube to fully extract the iron
- Centrifuge at 13,000 rpm to remove protein precipitate
- Take a sample aliquot (0.01–0.2 mL) so the final value is in the middle of the iron standard curve (25 nmol) and make the sample volume to 0.2 mL
- Using a 1 mM $FeCl_2$ solution prepare a standard curve (0–50 nmol) in a final volume of 0.2 mL
- To develop samples, add 0.55 mL Milli-Q water
- Add 0.2 mL saturated ammonium acetate
- Add 0.05 mL 0.1 M sodium metabisulfite and mix
- Add 0.01 mL 0.1 M bathophenanthroline disulfonic acid and mix
- After 10 min read the absorbance at 535 nm

3.3.3 Note

If too much protein is taken for extraction before analysis, it can solidify upon heating and it is impossible to remove a sample for analysis.

4. X-ray crystallography of RdhA proteins

Detailed accounts of the X-ray crystallographic studies of NpRdhA are available in Payne, Quezada, et al. (2015) and Halliwell et al. (2020). Initial studies of C-terminally His-tagged enzyme ($NpRdhA^C$) produced diffracting crystals in 0.2 M potassium thiocyanate, 0.1 M sodium cacodylate pH 6.5, 8%*w*/*v* PEG 20K plus 8%*v*/v PEG550MME at 4 °C (Payne, Quezada, et al., 2015). Subsequent studies of N-terminally His-tagged enzyme ($NpRdhA^N$) revealed a new crystallization regime, 0.1 M Bis-Tris propane (pH 8.5), 0.2 M sodium nitrate and 20%w/v PEG3350, again at 4 °C. $NpRdhA^N$ gave crystals diffracting to ~2.0 Å, rather than ~3.0 Å for $NpRdhA^C$, due to improved packing and reduced mobility of the three monomers in the asymmetrical unit.

4.1 Equipment

- Mosquito® LCP
- HD clear sealing tape
- Microscope
- Mounted cryoloops
- Universal storage pucks (Unipuck)

4.2 Buffers and reagents

- ~95% pure, 30–7.5 mg/mL, $NpRdhA^N$, in 200 mM NaCl, 50 mM Tris-HCl, pH 7.5
- SWISSCI SD3 crystallography trays containing a commercial screen (e.g., PACT premier™)
- 10% PEG 200
- 100 mM substrate stock (e.g., 3,5-dibromo-4-hydroxybenzoic acid) in crystal mother liquor

4.3 Procedure

- The Mosquito® LCP program for NpRdhA crystallogenesis by sitting drop involves mixing equal volumes of protein and mother liquor (200 nL). SWISSCI SD3 trays allow preparation of 3 drops per buffer condition allowing for screening of protein concentrations; this should be reflected in the program

- Seal completed trays with tape and store at 4 °C for 24–72 h, after which crystals should be identifiable by microscopy

Extract crystals using an appropriately sized cryoloop and either soak briefly in a 10% PEG 200 solution or substrate solution before submersion and storage in a Unipuck in liquid nitrogen.

4.4 Structure solution

Initial NpRdhA^N structure solution was performed by molecular replacement using the NpRdhA^C structure as a model (PDB: 4RAS). Manual inspection of the model: density fit can be performed in Coot, identifying and appropriately altering clashes and performing model refinement using Phenix.refine between iterations, until the R_{work}/R_{free} could no longer be improved. Identification of 3,5-dibromo-4-hydroxybenzoic acid bound structures can be determined by difference Fourier transform using the NpRdhA^N model, revealing substrate density in the active site. Ligand fit can be used to model substrate, generating ligand restraints via eLBOW.

5. In vitro NpRdhA enzyme catalysis

RdhA activity has been measured using chemical reductants such as reduced methyl viologen ($Mv^{+\bullet}$) or titanium (III) citrate as electron donors, or by use of a non-physiological system consisting of spinach ferredoxin and *E. coli* flavodoxin reductase, which can support NADPH-dependent organohalide reduction (Collins et al., 2018). NpRdhA activity has been determined for several ortho-halogenated phenolic substrates and shows a strong preference for substrates having a 2,6-dihalogenated phenolic moiety. Of the NpRdhA substrates discovered the fastest reaction is the debromination of 3,5-dibromo-4-hydroxybenzoic acid to 3-bromo-4-hydroxybenzoic acid and ultimately to hydroxybenzoic acid.

5.1 Real time spectrophotometric RdhA activity measurement with Mv^+

RdhA enzyme activity can be measured indirectly, under strict anaerobic conditions, by monitoring the anaerobic oxidation of the brilliant blue $Mv^{+\bullet}$ to colorless $Mv^{2+\bullet}$ during RdhA turnover. This technique provides simple, rapid, very sensitive, and accurate determination of kinetic data as well as screening of substrate libraries. Assays are usually initiated by the addition of RdhA after an initial background rate has been determined because $Mv^{+\bullet}$ reacts rapidly with any contaminating oxygen to form $Mv^{2+\bullet}$.

5.1.1 Equipment

- Anaerobic chamber (e.g., Belle Technology glovebox operating at <1 ppm oxygen)
- UV–vis spectrophotometer in anaerobic chamber—preferably thermostatically controlled (e.g., Cary Win-UV 60, Agilent)
- 10DG desalting column
- Cuvettes

5.1.2 Buffers and reagents

- Anaerobic buffer
- Methyl viologen (30 mM stock solution)
- Sodium dithionite (50 mM stock solution)
- Substrate for RdhA (10 mM solution)

5.1.3 Procedure

- All buffers are made anaerobic by sparging with nitrogen gas prior to being left to equilibrate in the anaerobic chamber overnight
- All chemical reductants and substrates are weighed outside the anaerobic chamber and gently flushed in Subaseal capped vials before being dissolved in anaerobic buffer inside the anaerobic chamber before use
- Protein samples are made anaerobic by buffer exchange through a 10DG desalting column (GE Healthcare Life Sciences, UK) prepared and washed with anaerobic buffer (50 mM Tris, pH 7.7, 100 mM NaCl)
- Anaerobic protein concentrations are determined from A280. A typical 1 mL methyl viologen reaction consists of 0.3 mM methyl viologen, 0.05 mM sodium dithionite, 0.5 mM substrate (e.g., 3,5-dibromo-4-hydroxy benzoic acid) and reactions started by the addition of 1 μM RdhA followed by mixing by inversion
- The RdhA dependent rate of methyl viologen oxidation is measured at 578 nm ($\varepsilon = 9.78\,\text{mM}^{-1}\,\text{cm}^{-1}$) using a Cary UV–vis spectrophotometer over 5 min and fit to a linear rate where appropriate

5.1.4 Note

Sodium dithionite powder is prone to breakdown and should be stored under an inert atmosphere when not in use. Stock substrate solutions can be sparged with oxygen-free nitrogen before being taken into the anaerobic chamber and then diluted to a working concentration with anaerobic buffer before use. Many substrates are sparingly soluble in water and must either be pH adjusted before use or dissolved in DMSO or an organic solvent before being added to an assay. We recommend keeping the concentration of organic solvent in an assay below 5% unless a biphasic assay is being performed. The rate of dithionite breakdown increases with acidity so solutions should be kept slightly alkaline. Reduced methyl viologen should be made fresh immediately prior to use and kept under strict anaerobic conditions.

5.2 Steady-state assay of NpRdhA

NpRdhA is exchanged into anaerobic 50 mM Tris, pH 7.7, 100 mM NaCl in an anaerobic chamber. Typically 500 μL reactions are set up under anaerobic conditions in crimp seal vials and include 20 mM NADPH, NpRdhA, spinach ferredoxin and ferredoxin oxidoreductase in an enzyme ratio of 1:10:2 or NpRdhA, spinach ferredoxin and *E. coli* flavodoxin reductase in a ratio of 1:40:400, where NpRdhA is 1 μM (Collins et al., 2018). Reactions are initiated by the addition of 10 mM organohalide substrate and incubated at 30 °C, 250 rpm for 15–60 min. Terminate assays by the addition of 50 μL of 30% TCA before centrifugation in a bench top microfuge at 13,200 rpm to sediment precipitated reaction components. Sample analysis is performed using an HPLC equipped with a UV detector, e.g., Agilent 1100 Series, and a 250 × 4.6 mm Kinetex 5 μ C18 100A column developed with a mobile phase of acetonitrile: water (50:50) with 0.1% TFA at a flow rate of 1 mL min^{-1}. Standard curves for each substrate and product (0–10 mM) should be used to compare product peaks and to calculate turnover rates (μM/min).

6. EPR spectroscopy as a tool to study RdhA

The Co^{2+} oxidation state of cobalamin (cob(*II*)alamin) and the $[4Fe\text{-}4S]^{1+}$ (reduced) oxidation state of low potential [4Fe-4S] clusters can be studied using electron paramagnetic resonance (EPR) spectroscopy, that measures the absorption of electromagnetic radiation (usually microwave radiation in the GHz range) by an unpaired electron that has been placed in a magnetic field (Hagen, 2008). EPR spectroscopy of transition metal ions is typically carried out at cryogenic temperatures (2 – 100 K) using frozen solutions of the protein under study. The EPR spectrum is an exquisitely sensitive probe of the environment of the unpaired electron and so provides a metallocentric view of a metalloprotein. EPR reports on the symmetry of the environment of the electron (anisotropy) and "hyperfine" interactions between the electron and nearby atomic nuclei that have a magnetic moment, which can also be studied using EPR's sister technique nuclear magnetic resonance (NMR) spectroscopy, including the nucleus of the transition metal ion itself (Bertini, Luchinat, Parigi, & Ravera, 2016).

6.1 Equipment

- X-band EPR spectrometer (Bruker E500/580, ER4122SHQ resonator)
- Liquid helium sample cooling system (Oxford ESR900 Cryostat and LLT 600 series transfer line, ITC 503 temperature controller)
- Quartz EPR sample tubes, 4 mm o.d. (Wilmad 707-SQ-160)
- Anaerobic chamber
- Subaseal stoppers, size 9
- Cryogens (liquid helium and liquid nitrogen)

6.2 Procedure

- NpRdhA from cryo-storage (Section 3.1.3 step 6) was thawed in the anaerobic chamber. Samples were made in quartz EPR tubes, 200 μL volume and [NpRdhA] = 100 μM, with the addition of substrates or products from concentrated anaerobic stock solutions. Reductants were added where necessary and illumination was effected within the chamber if required. Tubes were sealed with subaseal closures and taken out of the chamber for flash freezing in liquid nitrogen
- The EPR spectrometer sample compartment was precooled using liquid helium gas flow. Samples were transferred to the spectrometer without thawing from 77 K. Optimum experimental temperature, microwave power and field modulation amplitude were determined in preliminary experiments. Wide scans of 450 mT were made to ensure detection of all signals. Experimental data presented were recorded over field ranges optimized for resolution of the signal under study

6.3 Note

Aqueous samples should be frozen slowly (10–15 s) in liquid nitrogen because water expands on freezing and can crack the EPR tube. When samples are removed from liquid nitrogen for analysis, use a tissue to wipe condensation from the outside of the tube.

6.4 Analysis of spectra

The single unpaired electron of cob(*II*)alamin (formally $S = ½$, where S is the electron spin quantum number) is located in the cobalt d_z2 orbital (Fig. 2A) that is essentially perpendicular to the plane of the corrin ring.

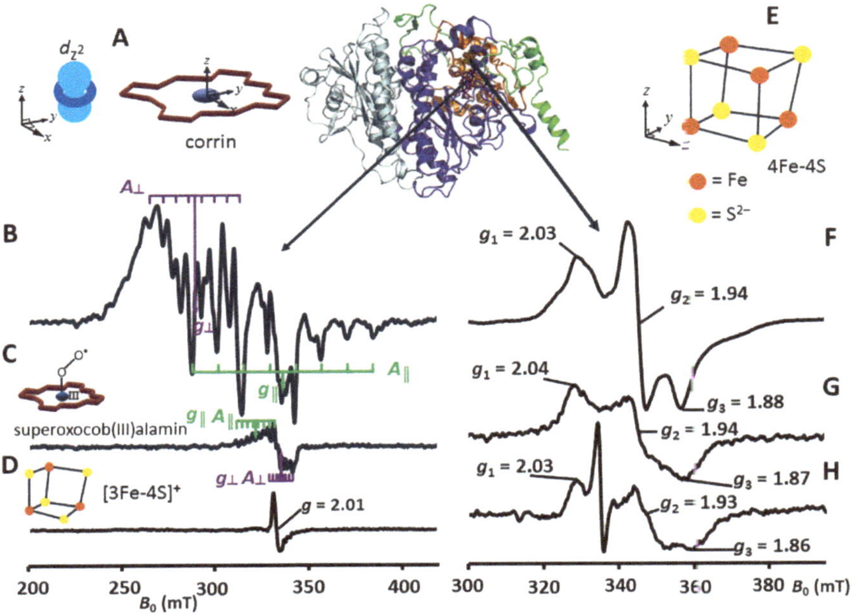

Fig. 2 (A) representation of the electron spin density distribution of the d_z2 electron orbital and its orientation within the cobalamin structure. X-band EPR spectra at 30 K of (B) 100 μM *Nitratireductor pacificus* reductive dehalogenase (NpRdhA), (C) the isolated superoxo-cob(*III*)alamin contamination, (D) the isolated $[3Fe\text{-}4S]^{1+}$ contamination. (E) a representation of the [4Fe-4S] cluster structure. X-band EPR spectra at of 79 μM NpRdhA (F) reduced using EDTA, deazaflavin and light at 10 K, (G) reduced using EDTA, deazaflavin and light at 40 K, (H) reduced using 5 mM sodium dithionite at 10 K. Experimental parameters: microwave power 0.5 mW, field modulation frequency 100 KHz, field modulation amplitude 0.5 mT (B, C and D) 0.7 mT (F, G and H). Equipment: Bruker ELEXSYS E500/E580 equipped with Bruker Super High Q (ER 4122SHQ) cavity resonator and Oxford Instruments ESR900 helium flow cryostat and ITC503 temperature controller.

As such, its environment shows cylindrical or "axial" symmetry being identical along the x and y directions and unique in the z direction. This leads to an axial EPR spectrum (EPR presents spectra as first derivatives of the absorption line) having the same *g* values (the spectrometer-independent measure of resonance condition) in the x and y directions $g_\perp$ ($g_{\text{perpendicular}}$) and a unique z value $g_\parallel$ (g_{parallel}), Fig. 2B. Furthermore, hyperfine interactions between the unpaired electron and the cobalt nuclear magnetic moment "split" the spectrum into eight lines, separated by the hyperfine coupling constant A, at each g value since there are eight spin energy levels available to that moment (formally, nuclear spin quantum number $I = 7/2$). The basic concepts underlying the origins of cob(*II*)alamin spectra were reported from the early 1960s (Hogenkamp, Barker, & Mason, 1963); (Bayston, Looney, Pilbrow, & Winfield, 1970). The EPR spectrum of cob(*II*)alamin in NpRdhA is typical of a five coordinate "base off" structure (i.e., with no cobalt coordination by either the intrinsic DMB ligand or a histidine sidechain of the protein) with water as the fifth ligand to the cobalt ion in addition to the four corrin nitrogens. Two small contaminating signals are present, from a cob(*III*)alamin-superoxide complex, Fig. 2C, and a $[3Fe\text{-}4S]^{1+}$ species, Fig. 2D, but both at less than 4% of the cob(*II*)alamin signal (Payne, Quezada, et al., 2015).

The origin of the apparently simpler spectrum arising from the $[4Fe\text{-}4S]^{1+}$ state, Fig. 2F, of the two [4Fe-4S] clusters is, in fact, more complex and beyond the scope of this chapter. The spectra derived from the individual $[4Fe\text{-}4S]^{1+}$ clusters are not easily separated in RdhA (unlike in the related QueG protein; (Payne et al., 2015), however, spectra obtained using different reducing agents and at different temperatures, Fig. 2G and H, provide evidence for two $[4Fe\text{-}4S]^{1+}$ signals with slightly different g values.

Binding substrates and products to NpRdhA affects the environment of the Co^{2+} ion (Collins et al., 2018). An excess of 3,5-dichloro-4-hydroxybenzoic acid produces the spectrum of Fig. 3B. This spectrum clearly consists of more than two groups of eight lines and can be rationalized as a rhombic spectrum having three g values and three cobalt hyperfine coupling constants (A). This suggests that the axial symmetry at the Co^{2+} ion has been broken by the substrate, i.e., the environment of the unpaired electron in the x and y directions is no longer equivalent; the X-ray structures of related complexes support this (Halliwell et al., 2020). The same rationale can be applied to the binding of the first product, 3-chloro-4-hydroxybenzoic acid, Fig. 3C. However, binding the ultimate product, 4-hydroxybenzoic acid, does not duplicate this effect on anisotropy, giving rise to an axial spectrum with very similar g and A values to H_2O coordination, Fig. 3D. This suggests that in these solution-state experiments the 4-hydroxybenzoic acid binds in one of the non-coordinating locations identified in X-ray crystal structures (Halliwell et al., 2020) with the closest approach of product carbon atom and Co^{2+} ion being ~7 Å. Chloride ion binding leads to a distinctive "quartet" splitting of the parallel

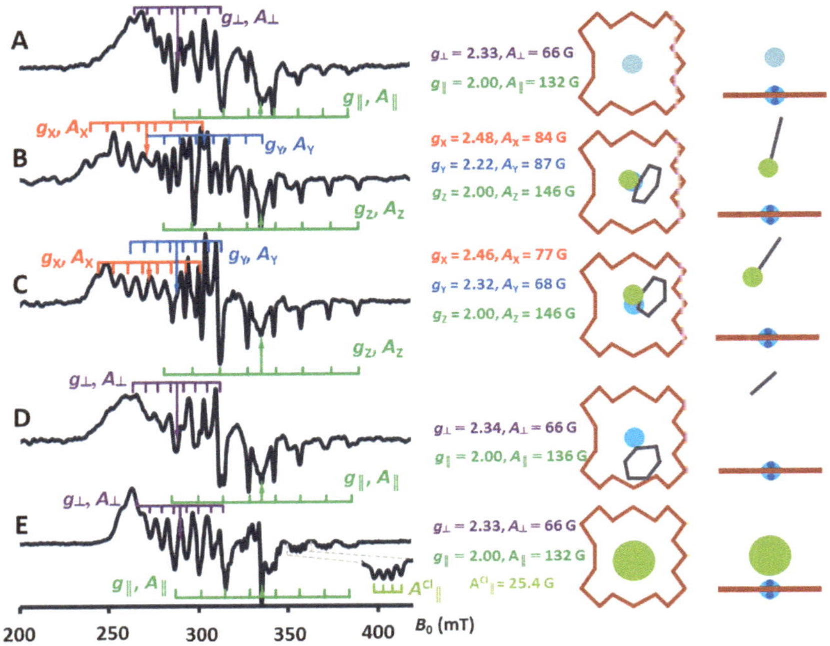

Fig. 3 X-band EPR spectra of the cob(*II*)alamin signal of 100 μM NpRdhA measured at 30 K in different binding states. Each spectrum is accompanied by the *g* and *A* values obtained from the data and diagrams showing binding of substrate/product relative to the cobalamin from both above and the edge of the cobalamin plane (derived from X-ray crystallographic structures). (A) NpRdhA alone, (B) NpRdhA and 20 mM 3,5-dichloro-4-hydroxybenzoic acid, (C) NpRdhA and 20 mM 3-chloro-4-hydroxybenzoic acid, and (D) NpRdhA and 20 mM 4-hydroxybenzoic acid, (E) NpRdhA and 200 mM NaCl. Experimental parameters as in Fig. 2B.

lines in the otherwise axial EPR spectrum, Fig. 3E (Payne, Quezada, et al., 2015). This superhyperfine splitting arises from interactions between the unpaired electron and the magnetic moment of the chloride nucleus ($I=3/2$) and was not observed for any cob(*II*)alamin spectra prior to the study of RdhA.

Having characterized the cob(*II*)alamin EPR spectra of several RdhA substrate/product binding complexes, such information can be used to study the enzyme under turnover conditions (Collins et al., 2018). The basic turnover system consisted of equimolar (100 μM) RdhA, spinach ferredoxin (SpFd) and *E. coli* ferredoxin reductase (EcFldr) with 20 mM NADPH as reductant and was incubated for 5 min at 30 °C, Fig. 4.

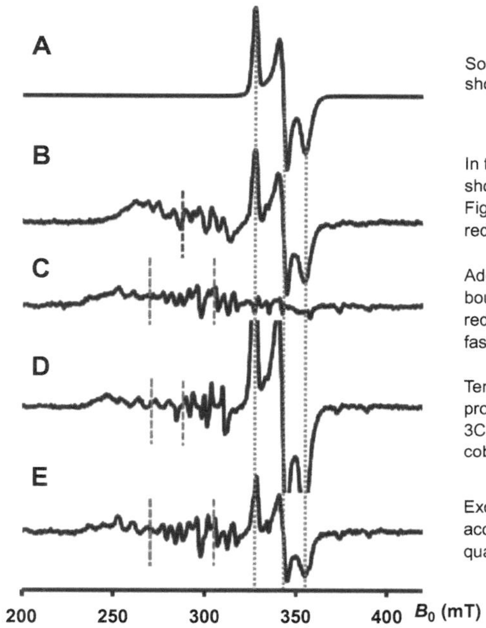

Fig. 4 (A) X-band continuous wave EPR spectrum of 1 mM spinach [2Fe-2S] ferredoxin (SpFd) reduced by an excess of sodium dithionite ($[2Fe\text{-}2S]^{1+}$ state), recorded at 30 K. X-band continuous wave EPR spectra of 100 μM NpRdhA incubated with the reduction system consisting of SpFd and *E. coli* flavodoxin reductase (EcFldr) and NADPH, measured at 30 K: (B) NpRdhA, 100 μM SpFd, 100 μM EcFldr, and 20 mM NADPH incubated for 5 min; (C) NpRdhA, 100 μM SpFd, 100 μM EcFldr, and 20 mM NADPH incubated for 5 min with 100 μM 3,5-dichloro-4-hydroxybenzoic acid; (D) NpRdhA, 1 mM SpFd, 100 μM EcFldr, and 20 mM NADPH incubated for 5 min with 100 μM 3,5-dichloro-4-hydroxybenzoic acid; and (E) NpRdhA, 1 mM SpFd, 100 μM EcFldr, and 20 mM NADPH incubated for 5 min with 20 mM 3,5-dichloro-4-hydroxybenzoic acid. Experimental parameters as in Fig. 2B.

7. Summary

Persistent organohalide pollutants pose a threat to both humans and the environment; therefore, it is crucial that further study of reductive dehalogenases continues for use as a tool in bioremediation. The methods described focus primarily on NpRdhA expression and characterization; however, they can, and have, been used with other RdhAs and related enzymes (e.g., epoxyqueuosine reductases) (Payne, Fisher, et al., 2015). Using these methods, research can now focus on other members of the catabolic reductive dehalogenases, including the "self-sufficient" reductive dehalogenases, which do not rely on external partner proteins for electron delivery and can use NADPH directly to perform dehalogenation.

References

Atashgahi, S., Häggblom, M. M., & Smidt, H. (2018). Organohalide respiration in pristine environments: Implications for the natural halogen cycle. *Environmental Microbiology*, *20*(3), 934–948.

Bayston, J. H., Looney, F., Pilbrow, J., & Winfield, M. E. (1970). Electron paramagnetic resonance studies of cob (II) alamin and cob (II) inamides. *Biochemistry*, *9*(10), 2164–2172.

Bertini, I., Luchinat, C., Parigi, G., & Ravera, E. (2016). *NMR of paramagnetic molecules: Applications to metallobiomolecules and models*. Elsevier.

Collins, F. A., Fisher, K., Payne, K. A., Gaytan Mondragon, S., Rigby, S. E., & Leys, D. (2018). NADPH-driven organohalide reduction by a nonrespiratory reductive dehalogenase. *Biochemistry*, *57*(25), 3493–3502.

Dobbek, H., & Leys, D. (2016). Insights into reductive dehalogenase function obtained from crystal structures. In *Organohalide-respiring bacteria* (pp. 485–495). Springer.

Genomics, S. (Producer). (2017). *VmaxTM express electrocompetent cells user guide*.

Gribble, G. W. (2015). A recent survey of naturally occurring organohalogen compounds. *Environmental Chemistry*, *12*(4), 396–405.

Hagen, W. R. (2008). *Biomolecular EPR spectroscopy*. CRC press.

Halliwell, T., Fisher, K., Payne, K. A., Rigby, S. E., & Leys, D. (2020). Catabolic reductive dehalogenase substrate complex structures underpin rational repurposing of substrate scope. *Microorganisms*, *8*(9), 1344.

Halliwell, T., Fisher, K., Payne, K. A., Rigby, S. E., & Leys, D. (2021). Heterologous expression of cobalamin dependent class-III enzymes. *Protein Expression and Purification*, *177*, 105743.

Hausjell, J., Weissensteiner, J., Molitor, C., Halbwirth, H., & Spadiut, O. (2018). E. coli HMS174 (DE3) is a sustainable alternative to BL21 (DE3). *Microbial Cell Factories*, *17*(1), 1–9.

Hogenkamp, H., Barker, H., & Mason, H. (1963). An electron-spin resonance study of coenzyme B12. *Archives of Biochemistry and Biophysics*, *100*(3), 353–359.

Inoue, H., Nojima, H., & Okayama, H. (1990). High efficiency transformation of Escherichia coli with plasmids. *Gene*, *96*(1), 23–28.

Jugder, B.-E., Ertan, H., Bohl, S., Lee, M., Marquis, C. P., & Manefield, M. (2016). Organohalide respiring bacteria and reductive dehalogenases: Key tools in organohalide bioremediation. *Frontiers in Microbiology*, 7, 249.

Jugder, B.-E., Ertan, H., Lee, M., Manefield, M., & Marquis, C. P. (2015). Reductive dehalogenases come of age in biological destruction of organohalides. *Trends in Biotechnology*, *33*(10), 595–610.

Kodavanti, P. R. S., & Loganathan, B. G. (2017). Organohalogen pollutants and human health. *5. International Encyclopedia of Human Health* (2nd, pp. 359–366). Oxford: Academic Press.

Leys, D., Adrian, L., & Smidt, H. (2013). *Organohalide respiration: Microbes breathing chlorinated molecules*. In: The Royal Society.

Mac Nelly, A., Kai, M., Svatoš, A., Diekert, G., & Schubert, T. (2014). Functional heterologous production of reductive dehalogenases from desulfitobacterium hafniense strains. *Applied and Environmental Microbiology*, *80*(14), 4313–4322.

Moore, S. J., Lawrence, A. D., Biedendieck, R., Deery, E., Frank, S., Howard, M. J., et al. (2013). Elucidation of the anaerobic pathway for the corrin component of cobalamin (vitamin B12). *Proceedings of the National Academy of Sciences*, *110*(37), 14906–14911.
Nou, X., & Kadner, R. J. (2000). Adenosylcobalamin inhibits ribosome binding to btuB RNA. *Proceedings of the National Academy of Sciences*, *97*(13), 7190–7195.
Payne, K. A., Fisher, K., Sjuts, H., Dunstan, M. S., Bellina, B., Johannissen, L., et al. (2015). Epoxyqueuosine reductase structure suggests a mechanism for cobalamin-dependent tRNA modification. *Journal of Biological Chemistry*, *290*(46), 27572–27581.
Payne, K. A., Quezada, C. P., Fisher, K., Dunstan, M. S., Collins, F. A., Sjuts, H., et al. (2015). Reductive dehalogenase structure suggests a mechanism for B12-dependent dehalogenation. *Nature*, *517*(7535), 513–516.
Vary, P. S., Biedendieck, R., Fuerch, T., Meinhardt, F., Rohde, M., Deckwer, W.-D., et al. (2007). Bacillus megaterium—From simple soil bacterium to industrial protein production host. *Applied Microbiology and Biotechnology*, *76*(5), 957–967.

CHAPTER FOURTEEN

An unusual light-sensing function for coenzyme B_{12} in bacterial transcription regulator CarH

Harshwardhan Poddar, Derren J. Heyes, Shaowei Zhang, Samantha J. Hardman, Michiyo Sakuma, and Nigel S. Scrutton*

Manchester Institute of Biotechnology, Department of Chemistry, University of Manchester, Manchester, United Kingdom

*Corresponding author: e-mail address: nigel.scrutton@manchester.ac.uk

Contents

Abstract

Coenzyme B_{12} is one of the most complex cofactors found in nature and synthesized *de novo* by certain groups of bacteria. Although its use in various enzymatic reactions is well characterized, only recently an unusual light-sensing function has been ascribed to coenzyme B_{12}. It has been reported that the coenzyme B_{12} binding protein CarH, found in the carotenoid biosynthesis pathway of several thermostable bacteria, binds to the promoter region of DNA and suppresses transcription. To overcome the harmful effects of light-induced damage in the cells, CarH releases DNA in the presence of light

Methods in Enzymology, Volume 668
ISSN 0076-6879
https://doi.org/10.1016/bs.mie.2021.12.010

and promotes transcription and synthesis of carotenoids, thereby working as a photoreceptor. CarH is able to achieve this by exploiting the photosensitive nature of the Co—C bond between the adenosyl moiety and the cobalt atom in the coenzyme B_{12} molecule. Extensive structural and spectroscopy studies provided a mechanistic understanding of the molecular basis of this unique light-sensitive reaction. Most studies on CarH have used the ortholog from the thermostable bacterium *Thermus thermophilus*, due to the ease with which it can be expressed and purified in high quantities. In this chapter we give an overview of this intriguing class of photoreceptors and report a step-by-step protocol for expression, purification and spectroscopy experiments (both static and time-resolved techniques) employed in our laboratory to study CarH from *T. thermophilus*. We hope the contents of this chapter will be of interest to the wider coenzyme B_{12} community and apprise them of the potential and possibilities of using coenzyme B_{12} as a light-sensing probe in a protein scaffold.

1. Introduction

The sun has played a significant role in the evolution of life on Earth. Without the energy supplied by the sun in the form of heat and light, Earth would have been a lifeless planet (Ruban, 2015). Most living forms on Earth have evolved to harness this unlimited source of transferable light energy by tailoring their important physiological functions to detect and respond to sunlight. The life sustaining energy from sunlight is indispensable for a multitude of critical life processes. The ability of vision in organisms from the animal kingdom (Palczewski, 2012; Wald, 1968), absorption of photons by special light harvesting complexes in plants and subsequent photolysis of water during photosynthesis (Croce & van Amerongen, 2014), control of circadian rhythms (Cohen & Golden, 2015; Nagel & Kay, 2012), crucial photoenzyme catalyzed reactions (Brettel & Byrdin, 2010; Heyes et al., 2021; Sorigué et al., 2017; Yang & Cheng, 2004) and tactic movement in different organisms (Armitage, 1999) are among some of the important examples of light-dependent biochemical processes. Several organisms also employ specialized proteins called photoreceptors to detect light and trigger an appropriate cellular response, either to mitigate the harmful effects of light-induced damage (Fasnacht & Polacek, 2021; Purcell & Crosson, 2008) or to affect an important biological process (Elías-Arnanz, Padmanabhan, & Murillo, 2011). Photoreceptors utilize special chemical moieties called chromophores to sense light at different wavelengths comprising the electromagnetic spectrum (380–760 nm) (Poddar et al., 2021). The study of different photoreceptors and the underlying molecular details of their photochemistry have gained a lot of traction in recent years due to the possibility of recruiting photoreceptors in a myriad of optogenetic applications (Banerjee & Mitra, 2020; Krueger et al., 2019; Losi, Gardner, & Möglich, 2018; Shcherbakova, Shemetov, Kaberniuk, & Verkhusha, 2015). The most commonly used chromophores found in different proteins that have been characterized in detail include retinal (Wickstrand et al., 2019; Wickstrand, Dods, Royant, & Neutze, 2015), flavin mononucleotide (FMN) (Hart & Gardner, 2021), flavin adenine dinucleotide (FAD) (Park & Tame, 2017), carotenoids (Konold et al., 2019), *p*-coumaric acid (Pande et al., 2016) and linear tetrapyrroles (Takala et al., 2014).

In recent times a new class of coenzyme B_{12}-dependent photoreceptor, involved in transcriptional regulation of the carotenoid biosynthesis pathway genes in various bacteria, has been identified (Padmanabhan, Pérez-Castaño, & Elías-Arnanz, 2019). The ability of coenzyme B_{12} to sense light in the green region of the visible spectrum and work as a chromophore is remarkable and expands the already diverse chemical repertoire of photoreceptor families (Cheng, Yamamoto, & Bauer, 2016; Padmanabhan, Jost, Drennan, & Elías-Arnanz, 2017). This chapter highlights the recent progress made in our mechanistic understanding of these unique B_{12}-dependent photoreceptors using the CarH protein from *Thermus thermophilus* (TtCarH) as an exemplar model (Ortiz-Guerrero, Polanco, Murillo, Padmanabhan, & Elías-Arnanz, 2011). We aim to provide a comprehensive step-by-step protocol for expression, purification and spectroscopic analysis of TtCarH and also give a prospective view on the scope and potential for future studies on this intriguing class of photoreceptors.

2. Background

2.1 Discovery of CarH photoreceptors

Light-induced photooxidative damage of DNA and proteins caused by reactive oxygen species (ROS) is a major threat for microorganisms as it leads to malfunctioning of cellular processes (Glaeser, Nuss, Berghoff, & Klug, 2011; Li, Wakao, Fischer, & Niyogi, 2009). Many bacteria produce carotenoids to protect themselves against the harmful effects of singlet oxygen and other ROS (Leverenz et al., 2015; Padmanabhan et al., 2021). Light-dependent regulation of carotenoid producing genes facilitated by CarH was first reported in the soil dwelling bacterium *Myxococcus xanthus* (Pérez-Marín, Padmanabhan, Polanco, Murillo, & Elías-Arnanz, 2008). The gene encoding CarH protein was found to be located downstream of structural genes comprising the carotenoid biosynthesis pathway. Sequence comparison suggested that the N-terminal region of these proteins closely resembles the DNA binding domain (DBD) of the MerR (mercury-sensing regulator) family of transcription factors (Cervantes & Murillo, 2002; Pérez-Marín et al., 2008), whereas the C-terminal region (also referred to as cobalamin binding domain—CBD) had remarkable similarity to the methylcobalamin-binding domain of the B_{12}-dependent enzyme, methionine synthase (MetH) (Drennan, Huang, Drummond, Matthews, & Ludwig, 1994). More importantly, the B_{12}-binding motif was also found to be conserved. In vivo studies in *M. xanthus* (Cervantes & Murillo, 2002; Ortiz-Guerrero et al., 2011; Pérez-Marín et al., 2008), *T. thermophilus* (Díez et al., 2013; Ortiz-Guerrero et al., 2011) and *Bacillus megaterium* (Fernández-Zapata et al., 2018; Takano et al., 2015) confirmed the role of CarH in gene regulation and established a role for coenzyme B_{12} as a light sensitive chromophore. Subsequent bioinformatic analysis has revealed that these proteins are found in hundreds of different bacterial species belonging to multiple phyla (Padmanabhan et al., 2017).

In the following sections, we delve deeper into the chemistry of coenzyme B_{12} and its important role in the physiological function of CarH. We also discuss the crystal structures of CarH which have given unprecedented insights into how light-induced structural dynamics control transcriptional regulation.

2.2 Different forms of vitamin B_{12}

Vitamin B_{12} is one of the most complex cofactors found in nature (Hodgkin et al., 1956), and is synthesized only by a few prokaryotic organisms *de novo* (Rodionov, Vitreschak, Mironov, & Gelfand, 2003; Zhang, Rodionov, Gelfand, & Gladyshev, 2009). Briefly, the structure comprises a central cobalt atom (in Co^{3+} oxidation state) which is coordinated by four nitrogen atoms from the corrin ring. A 'tail' comprising 5,6-dimethylbenzimidazole (DMB) ribonucleotide is attached to the corrin via a propionamide group. A nitrogen atom from this DMB tail makes a bond with Co to serve as the lower axial ligand in free B_{12}, a conformation referred to as the base-on state. The corrin ring and DMB tail together are also referred to as cobalamin (Cbl). The upper axial ligand, which may differ, completes the octahedral hexacoordinated geometry of Co^{3+}. Depending on the upper axial ligand, the different naturally available forms of cobalamin are adenosylcobalamin (AdoCbl, also called coenzyme B_{12}), methylcobalmin (MeCbl), cyanocobalamin (CNCbl) and hydroxocobalamin/aquo (OH/AqCbl) (Gruber, Puffer, & Kräutler, 2011) (Fig. 1).

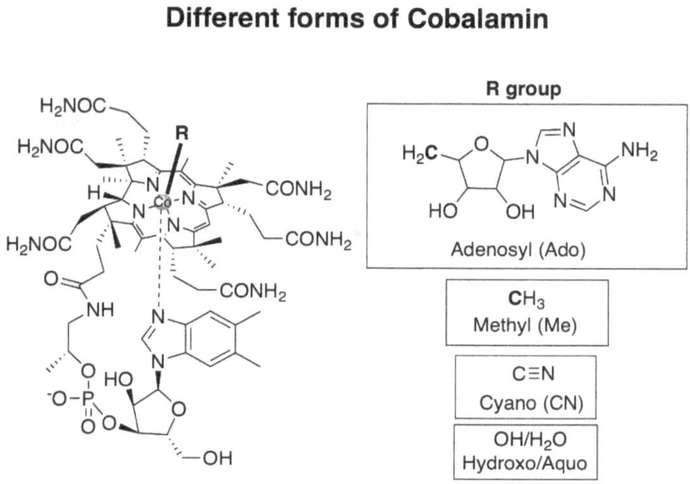

Fig. 1 Different forms of naturally found cobalamin. The carbon atom involved in forming the photosensitive Co—C bond in AdoCbl and MeCbl is shown in bold.

Of these four, AdoCbl and MeCbl have been shown to undergo photolysis upon illumination (Kozlowski, Garabato, Lodowski, & Jaworska, 2016). The Co—C bond between the methyl carbon atom or the C5′ carbon atom of the adenosyl moiety in MeCbl and AdoCbl, respectively, is a photosensitive bond with low bond dissociation energies that enables them to cleave easily following photon absorption (Rury, Wiley, & Sension, 2015). The MeCbl-binding enzyme, MetH, and other AdoCbl-dependent enzymes, such as mutases, dehydratases, deaminases and reductases, which have been studied in detail, do not use light to photolyze the Co—C bond and afford catalysis (Bridwell-Rabb & Drennan, 2017). CarH, on the other hand, with its CBD resembling that of MeCbl-binding MetH enzyme, was unexpectedly found to produce a stable tetramer with AdoCbl, which was capable of binding to the promoter region and suppressing transcription in dark conditions (Ortiz-Guerrero et al., 2011). In the presence of light, however, the Co—C bond of AdoCbl was photolyzed which promoted tetramer disassembly and release of DNA to induce carotenoid biosynthesis.

2.3 Crystal structure of CarH in dark and light state

The crystal structures of TtCarH in dark, DNA-bound and light-adapted states reported by Jost and coworkers have given us unparalleled insights into the structural intricacies of the CarH mechanism (Jost, Fernández-Zapata, et al., 2015) (Fig. 2). Briefly, the active tetrameric form of CarH was shown

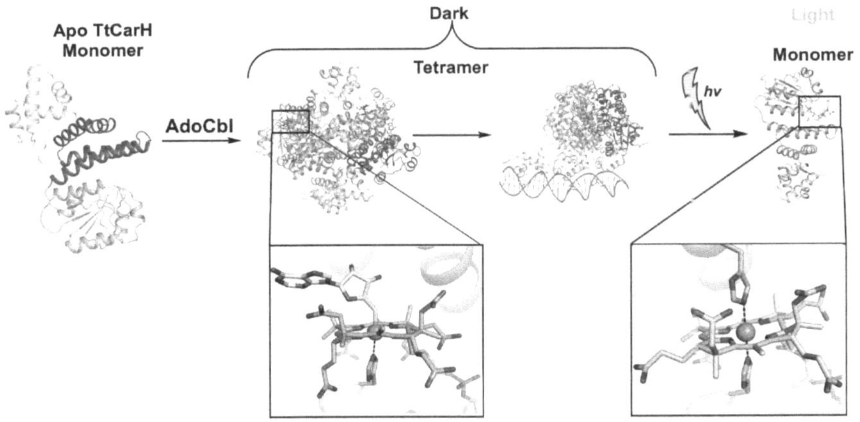

Fig. 2 Cartoon representation of CarH structures in dark and light states. The Apo monomeric structure consists of three domains—DNA-binding domain (*gray*), four helix bundle (*red*) and Rossmann fold (*orange*). Addition of AdoCbl promotes tetramerization (CBD shown in dark colors and DBD in *gray*) (PDB id 5C8D), which is capable of binding DNA (PDB id 5C8E). Following light illumination, the tetramer dissembles into monomers, which leads to release of DNA (PDB id 5C8F). The bound AdoCbl in the dark structure and bis-His ligated light state are shown as sticks in the insets (Jost, Fernández-Zapata, et al., 2015).

to be a dimer of dimers with residues belonging only to the CBD region involved in dimeric and tetrameric interactions. The dimer itself is formed by a head-to-tail arrangement of individual monomeric units. The N-terminal region (residues 1–74) of each monomer forms the DBD, which is tethered to the CBD by a short linker of six residues. This flexible linker allows the DBD to adopt different spatial orientations with respect to the CBD region. The DBD consists of a conserved RxWERRY motif (where x can be any residue). The CBD (residues 80–285) comprises two distinct structural regions—a four-helix bundle with a conserved WH-(8)x-EH motif and a Rossmann fold accommodating a ExHxxG motif (Marsh & Holloway, 1992). AdoCbl binds at the interface of these two domains and residues from the two conserved motifs, which collectively form the B_{12}-binding motif, are critical in its binding. His177 from the ExHxxG motif makes a direct interaction with the Co atom in a state referred to as base-off/His-on mode. The four-helix bundle domain closes over the bound AdoCbl and residues from this region make direct contact with the adenosyl (Ado) moiety of AdoCbl. Important interactions include those mediated by Trp131, Glu141 and His142, all belonging to the conserved WH-(8)x-EH motif. It has been shown that mutating these residues results in poor AdoCbl binding (Fernández-Zapata et al., 2018; Jost, Fernández-Zapata, et al., 2015). This in turn affects the oligomerization as binding of AdoCbl allows the monomers to assume an extended conformation, which is imperative for the head-to-tail arrangement in dimers.

The DNA-bound structure of TtCarH shed light on its unusual binding mode to the promoter region. Only three of the four DBDs from the tetramer were involved in binding DNA. Residues from the RxWERRY motif were found to form H-bonds and electrostatic interactions. Importantly, binding to DNA via DBD did not elicit any conformational change in the CBD region and the tetramer assembly assumed identical arrangement as the DNA-free structure. The light-adapted structure further provided a rationale for the tetramer disassembly following light-activation of the Co—C bond in AdoCbl. The light structure showed a bis-His ligated cobalamin, where the upper His ligand is contributed by His132 from the WH motif in the four-helix region. It was proposed that following Co—C bond cleavage the Ado moiety leaves the B_{12}-binding pocket. Trp131, which was making stacking interactions with the Ado group, swings in to occupy the space left vacant by the leaving group, therefore bringing H132 closer to the Co atom. This dynamic movement disrupts the head-to-tail dimeric organization eventually leading to tetramer disassembly and DNA release.

2.4 Photochemical mechanism of Co—C bond cleavage

Although structural dynamics at the global level addressed the mechanism of CarH-DNA complex disintegration, understanding of the molecular events in the B_{12} binding pocket immediately following light activation remained unclear. Jost and coworkers reported the formation of 4′,5′-anhydroadenosine as a product following photolysis of the Co—C bond using liquid chromatography and spectroscopic techniques (Jost, Simpson, & Drennan, 2015). This observation suggested that CarH was capable of altering the photochemistry of AdoCbl and prevented the escape of the highly reactive Ado• radical or Ado^- anion in solution. Formation of Co^{2+} was also detected during the anaerobic photoconversion of CarH using EPR spectroscopy (Jost, Simpson, & Drennan, 2015). Based on these observations Jost et al. proposed a photochemical mechanism, which suggested that the Co—C bond cleavage pathway could proceed via either a heterolytic or homolytic route (Fig. 3A). Kutta et al. proposed that CarH largely prefers the heterolytic cleavage route (Kutta et al., 2015). They also

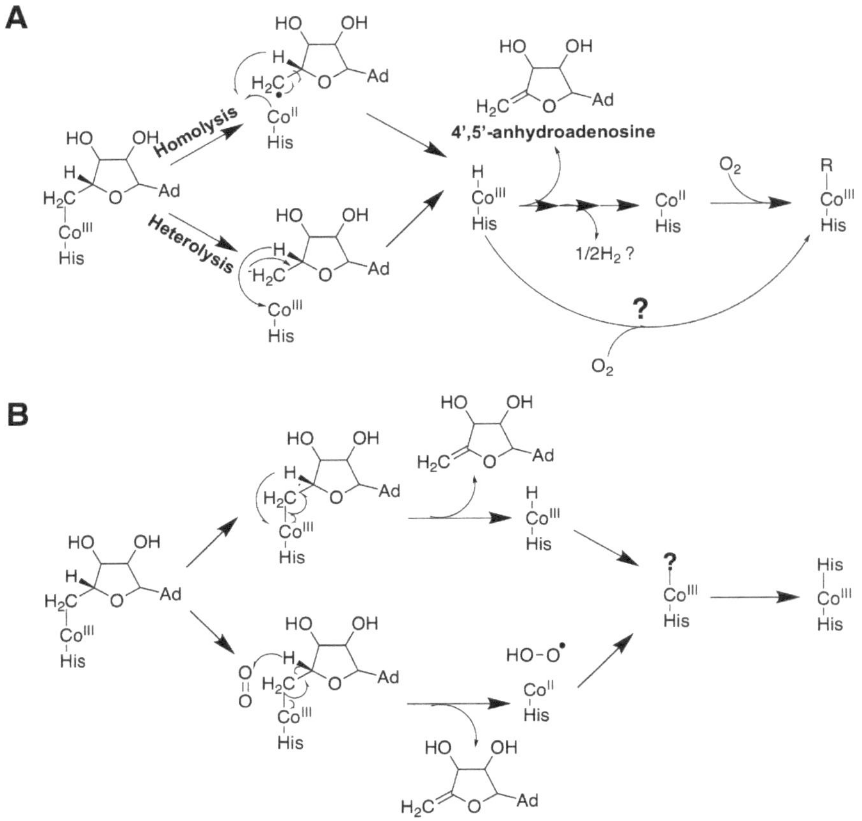

Fig. 3 Proposed photochemical mechanism of CarH by Jost and coworkers (A) and Miller et al. (B).

identified several reaction intermediates during the photocycle using time-resolved spectroscopy over femtoseconds through to seconds timescale, although the precise chemical nature of these intermediates remains poorly defined. Whereas Kutta and coworkers used excitation at 370 nm in their studies, Miller et al. reported similar studies, using excitation pulses in the range of 550–565 nm, to probe the excited electronic state dynamics (Miller et al., 2020). They propose the formation of an excited triplet state during the photocycle, which results from the ability of CarH to control the reactivity of AdoCbl. CarH appears to efficiently stabilize the excited state and slow down the reaction, therefore, allowing the formation of 4′,5′-anhydroadenosine (Fig. 3B).

More recently, computational methods such as molecular dynamic simulations and QM/MM calculations have been increasingly employed to gain further insights into the photochemical mechanism of CarH (Cooper, Panitz, Edwards, & Goyal, 2021; Toda, Lodowski, Mamun, & Kozlowski, 2021; Toda, Mamun, Lodowski, & Kozlowski, 2020). More studies are required to better understand the unique photochemical mechanism of CarH and explain the altered photochemistry of AdoCbl in its B_{12} binding pocket vis-à-vis AdoCbl binding enzymes.

3. Protocols

Although CarH was first characterized in *M. xanthus*, its comprehensive analysis remained elusive due to the inability to express and purify this protein in sufficient quantities (Padmanabhan et al., 2017). TtCarH, an ortholog from *Thermus thermophilus*, was found to express in decent quantities in *E. coli*-based expression systems. This enabled the detailed characterization of CarH proteins as it was possible to produce purified TtCarH for crystallography and spectroscopy experiments (Jost, Fernández-Zapata, et al., 2015; Kutta et al., 2015; Miller et al., 2020). In the past, several publications have reported brief purification protocols for TtCarH (Díez et al., 2013; Jost, Fernández-Zapata, et al., 2015; Kutta et al., 2015; Ortiz-Guerrero et al., 2011). Holo TtCarH is light sensitive and therefore requires extra precautions during the purification stage. In the following sections we provide a detailed step-by-step protocol adapted in our laboratory for the successful expression and purification of TtCarH and subsequent analysis by (time-resolved) UV–vis absorbance spectroscopy.

3.1 Expression protocol for TtCarH

Here we describe the expression protocol for an N-terminal 6xHistag containing TtCarH (UniProt id Q746J7) cloned in pET15b vector.

3.1.1 Equipment

- Autoclave to prepare sterile cell culture media
- Variable temperature incubation shaker (INFORS HT)
- High speed centrifuge with refrigeration (Beckman Coulter Avanti J-30I, rotor JLA 8.1000)

3.1.2 Materials and reagents

- Competent *E. coli* BL21(DE3) cells (New England Biolabs)
- Ampicillin sodium salt (Formedium)—50 mg/mL stock filter sterilized by passing through a 0.2 m filter
- LB Broth Miller (Formedium)
- IPTG-Dioxan free (Formedium)—1 mM stock filter sterilized by passing through a 0.2 m filter
- LB agar plates containing Ampicillin (50 g/mL)
- Glass conical flasks of 2 L volume

3.1.3 Procedure

1. Obtain fresh transformants in *E. coli* BL21(DE3) cells using standard transformation protocol
2. Use a single colony from a freshly transformed plate to inoculate an overnight starter culture (5 mL sterile LB media, 37 °C, 50 μg/mL Amp).
3. After overnight growth, use 1 mL of this starter culture to inoculate a day culture (100 mL LB media, 37 °C, 50 μg/mL Amp).
4. After 3 h of growth at 37 °C with shaking at 180 rpm, use 5 mL of this fresh day culture to inoculate 500 mL sterile LB media in conical flasks
5. Grow cells at 37 °C with shaking (180 rpm) until an OD of ~0.4–0.5 is reached and induce the culture by adding 0.4 mM IPTG (final concentration in 500 mL volume)
6. Reduce the temperature to 25 °C and allow the cultures to grow for another 20 h
7. Harvest the cells by centrifuging at 5000 × *g* for 15 min at 4 °C
8. The harvested cells can be stored at −20 °C until needed or used immediately to start purification. Roughly ~15 g of wet cell weight is obtained from 6 × 500 mL culture

3.2 Cell lysis protocol

In a typical purification experiment cells from 3 L culture is used to carry out a large-scale purification.

3.2.1 Equipment

- Cell disruptor (Constant Systems Limited)
- High speed centrifuge with refrigeration (Beckman Coulter Avanti J-30I, rotor JA 25.50)

3.2.2 Materials and reagents

- Tris (Fisher)
- Sodium chloride (Fisher)
- Imidazole (Sigma)
- SigmaFast Protease Inhibitor EDTA free tablet (Sigma S8820)
- DNase I (Sigma DN25) 5 mg/mL stock prepared in 50 mM sodium acetate, 5 mM $CaCl_2$, 0.1 mM PMSF and 50% glycerol at pH 6.5

3.2.3 Procedure

1. Prepare the following buffer solutions
- 50 mM Tris-HCl, pH 7.5, 300 mM NaCl, 10 mM imidazole (Buffer A)
- 50 mM Tris-HCl, pH 7.5, 300 mM NaCl, 250 mM imidazole (Buffer B)

Note: It is a good practice to filter and degas the buffers to be used in protein purification experiments.

2. Resuspend cells in buffer A using a magnetic stirrer. For every g of wet cell weight use 5 mL of buffer to prepare a cell suspension
3. Ensure all cell clumps have dissolved and add one complete tablet of protease inhibitor and DNase (to a final concentration of 5 μg/mL) to the cell suspension

Tip: adding $MgCl_2$ (5 mM final concentration) helps in efficient DNase activity

4. Maintaining the resuspended cells at 4 °C, pass it once through a cell disruptor at 20 kpsi pressure setting

Note: Alternatively, a sonicator with a setting of 35% power output can also be used. 30 cycles with 10 s pulse on and 50 s pulse off is sufficient to lyse the cells.

5. Centrifuge the lysed cells at 50,000 × *g* for 60 min at 4 °C to separate the cell debris from the cell lysate
6. Carefully collect the supernatant from the centrifuge tubes and pass it through a 0.45 μm filter to remove any unwanted aggregates

Critical: If the supernatant after filtration appears to be cloudy then dilute the cell lysate with buffer A. Apo TtCarH tends to form aggregates if the protein concentration is above ~25 μM.

3.3 Affinity chromatography

The cell lysate obtained in the previous step is purified using immobilized metal affinity chromatography.

3.3.1 Equipment and materials

- ÄKTA Pure chromatography system (Cytiva) and related accessories (UV detector, sample collector, superloop)
- TALON metal affinity resin (TakaraBio)

3.3.2 Procedure

1. Equilibrate a 5 mL column volume (cv) TALON column with 5 cv of buffer A using a ÄKTA Pure chromatography system

Note: Ni-NTA resin can also be used but TALON provides better specificity and better yield of purified protein. In our laboratory we keep the chromatography system in a cooling cabinet set at 10 °C. However, purification can also be carried out at room temperature.

2. Apply the cell lysate to the equilibrated column at 1.0 mL/min flow rate and collect the flow through fraction

3. Remove any unbound protein by washing the column with 5 cv of buffer A at flow rate 2.0 mL/min and collect samples in 10 mL fraction size

4. Wash the column with 5 cv of 50 mM imidazole containing buffer

Note: Most of the impurities together with some loosely bound TtCarH elutes out during this washing step.

5. Apply a linear elution gradient of 50–250 mM imidazole concentration over 40 cv using buffer A and B at a flow rate of 2.0 mL/min. Collect elution in fraction size of 5 mL

6. Check for presence of TtCarH in different fractions by running an SDS-PAGE gel

Note: TtCarH mostly elutes out in the 150–200 mM segment of the linear gradient.

7. Pool the fractions containing TtCarH together

3.4 Adding AdoCbl to form holo TtCarH

To obtain holo TtCarH AdoCbl needs to be added to the sample pooled together after TALON purification.

3.4.1 Equipment and materials

- Coenzyme B_{12} or adenosylcobalamin (Sigma Aldrich C0884)
- Opaque containers-1.5 mL microtubes (Heathrow Scientific, HS4323K), 50 mL Black tubes (Heathrow Scientific, HS4429), Amber HDPE bottles (Nalgene, DS2185–0004)
- Rolling platform

3.4.2 Procedure

1. Check the protein concentration of the pooled sample using the published extinction coefficient of 37.9 mM^{-1} cm^{-1} at 280 nm for CarH (Jost, Fernández-Zapata, et al., 2015).
2. Add a three-fold molar excess of AdoCbl to the sample and store in opaque containers

Important: From this point onwards, the sample should not be exposed to light to prevent any photoconversion of the holo TtCarH. Use of red light is allowed as AdoCbl is not susceptible to it.

3. Leave the containers at 4 °C on a rolling platform set at a slow speed to ensure gentle mixing of protein and AdoCbl for at least 1 h

Note: This step can also be a pause point and the sample can be left for a prolonged period before continuing with the subsequent steps.

3.5 Size-exclusion chromatography

The final step in the purification protocol of TtCarH is a size-exclusion chromatography step which helps in obtaining a homogeneous dark-adapted tetrameric sample with greater than 95% purity.

3.5.1 Equipment and materials

- ÄKTA Pure chromatography system (Cytiva)
- HiLoad 26/60 Superdex 200 size-exclusion column (Cytiva)
- Vivaspin 20 centrifugal concentrator—MWCO 30000 Da (Sartorius)

3.5.2 Procedure

1. Prepare the elution buffer—50 mM Tris-HCl, pH 7.5, 150 mM NaCl
2. Equilibrate the HiLoad 26/60 column with 2 cv of elution buffer.
3. Concentrate AdoCbl bound CarH down to 10 mL volume using Vivaspin concentrators under dim red light.

Tip: Pass the mixture through a 0.45 μm filter prior to concentrating to remove any aggregates that can block the filters on Vivaspin concentrators. This can save a lot of time.

Note: Excess AdoCbl will be seen in the flow through compartment of concentrators during centrifugation.

4. Inject the concentrated sample on the equilibrated size-exclusion column. Ensure that the equipment is covered and not exposed to any form of ambient light

Note: In our laboratory we cover the cooling cabinet that contains the chromatography equipment with blackout curtains while the sample is on the column.

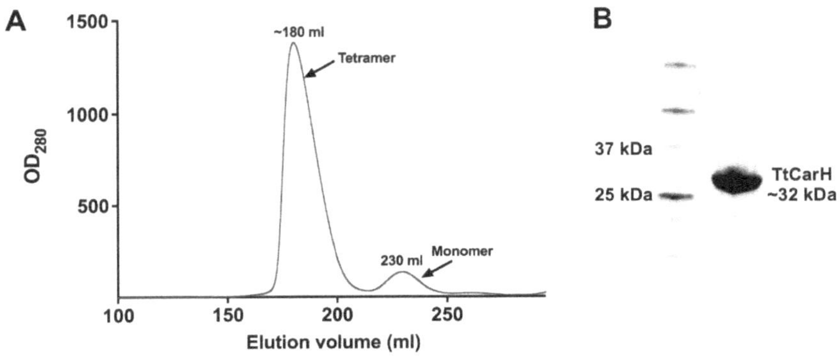

Fig. 4 (A) Elution profile of TtCarH from Superdex 200 Hiload 26/600 pg column. (B) SDS-PAGE gel showing purified TtCarH.

5. Using a flow rate of 1.5 mL/min collect all the fractions (2 mL each) over an elution volume of 1.5 cv

Note: Holo TtCarH (dark-adapted tetramer) elutes at ~180 mL during elution (Fig. 4A).

6. Check the fractions on an SDS-PAGE gel and pool the relevant fractions containing tetrameric TtCarH (Fig. 4B).

Note: Roughly 30 mg of pure protein can be obtained from 3 L culture.

7. The buffer-exchanged protein can be flash frozen in liquid nitrogen and stored at −80 °C for long-term storage

3.6 Static UV–vis absorbance spectroscopy

3.6.1 Equipment and materials

- Cary 60 UV–Vis spectrophotometer (Agilent Technologies)
- Quartz cuvette (10 mm × 2 mm pathlength)
- High-power 530 nm light-emitting diode (LED; M530L3, Thorlabs)
- TGP110 pulse generator (Thurlby Thandar)
- Anti-reflection-coated aspheric lens (COP1A, Thorlabs)

3.6.2 Procedure

1. Static UV–vis absorbance spectra, recorded using a spectrophotometer, can be used to determine the protein concentration and to monitor the conversion of the dark state to the light-adapted state.
2. Pipette 400 μL CarH sample into a cuvette (10 mm × 2 mm) and measure spectrum from 300 to 800 nm. The concentration of AdoCbl-bound CarH is determined using the published molar extinction coefficient of 8.0×10^3 M^{-1} cm^{-1} for AdoCbl at 522 nm (Hogenkamp & Holmes, 1970).

3. To monitor the photoconversion from the dark state to the light-adapted state, samples of CarH (typically between 10 and 50 μM) in a cuvette can be illuminated using the pulsed LED with λ_{max} of 530 nm. The duration of a single pulse should be set to ~100 ms, using the pulse generator, with an energy of ~2 mJ. This excitation light is collimated using the lens and focused onto the 2 mm pathlength of the cuvette at right-angles to the detection beam.

Note: the sample volume should be minimized to ensure that the entire sample is illuminated homogeneously.

4. Record an absorbance spectrum after each excitation pulse to produce a sequence of spectra as a function of illumination time

Note: It is important to wait more than 2 min after each excitation pulse before recording each spectrum to ensure proteins have had sufficient time to convert to the light-activated state—see below).

5. The degree of photoconversion of the sample can be determined by following the increase in absorbance associated with the light-adapted state (e.g., increase at ~360 nm) or the decrease in signal from the dark state (e.g., decrease at ~560 nm) (Fig. 5).

3.7 Time-resolved UV–vis absorbance spectroscopy

A detailed description of the current advances in time-resolved spectroscopy is provided elsewhere in this edition. Here, we provide a protocol for how these techniques can be used across multiple timescales from ps through to seconds to study the mechanism of CarH.

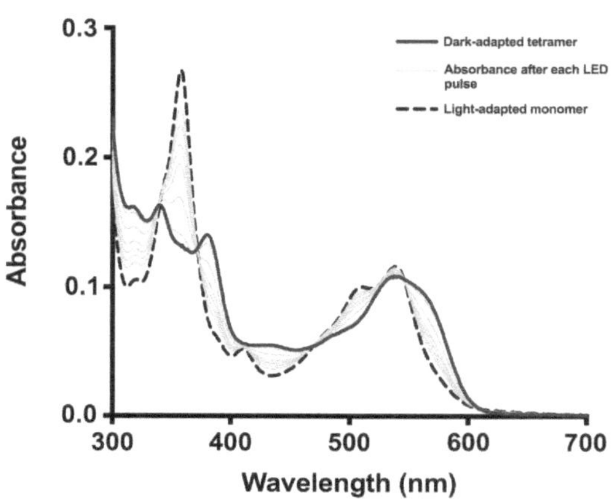

Fig. 5 Photoconversion of TtCarH followed using a UV–Vis spectrophotometer.

3.7.1 Equipment and materials

- Helios broadband pump-probe transient absorbance spectrometer (Ultrafast Systems LLC)
- 2 mm pathlength quartz cuvette (or flow cuvette)
- LP980 laser flash photolysis instrument (Edinburgh Instruments)
- Quartz cuvette (10 mm × 2 mm pathlength)
- Visible bandpass filters (Thorlabs)
- Cary 60 UV–Vis spectrophotometer (Agilent Technologies)
- High-power 530 nm light-emitting diode (LED; M530L3,Thorlabs)
- TGP110 pulse generator (Thurlby Thandar)
- Anti-reflection-coated aspheric lens (COP1A, Thorlabs)

3.7.2 Procedure for ultrafast transient absorbance (picosecond–nanosecond timescales)

1. Measurements should be performed at room temperature using a pump-probe transient absorbance spectrometer, which has an instrument response function of ~170 fs. This can be used to probe changes in the 350–750 nm spectral region after photoexcitation from ~200 fs to 3 ns
2. Briefly, the instrument consists of using a Ti:Sapphire amplifier system (Spectra-Physics Solstice Ace, 1 kHz, 6 W, 100 fs pulses), which is split to provide pump and probe beams (Fig. 6A). The pump beam is tuned using an optical parametric amplifiers (OPAs; TOPAS Prime) to the corresponding excitation wavelength (either ~370 nm for blue-light excitation or ~530 nm for green-light excitation) with an energy of 0.5 μJ per pulse. The white-light continuum probe is generated in either sapphire or CaF_2, along with an optical delay line to generate difference spectra over a range of time points.
3. The concentration of CarH samples used in these experiments should be ~200 μM in a 2 mm pathlength cuvette to give an optical density between 0.3 and 0.4 at the excitation wavelength.
4. Data should be recorded in a stirred quartz cuvette at time delays between −1.5 ps and 3 ns after excitation, the order of which is to be randomized.

Tip: It might be preferable to flow the sample through a flow cuvette using a peristaltic pump. In this case it is important to flow the sample fast enough to ensure that a fresh part of the sample is excited by each pulse.

5. At each time delay, 1000 single shots of pump and probe cycles should be collected and averaged to yield data with good signal-to-noise.

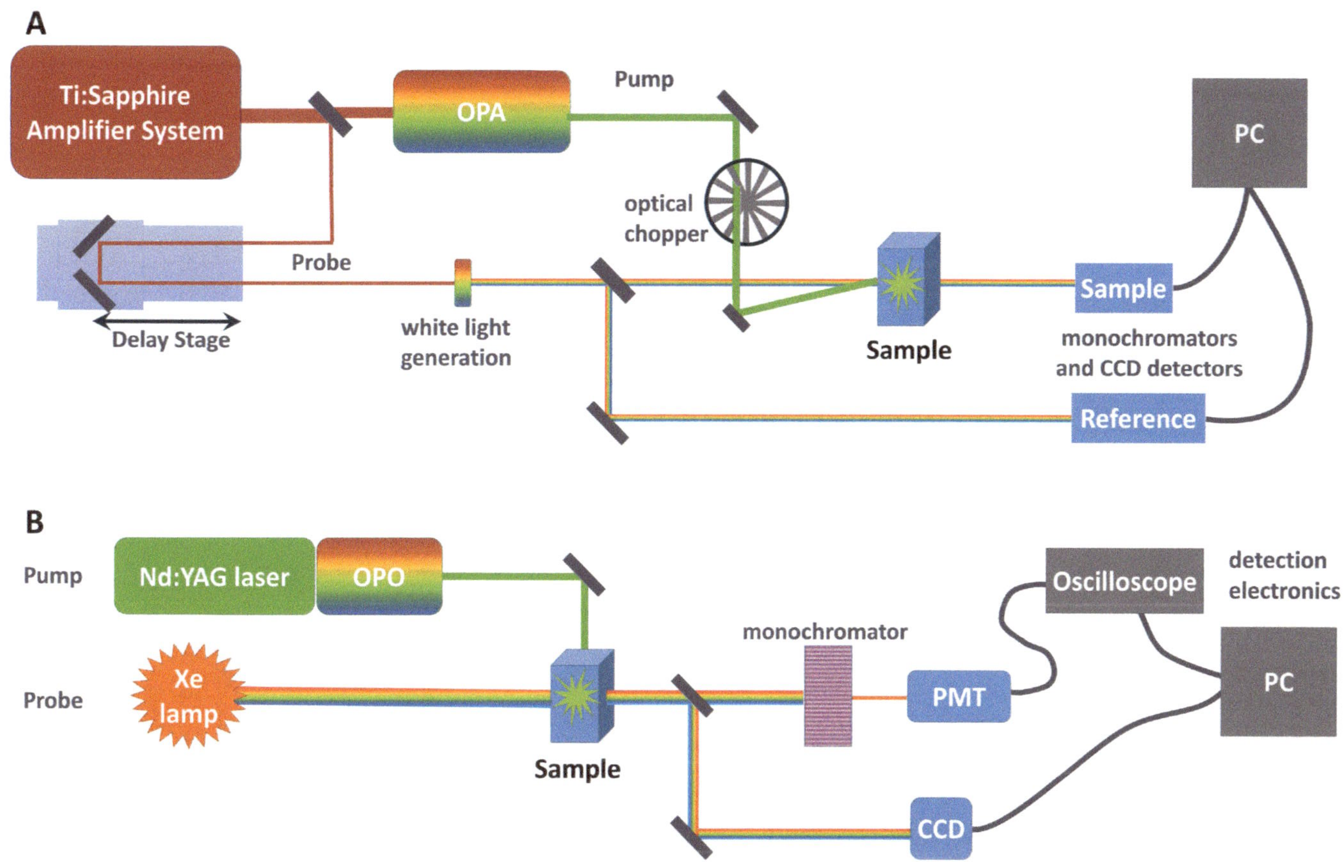

Fig. 6 Scheme of experimental layout for ultrafast transient absorbance (A) and laser flash photolysis (B).

6. Global analysis of the time-resolved spectral changes can be performed using the open-source software Glotaran to obtain an evolution associated difference spectra (Snellenburg, Laptenok, Seger, Mullen, & van Stokkum, 2012).

3.7.3 Procedure for laser flash photolysis (nanosecond to millisecond timescales)

1. Slower spectral changes on ns–s timescales can be followed using a laser flash photolysis instrument. With this set up data can either be acquired in kinetic mode (change in absorbance at single wavelength versus time) or spectral mode (absorbance difference spectrum at a single time point after excitation).
2. Briefly, photoconversion of CarH is initiated by excitation at 530 nm, using an optical parametric oscillator of a Q-switched Nd-YAG laser (EKSPLA N432). The laser pulses, which are between 6 and 8 ns in duration, should have an energy of ~15 mJ (Fig. 6B).
3. The detection system, which comprises probe light, sample, monochromator and either photomultiplier (for kinetic mode) or CCD camera (for spectral mode), is at right angles to the incident laser beam (Fig. 6B).
4. Prepare 400 μl samples of 50 μM CarH in a cuvette of 10 × 2 mm pathlength. For spectral mode acquisition 1.5 mL 50 μM CarH should be flowed through a 10 × 2 mm pathlength flow cuvette using a peristaltic pump.

Note: Excitation by the laser pulse should always be through the 2 mm path of the cuvette and the detection probe should be through the 10 mm path.

5. Initially, use the spectral mode to acquire absorbance difference spectra at various time points from 100 ns through to ~100 ms to gain insight into the spectral changes that occur during photoconversion.
6. In kinetic mode, rate constants can be measured by recording absorption transients at a range of wavelengths between 300 and 700 nm, using a corresponding bandpass filter between the probe light and the sample. Rate constants should be calculated from the average of at least three time-dependent absorption measurements by fitting to a single exponential function either by using the instrument software or by using Origin Pro 9.1 software.

3.7.4 Procedure for millisecond to seconds timescales

1. For measurements >~100 ms the intensity of the probe light of the laser flash photolysis instrument becomes problematic and triggers

photoconversion of the CarH sample. These slower measurements should be performed using a UV–Vis spectrophotometer in a similar way to the static UV–vis measurements described in Section 3.6.

2. 400 μL samples of 50 μM CarH in a cuvette (10 mm × 2 mm) are illuminated using a pulsed LED with λ_{max} of 530 nm. The duration of a single pulse should be set to ~10 ms using the pulse generator with an energy of ~2 mJ. This excitation light is collimated using the lens and focused onto the 2 mm pathlength of the cuvette at right-angles to the detection beam.
3. Record absorbance changes at single wavelengths ranging from 300 to 700 nm in 5 nm steps in 12.5 ms steps over a time range of ~50 s.
4. The LED flash should be manually triggered for each time trace using the pulse generator, which also triggers data collection via the spectrophotometer.
5. Kinetic data at single wavelengths can be analyzed by fitting to a single exponential function either by using Origin Pro 9.1 software data sets corrected to the same t¼0 during data processing. Alternatively, the raw data should be reduced to 500 points by averaging prior to global analysis of the entire dataset using the open-source software Glotaran (see Section 3.7.2).

4. Future directions and opportunities

The light-sensitive property of the CarH photoreceptor has been successfully exploited in a wide variety of applications, ranging from control of gene expression using green light (Chatelle et al., 2018) to preparation of hydrogels that can be used as light responsive carriers for drug delivery (Jiang et al., 2020; Narayan, Mu, Hasturk, & Kaplan, 2021; Wang, Yang, Luo, Hsing, & Sun, 2017). More recently, it was demonstrated that the CBD from TtCarH can be used in conjunction with smart watches to regulate the expression of mammalian genes using green light (Mansouri et al., 2021). This study has opened up a host of possibilities for CarH photoreceptors to be applied in smart wearable technology devices, which may have application in future personalized therapies for disease treatment.

To further realize the full potential of CarH proteins we need a better understanding of the photochemistry of the light-induced reaction. Crystal structure determination of the intermediates formed at different time points following light illumination will go a long way in our quest to identify the rate-limiting step in the photocycle. This information would be important to improve the quantum yield of the reaction. To this end,

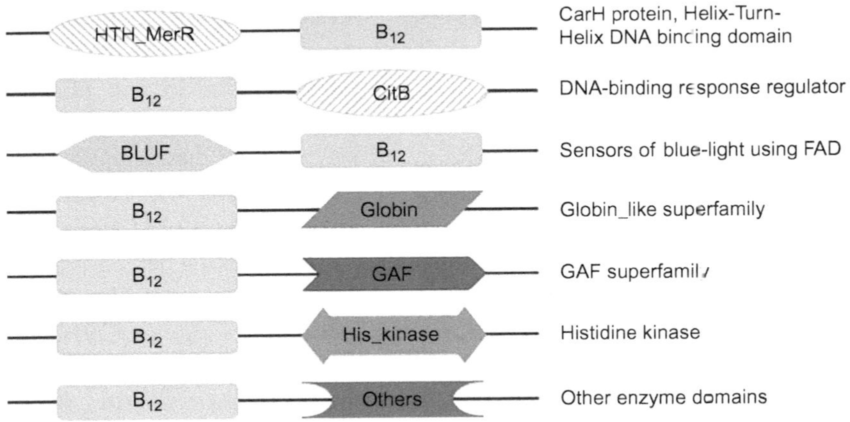

Fig. 7 Potential B_{12}-binding proteins with possible light-driven regulatory functions.

time-resolved serial crystallography at synchrotron and XFEL sources will be most beneficial. It has been suggested that CarH proteins are ideal targets for pump-probe based serial crystallography experiments (Poddar et al., 2021). Other time-resolved techniques, such as transient absorption spectroscopy, time-resolved X-ray scattering and mass spectroscopy, which can be conducted with protein in solution can also be further probed

Most of the studies summarized in the preceding sections have been carried out with TtCarH as the model protein. Bioinformatic analysis has revealed the presence of hundreds of CarH-like proteins found in different bacterial species (Padmanabhan et al., 2017; Padmanabhan et al., 2019). Many proteins have been found to contain a CarH like B_{12}-binding domain, fused with different functional domains. As shown in Fig. 7, different DNA binding domains and several potential photoreceptor domains such as, FAD-binding BLUF and tetrapyrrole-binding GAF proteins, have been found fused to the B_{12} domain (Cheng et al., 2016). Some enzyme domains, such as kinases have been also reported with putative B_{12}-binding domains which show high levels of identity to CarH CBD. These potential light-regulated proteins and enzymes remain uncharacterized.

5. Summary

CarH proteins are a unique class of coenzyme B_{12}-binding photoreceptors. The remarkable light-sensing function of coenzyme B_{12} in CarH, shows how cleverly nature has repurposed the light-sensitivity of Co—C bond in AdoCbl into regulating an important cellular process. With so many potential light-sensitive/regulated B_{12}-binding proteins awaiting characterization this is an exciting time for researchers in coenzyme B_{12} community. We hope the topics we have covered in this chapter will help researchers better understand the non-enzymatic photosensitive role of coenzyme B_{12} and its associated photochemistry. We further hope that the protocols described in this chapter would benefit the research community as they can be adapted to study other CarH proteins or any other coenzyme B_{12}-binding protein.

Acknowledgment

The research on B_{12}-binding proteins in the authors' laboratory is supported by an EPSRC Centre-to-Centre grant no. EP/S030336/1.

References

Armitage, J. P. (1999). Bacterial tactic responses. *Advances in Microbial Physiology*, *41*, 229–289. https://doi.org/10.1016/s0065-2911(08)60168-x.

Banerjee, S., & Mitra, D. (2020). Structural basis of design and engineering for advanced plant Optogenetics. *Trends in Plant Science*, *25*(1), 35–65. https://doi.org/10.1016/j.tplants.2019.10.002.

Brettel, K., & Byrdin, M. (2010). Reaction mechanisms of DNA photolyase. *Current Opinion in Structural Biology*, *20*(6), 693–701. https://doi.org/10.1016/j.sbi.2010.07.003.

Bridwell-Rabb, J., & Drennan, C. L. (2017). Vitamin B. *Current Opinion in Chemical Biology*, *37*, 63–70. https://doi.org/10.1016/j.cbpa.2017.01.013.

Cervantes, M., & Murillo, F. J. (2002). Role for vitamin B(12) in light induction of gene expression in the bacterium Myxococcus xanthus. *Journal of Bacteriology*, *184*(8), 2215–2224. https://doi.org/10.1128/JB.184.8.2215-2224.2002.

Chatelle, C., Ochoa-Fernandez, R., Engesser, R., Schneider, N., Beyer, H. M., Jones, A. R., et al. (2018). A green-light-responsive system for the control of transgene expression in mammalian and plant cells. *ACS Synthetic Biology*, 7(5), 1349–1358. https://doi.org/10.1021/acssynbio.7b00450.

Cheng, Z., Yamamoto, H., & Bauer, C. E. (2016). Cobalamin's (vitamin B12) surprising function as a photoreceptor. *Trends in Biochemical Sciences*, *41*(8), 647–650. https://doi.org/10.1016/j.tibs.2016.05.002.

Cohen, S. E., & Golden, S. S. (2015). Circadian rhythms in Cyanobacteria. *Microbiology and Molecular Biology Reviews*, *79*(4), 373–385. https://doi.org/10.1128/MMBR.00036-15.

Cooper, C. L., Panitz, N., Edwards, T. A., & Goyal, P. (2021). Role of the CarH photoreceptor protein environment in the modulation of cobalamin photochemistry. *Biophysical Journal*, *120*(17), 3688–3696. https://doi.org/10.1016/j.bpj.2021.07.020.

Croce, R., & van Amerongen, H. (2014). Natural strategies for photosynthetic light harvesting. *Nature Chemical Biology*, *10*(7), 492–501. https://doi.org/10.1038/nchembio.1555.

Díez, A. I., Ortiz-Guerrero, J. M., Ortega, A., Elías-Arnanz, M., Padmanabhan, S., & García de la Torre, J. (2013). Analytical ultracentrifugation studies of oligomerization and DNA-binding of TtCarH, a Thermus thermophilus coenzyme B12-based photosensory regulator. *European Biophysics Journal*, *42*(6), 463–476. https://doi.org/10.1007/s00249-013-0897-x.

Drennan, C. L., Huang, S., Drummond, J. T., Matthews, R. G., & Ludwig, M. L. (1994). How a protein binds B12: A 3.0 a X-ray structure of B12-binding domains of methionine synthase. *Science*, *266*(5191), 1669–1674. https://doi.org/10.1126/science.7992050.

Elías-Arnanz, M., Padmanabhan, S., & Murillo, F. J. (2011). Light-dependent gene regulation in nonphototrophic bacteria. *Current Opinion in Microbiology*, *14*(2), 128–135. https://doi.org/10.1016/j.mib.2010.12.009.

Fasnacht, M., & Polacek, N. (2021). Oxidative stress in Bacteria and the central dogma of molecular biology. *Frontiers in Molecular Biosciences*, *8*. https://doi.org/10.3389/fmolb.2021.671037, 671037.

Fernández-Zapata, J., Pérez-Castaño, R., Aranda, J., Colizzi, F., Polanco, M. C., Orozco, M., et al. (2018). Plasticity in oligomerization, operator architecture, and DNA binding in the mode of action of a bacterial B. *The Journal of Biological Chemistry*, *293*(46), 17888–17905. https://doi.org/10.1074/jbc.RA118.004838.

Glaeser, J., Nuss, A. M., Berghoff, B. A., & Klug, G. (2011). Singlet oxygen stress in microorganisms. *Advances in Microbial Physiology*, *58*, 141–173. https://doi.org/10.1016/B978-0-12-381043-4.00004-0.

Gruber, K., Puffer, B., & Kräutler, B. (2011). Vitamin B12-derivatives-enzyme cofactors and ligands of proteins and nucleic acids. *Chemical Society Reviews*, *40*(8), 4346–4363. https://doi.org/10.1039/c1cs15118e.

Hart, J. E., & Gardner, K. H. (2021). Lighting the way: Recent insights into the structure and regulation of phototropin blue light receptors. *The Journal of Biological Chemistry*, *296*. https://doi.org/10.1016/j.jbc.2021.100594, 100594.

Heyes, D. J., Zhang, S., Taylor, A., Johannissen, L. O., Hardman, S. J. O., Hay, S., et al. (2021). Photocatalysis as the 'master switch' of photomorphogenesis in early plant development. *Nature Plants*, 7(3), 268–276. https://doi.org/10.1038/s41477-021-00866-5.

Hodgkin, D. C., Kamper, J., Mackay, M., Pickworth, J., Trueblood, K. N., & White, J. G. (1956). Structure of vitamin B12. *Nature*, *178*(4524), 64–66. https://doi.org/10.1038/178064a0.

Hogenkamp, H. P., & Holmes, S. (1970). Polarography of cobalamins and cobinamides. *Biochemistry*, *9*(9), 1886–1892. https://doi.org/10.1021/bi00811a004.

Jiang, B., Liu, X., Yang, C., Yang, Z., Luo, J., Kou, S., et al. (2020). Injectable, photoresponsive hydrogels for delivering neuroprotective proteins enabled by metal-directed protein assembly. *Science Advances*, *6*(41). https://doi.org/10.1126/sciadv.abc4824.

Jost, M., Fernández-Zapata, J., Polanco, M. C., Ortiz-Guerrero, J. M., Chen, P. Y., Kang, G., et al. (2015). Structural basis for gene regulation by a B12-dependent photoreceptor. *Nature*, *526*(7574), 536–541. https://doi.org/10.1038/nature14950.

Jost, M., Simpson, J. H., & Drennan, C. L. (2015). The transcription factor CarH safeguards use of Adenosylcobalamin as a light sensor by altering the photolysis products. *Biochemistry*, *54*(21), 3231–3234. https://doi.org/10.1021/acs.biochem.5b00416.

Konold, P. E., van Stokkum, I. H. M., Muzzopappa, F., Wilson, A., Groot, M. L., Kirilovsky, D., et al. (2019). Photoactivation mechanism, timing of protein secondary structure dynamics and carotenoid translocation in the Orange carotenoid protein. *Journal of the American Chemical Society*, *141*(1), 520–530. https://doi.org/10.1021/jacs.8b11373.

Kozlowski, P. M., Garabato, B. D., Lodowski, P., & Jaworska, M. (2016). Photolytic properties of cobalamins: A theoretical perspective. *Dalton Transactions*, *45*(11), 4457–4470. https://doi.org/10.1039/c5dt04286k.

Krueger, D., Izquierdo, E., Viswanathan, R., Hartmann, J., Pallares Cartes, C., & De Renzis, S. (2019). Principles and applications of optogenetics in developmental biology. *Development*, *146*(20). https://doi.org/10.1242/dev.175067.

Kutta, R. J., Hardman, S. J. O., Johannissen, L. O., Bellina, B., Messiha, H. L., Ortiz-Guerrero, J. M., et al. (2015). The photochemical mechanism of a

B12-dependent photoreceptor protein. *Nature Communications*, *6*, 7907. https://doi.org/10.1038/ncomms8907.

Leverenz, R. L., Sutter, M., Wilson, A., Gupta, S., Thurotte, A., Bourcier de Carbon, C., et al. (2015). PHOTOSYNTHESIS. A 12 Å carotenoid translocation in a photoswitch associated with cyanobacterial photoprotection. *Science*, *348*(6242), 1463–1466. https://doi.org/10.1126/science.aaa7234.

Li, Z., Wakao, S., Fischer, B. B., & Niyogi, K. K. (2009). Sensing and responding to excess light. *Annual Review of Plant Biology*, *60*, 239–260. https://doi.org/10.1146/annurev.arplant.58.032806.103844.

Losi, A., Gardner, K. H., & Möglich, A. (2018). Blue-light receptors for Optogenetics. *Chemical Reviews*, *118*(21), 10659–10709. https://doi.org/10.1021/acs.chemrev.8b00163.

Mansouri, M., Hussherr, M. D., Strittmatter, T., Buchmann, P., Xue, S., Camenisch, G., et al. (2021). Smart-watch-programmed green-light-operated percutaneous control of therapeutic transgenes. *Nature Communications*, *12*(1), 3388. https://doi.org/10.1038/s41467-021-23572-4.

Marsh, E. N. G., & Holloway, D. E. (1992). Cloning and sequencing of glutamate mutase component S from *Clostridium tetanomorphum*: Homologies with other cobalamin-dependent enzymes. *FEBS Letters*, *310*, 147–149.

Miller, N. A., Kaneshiro, A. K., Konar, A., Alonso-Mori, R., Britz, A., Deb, A., et al. (2020). The photoactive excited state of the B. *The Journal of Physical Chemistry. B*, *124*(47), 10732–10738. https://doi.org/10.1021/acs.jpcb.0c09428.

Nagel, D. H., & Kay, S. A. (2012). Complexity in the wiring and regulation of plant circadian networks. *Current Biology*, *22*(16), R648–R657. https://doi.org/10.1016/j.cub.2012.07.025.

Narayan, O. P., Mu, X., Hasturk, O., & Kaplan, D. L. (2021). Dynamically tunable light responsive silk-elastin-like proteins. *Acta Biomaterialia*, *121*, 214–223. https://doi.org/10.1016/j.actbio.2020.12.018.

Ortiz-Guerrero, J. M., Polanco, M. C., Murillo, F. J., Padmanabhan, S., & Elías-Arnanz, M. (2011). Light-dependent gene regulation by a coenzyme B12-based photoreceptor. *Proceedings of the National Academy of Sciences of the United States of America*, *108*(18), 7565–7570. https://doi.org/10.1073/pnas.1018972108.

Padmanabhan, S., Jost, M., Drennan, C. L., & Elías-Arnanz, M. (2017). A new facet of vitamin B. *Annual Review of Biochemistry*, *86*, 485–514. https://doi.org/10.1146/annurev-biochem-061516-044500.

Padmanabhan, S., Monera-Girona, A. J., Pérez-Castaño, R., Bastida-Martínez, E., Pajares-Martínez, E., Bernal-Bernal, D., et al. (2021). Light-Triggered Carotenogenesis in. *Microorganisms*, *9*(5). https://doi.org/10.3390/microorganisms9051067.

Padmanabhan, S., Pérez-Castaño, R., & Elías-Arnanz, M. (2019). B12-based photoreceptors: From structure and function to applications in optogenetics and synthetic biology. *Current Opinion in Structural Biology*, *57*, 47–55. https://doi.org/10.1016/j.sbi.2019.01.020.

Palczewski, K. (2012). Chemistry and biology of vision. *The Journal of Biological Chemistry*, *287*(3), 1612–1619. https://doi.org/10.1074/jbc.R111.301150.

Pande, K., Hutchison, C. D., Groenhof, G., Aquila, A., Robinson, J. S., Tenboer, J., et al. (2016). Femtosecond structural dynamics drives the trans/cis isomerization in photoactive yellow protein. *Science*, *352*(6286), 725–729. https://doi.org/10.1126/science.aad5081.

Park, S. Y., & Tame, J. R. H. (2017). Seeing the light with BLUF proteins. *Biophysical Reviews*, *9*(2), 169–176. https://doi.org/10.1007/s12551-017-0258-6.

Pérez-Marín, M. C., Padmanabhan, S., Polanco, M. C., Murillo, F. J., & Elías-Arnanz, M. (2008). Vitamin B12 partners the CarH repressor to downregulate a photoinducible promoter in Myxococcus xanthus. *Molecular Microbiology*, *67*(4), 804–819. https://doi.org/10.1111/j.1365-2958.2007.06086.x.

Poddar, H., Heyes, D. J., Schirò, G., Weik, M., Leys, D., & Scrutton, N. S. (2021). A guide to time-resolved structural analysis of light-activated proteins. *The FEBS Journal*. https://doi.org/10.1111/febs.15880.

Purcell, E. B., & Crosson, S. (2008). Photoregulation in prokaryotes. *Current Opinion in Microbiology*, *11*(2), 168–178. https://doi.org/10.1016/j.mib.2008.02.014.

Rodionov, D. A., Vitreschak, A. G., Mironov, A. A., & Gelfand, M. S. (2003). Comparative genomics of the vitamin B12 metabolism and regulation in prokaryotes. *The Journal of Biological Chemistry*, *278*(42), 41148–41159. https://doi.org/10.1074/jbc.M305837200.

Ruban, A. V. (2015). Evolution under the sun: Optimizing light harvesting in photosynthesis. *Journal of Experimental Botany*, *66*(1), 7–23. https://doi.org/10.1093/jxb/eru400.

Rury, A. S., Wiley, T. E., & Sension, R. J. (2015). Energy cascades, excited state dynamics, and photochemistry in cob(III)alamins and ferric porphyrins. *Accounts of Chemical Research*, *48*(3), 860–867. https://doi.org/10.1021/ar5004016.

Shcherbakova, D. M., Shemetov, A. A., Kaberniuk, A. A., & Verkhusha, V. V. (2015). Natural photoreceptors as a source of fluorescent proteins, biosensors, and optogenetic tools. *Annual Review of Biochemistry*, *84*, 519–550. https://doi.org/10.1146/annurev-biochem-060614-034411.

Snellenburg, J. J., Laptenok, S., Seger, R., Mullen, K. M., & van Stokkum, I. H. M. (2012). Glotaran: A Java-based graphical user interface for the R package TIMP. *Journal of Statistical Software*, *49*(3), 1–22. https://doi.org/10.18637/jss.v049.i03.

Sorigué, D., Légeret, B., Cuiné, S., Blangy, S., Moulin, S., Billon, E., et al. (2017). An algal photoenzyme converts fatty acids to hydrocarbons. *Science*, *357*(6354), 903–907. https://doi.org/10.1126/science.aan6349.

Takala, H., Björling, A., Berntsson, O., Lehtivuori, H., Niebling, S., Hoernke, M., et al. (2014). Signal amplification and transduction in phytochrome photosensors. *Nature*, *509*(7499), 245–248. https://doi.org/10.1038/nature13310.

Takano, H., Mise, K., Hagiwara, K., Hirata, N., Watanabe, S., Toriyabe, M., et al. (2015). Role and function of LitR, an Adenosyl B12-bound light-sensitive regulator of bacillus megaterium QM B1551 in, Regulation of Carotenoid Production. *Journal of Bacteriology*, *197*(14), 2301–2315. https://doi.org/10.1128/JB.02528-14.

Toda, M. J., Lodowski, P., Mamun, A. A., & Kozlowski, P. M. (2021). Electronic and photolytic properties of hydridocobalamin. *Journal of Photochemistry and Photobiology. B*, *224*. https://doi.org/10.1016/j.jphotobiol.2021.112295, 112295.

Toda, M. J., Mamun, A. A., Lodowski, P., & Kozlowski, P. M. (2020). Why is CarH photolytically active in comparison to other B. *Journal of Photochemistry and Photobiology. B*, *209*. https://doi.org/10.1016/j.jphotobiol.2020.111919, 111919.

Wald, G. (1968). The molecular basis of visual excitation. *Nature*, *219*(5156), 800–807. https://doi.org/10.1038/219800a0.

Wang, R., Yang, Z., Luo, J., Hsing, I. M., & Sun, F. (2017). *Proceedings of the National Academy of Sciences of the United States of America*, *114*(23), 5912–5917. https://doi.org/10.1073/pnas.1621350114.

Wickstrand, C., Dods, R., Royant, A., & Neutze, R. (2015). Bacteriorhodopsin: Would the real structural intermediates please stand up? *Biochimica et Biophysica Acta*, *1850*(3), 536–553. https://doi.org/10.1016/j.bbagen.2014.05.021.

Wickstrand, C., Nogly, P., Nango, E., Iwata, S., Standfuss, J., & Neutze, R. (2019). Bacteriorhodopsin: Structural insights revealed using X-ray lasers and synchrotron radiation. *Annual Review of Biochemistry*, *88*, 59–83. https://doi.org/10.1146/annurev-biochem-013118-111327.

Yang, J., & Cheng, Q. (2004). Origin and evolution of the light-dependent protochlorophyllide oxidoreductase (LPOR) genes. *Plant Biology (Stuttgart, Germany)*, *6*(5), 537–544. https://doi.org/10.1055/s-2004-821270.

Zhang, Y., Rodionov, D. A., Gelfand, M. S., & Gladyshev, V. N. (2009). Comparative genomic analyses of nickel, cobalt and vitamin B12 utilization. *BMC Genomics*, *10*, 78. https://doi.org/10.1186/1471-2164-10-78.

Kolibri 360 GmbH
Ferdinand-Jühlke-Straße 7
99095 Erfurt, Deutschland
produktsicherheit@kolibri360.de